ELEMENTS OF STATISTICAL MECHANICS

With an Introduction to Quantum Field Theory and Numerical Simulation

Elements of Statistical Mechanics provides a concise, self-contained introduction to the key concepts and tools of statistical mechanics including advanced topics such as numerical methods and non-relativistic quantum field theory.

Beginning with an introduction to classical thermodynamics and statistical mechanics the reader is exposed to simple, exactly soluble models and their application to biological systems. Analytic and numerical tools are then developed and applied to realistic systems such as magnetism and fluids.

The authors discuss quantum statistical mechanics in detail with applications to problems in condensed matter as well as a selection of topics in astrophysics and cosmology. The book concludes with a presentation of emergent phenomena such as phase transitions, their critical exponents and the renormalization group.

Combining their extensive experience in research and teaching in this field, the authors have produced a comprehensive textbook accessible to advanced undergraduate and graduate students in physics, chemistry and mathematics.

IVO SACHS is Professor of Physics at the Department of Physics, Ludwig-Maximilians-Universität (LMU), Munich, Germany. After gaining his Ph.D. from Eidgenössische Technische Hochschule (ETH), Zurich in 1995, he held research positions at the Dublin Institute for Advanced Studies, the University of Durham and LMU, before his appointment as Lecturer in Mathematics at Trinity College Dublin. He has held his current position since 2003. His main areas of research include quantum field theory, string theory and black holes.

SIDDHARTHA SEN is the Deputy Director of the Hamilton Mathematics Institute and an emeritus fellow of Trinity College Dublin. Following his education in Calcutta and Sc.D. at the Massachusetts Institute of Technology, he has held visiting positions at CERN, Fermilab, Brookhaven National Laboratory, NY; the TATA Institute, Bombay, and currently at the Indian Association for the Cultivation of Science, Calcutta. He has published over 80 papers and co-authored a book in mathematical physics; his other research interests include black hole physics and string theory.

JAMES C. SEXTON is Director of High Perfomance Computing and also Associate Professor at the Department of Pure and Applied Mathematics at Trinity College Dublin. Following his education there, he gained an M.Sc. and D.Phil. from Columbia University, New York. He has conducted research at Fermilab, the Institute for Advanced Study, Princeton, IBM and Yorktown. His main research areas are lattice QCD, the numerical simulation of complex systems and supercomputer coding.

ELEMENTS OF STATISTICAL MECHANICS

With an Introduction to Quantum Field Theory and Numerical Simulation

IVO SACHS
Ludwig-Maximilians-Universität, Munich

SIDDHARTHA SEN
Trinity College Dublin

JAMES SEXTON
Trinity College Dublin

CAMBRIDGE UNIVERSITY PRESS
Cambridge, New York, Melbourne, Madrid, Cape Town, Singapore,
São Paulo, Delhi, Dubai, Tokyo

Cambridge University Press
The Edinburgh Building, Cambridge CB2 8RU, UK

Published in the United States of America by Cambridge University Press, New York

www.cambridge.org
Information on this title: www.cambridge.org/9780521143646

First published 2006
This digitally printed version 2010

A catalogue record for this publication is available from the British Library

ISBN 978-0-521-84198-6 Hardback
ISBN 978-0-521-14364-6 Paperback

Contents

Preface

Statistical mechanics is a fundamental part of theoretical physics. Not only does it provide the basic tools for analyzing the behavior of complex systems in thermal equilibrium, but also hints at, and is fully compatible with, quantum mechanics as the theory underlying the laws of nature. In the process one encounters such complex emergent phenomena as phase transitions, superfluidity, and superconductivity which are highly non-trivial consequences of the microscopic dynamics. At the same time statistical mechanics poses conceptual problems such as how irreversibilty can appear from an underlying microscopic system governed by reversible laws.

Historically, statistical mechanics grew out of classical thermodynamics with the aim of providing a dynamical foundation for this phenomenological theory. It thus deals with many-body problems starting from a microscopic model which is typically described by a simple Hamiltonian. The power of statistical mechanics lies in both its simplicity and universality. Indeed the same concept can be applied to a wide variety of systems both classical and quantum mechanical. These include non-interacting and interacting gases, chemical interactions, paramagnetic and spin systems, astrophysics, and solids. On the other hand statistical mechanics brings together a variety of different tools and methods used in theoretical physics, chemistry, and mathematics. Indeed while the basic concepts are easily explained in simple terms a quantitative analysis will quickly involve sophisticated methods.

The purpose of this book is twofold: to provide a concise and self-contained introduction to the key concepts of statistical mechanics and to present the important results from a modern perspective. The book is introductory in character, and should be accessible to advanced undergraduate and graduate students in physics, chemistry, and mathematics. It is a synthesis of a number of distinct undergraduate and graduate courses taught by the authors at the University of Dublin, Trinity College over a number of years.

Chapters 1 to 4 provide an introduction to classical thermodynamics and statistical mechanics beginning with basic concepts such as Carnot cycles and thermodynamic potentials. We then introduce the basic postulates of statistical mechanics which are applied to simple systems. This part then ends with a classical treatment of interactions in terms of a cluster expansion.

Although these techniques can provide us with considerable information about many systems they are in general not sufficient for all purposes. Therefore, in Chapters 5 and 6 we present a short self-contained account of the techniques used in the numerical approach to the statistical mechanics of interacting systems. In particular, Monte Carlo integration is reviewed in Chapter 5, while the powerful method of symplectic integrators is described in Chapter 6.

The second part of the book is devoted mostly to quantum statistical mechanics. In particular, in Chapter 7 Bose–Einstein and Fermi–Dirac systems are explained in detail, including high- and low-temperature expansions, Bose–Einstein condensation as well as blackbody radiation and phonon excitations in solids.

After introducing the basic concepts of classical and quantum statistical mechanics we make a short excursion into astrophysics and cosmology in Chapter 8, where we illustrate the importance of this formalism for a quantitative understanding of important problems such as determining the surface temperature of stars, the stability of compact objects such as white dwarfs and neutron stars, as well as the cosmic background radiation.

We then return to the main focus by systematically including interactions into quantum statistical mechanics. For this, the framework of non-relativistic quantum field theory is developed in Chapter 9, leading to a systematic, perturbative formalism for the evaluation of the partition function for interacting systems. In addition, in Chapter 10, we develop non-perturbative techniques which are used to give a qualitative derivation of the phenomenon of superfluidity.

In Chapter 11 the path integral formulation of quantum mechanics and field theory is described. The purpose of this chapter is to establish an intuitive link between classical and quantum statistical mechanics and also to establish some tools required in the treatment of critical phenomena in the last chapter.

Before that, however, we take another break in Chapter 12 with a critical review of the material presented thus far. Among the issues analyzed in some detail are the question of ergodicity, Poincaré recurrence, negative temperatures, and surface effects.

The final chapter is then devoted to the important subject of phase transitions and critical phenomena. After some general comments, Landau's phenomenological theory for phase transitions is introduced and then generalized, using the path integral formalism, to compute critical exponents which, in turn, play a central role in any quantitative discussion of phase transitions.

For the first half of the book a good knowledge of classical mechanics is assumed. For the second part some practice with quantum mechanics will be necessary. The aim is to present the concepts and methods of statistical mechanics with the help of simple examples. These examples illustrate a variety of phenomena that can be analyzed within statistical mechanics. We have tried, wherever possible, to physically motivate each step and to point out interconnections between different concepts. When appropriate we have included a short paragraph with some historical comments to place the various developments in this context.

We would like to thank Samik Sen for help in preparing the manuscript and Seamus Murphy for help with the figures. Many thanks to Matthew Parry, Mike Peardon, Fabian Sievers and Wolfgang Wieser for their help with the numerical algorithms. Further thanks to Herbert Wagner for helpful comments and to Sean Keating for a careful reading of the manuscript.

Fundamental physical constants

Planck constant	$h = 6.6260755 \cdot 10^{-34}$ J s
	$\hbar = h/2\pi = 1.05457266 \cdot 10^{-34}$ J s
Boltzmann constant	$k = 1.380658 \cdot 10^{-23}$ J/K
Avogadro number	$N_A = 6.0221367 \cdot 10^{23}$ particles/mol
Gas constant	$R = N_A k = 8.31451$ J/K per mol
Speed of light	$c = 2.99792458 \cdot 10^{8}$ m/s
Elementary charge	$e = 1.60217733 \cdot 10^{-19}$ J
Electron rest mass	$m_e = 9.1093897 \cdot 10^{-31}$ kg
Neutron rest mass	$m_n = 1.6749286 \cdot 10^{-27}$ kg
Proton rest mass	$m_p = 1.6726231 \cdot 10^{-27}$ kg
Compton wavelength of the electron	$\lambda_C = \frac{h}{m_e c} = 2.42631 \cdot 10^{-12}$ m
Acceleration due to gravity	$g = 9.80665$ m/s^2
Newton constant	$G = 6.67 \cdot 10^{-11}$m^3/kg $\cdot$ s^2
Bohr radius	$a_0 = 5.29177 \cdot 10^{-11}$ m
Stefan–Boltzmann constant	$\sigma = \frac{\pi^2 k^4}{60\hbar^3 c^2} = 5.6703 \cdot 10^{-8}W/m^2 \cdot$ K^4

1
The problem

We shall be concerned in this book with the modeling of physical systems containing a large number of interacting particles. These physical systems include volumes of gases, liquids, and solids, which may or may not be isolated from their surroundings. The constituents making up these systems can be molecules, atoms, or elementary particles like electrons and protons. The techniques we will develop in this book, however, will be applicable to much broader classes of problems than those which serve as our initial focus.

1.1 A molecular model

Let us begin with a concrete physical system containing N non-relativistic, identical particles which interact pairwise. The Hamiltonian for such a system is

$$H = \frac{1}{2m}\sum_i |\mathbf{p}_i|^2 + V(\mathbf{x}_1, \ldots, \mathbf{x}_N),$$

where $\mathbf{x}_i$ and $\mathbf{p}_i$ are the positions and momenta of the particles in the system, and $i = 1, \ldots, N$. The classical approach to solving this system involves integrating Hamilton's equations of motion. A full solution, either analytic or numerical, requires that the positions and momenta of all N particles be specified at some particular time t_0. The solution must then solve the equations to trace out the full trajectories followed by all N particles in the system for some finite time interval beyond t_0.

To grasp the scope of the problem when addressed in this way, let us consider some of the numbers involved. One gram of O_2 gas contains about 2×10^{22} molecules. At typical atmospheric pressure and temperature this amount of gas will occupy a volume of about 10 cm^3. Clearly this represents a reasonably small amount of gas. We see immediately that any analysis of this system which depends on integrating the equations of motion is doomed to fail. There are just too many molecules involved. Modern computers can execute individual calculations at a rate of approximately one per nanosecond. Solving equations of motion takes many operations per molecule. But as a lower bound, suppose it took a computer a nanosecond to plot the motion of a single molecule in this gas for some period of its evolution. The same computer would then require over 300 000 years to calculate the motion of all 10^{22} molecules over the same time period!

The statistical approach to modeling systems with large numbers of degrees of freedom involves describing the system's properties using averages from a small set of appropriate and relevant observables. The thought behind the statistical approach is that the exact details of the behavior of any given degree of freedom has no significant observable effect since there are so many degrees of freedom involved. The only important variables of interest must involve averaging over many of the degrees of freedom. Statistical mechanics is the formalization of this intuitive concept. The problems to be addressed in describing statistical mechanics are threefold: under what circumstances can the properties of a physical system be defined by the behavior of an appropriate small set of variables, what are the appropriate sets of relevant variables, and how can one calculate the properties of the system in terms of these variables.

In what is to follow we will answer these questions in considerable detail. In the process we will furthermore discover a number of so-called emergent phenomena, that is physical properties, such as phase transitions or superfluidity of certain N-particle systems which are not inherent in the microscopic 1-particle Hamiltonian, but emerge in the limit when the number of particles and the size of the systems tends to infinity. The framework for our answers will be built initially from the body of work of Maxwell, Boltzmann, and Gibbs. The formal treatment of this framework begins in Chapter 2. Before beginning that formal treatment, it is necessary to describe the physical concept which is the precursor to statistical mechanics, that is classical thermodynamics. It was developed early in the nineteenth century, and is based wholly on experimental observations. What is particularly impressive about classical thermodynamics is that most of the significant results in it were developed independently of any detailed picture of the atomic nature of matter. Classical thermodynamics axiomatizes the modeling of physical systems in four basic laws. The following sections will present these laws of thermodynamics, provide modern qualitative justifications for them, and discuss some of their simple consequences.

1.2 The basics of classical thermodynamics

Consider a gas of N identical molecules contained in a volume V. The modern view of this gas is that the molecules are all in motion. For simplicity in the following discussion we will assume that the molecules all behave like hard spheres, so that molecules move in straight lines with constant velocity except during the process of collision. This motion results in constant collisions between molecules themselves, and also between the molecules and the walls which contain them in the volume V. At any given instant, one would expect that molecules will be scattered throughout the volume. Different molecules will be moving in different directions. Some molecules will be moving fast, and some will be moving slowly.

Suppose, at some particular instant, we focus on a molecule which is moving very much faster than any of its neighbors, and follow the path of this molecule for a short time. This molecule will collide with its neighbors as it moves, and because its neighbors are moving more slowly, it will tend to slow down and lose energy in these collisions. Its neighbors in turn will tend to speed up and gain energy. After a number of collisions, therefore, we can expect that our molecule will have slowed sufficiently that its momentum will become comparable to that of its neighbors, and it will no longer stand out as a fast-moving molecule.

If we focus instead on a molecule which is moving slowly relative to its neighbors then the opposite effect will occur. In collisions with its faster-moving neighbors, the slowly moving molecule will tend to pick up energy. After a number of collisions the molecule will have speeded up sufficiently that it also no longer appears exceptional when compared with its neighbors.

Thought experiments of this kind can be applied to more general situations also. Suppose for example that molecules in one half of the volume V are all moving relatively fast, while molecules in the other half of the volume are all moving relatively slowly. At the interface between the two halves of V, fast-moving molecules will be colliding with slow-moving molecules. The arguments of the previous paragraphs apply here, and we would expect that the slow-moving molecules at the interface will speed up as a result of these collisions while the fast-moving molecules will slow down. After a sufficient number of collisions, the distinction between fast- and slow-moving molecules will wash out, and the interface region will contain molecules which are moving at some common speed which is faster than the original slow-moving half volume and slower than the original fast-moving half volume. In effect, a third region will have been introduced into our system, and the original interface will have become two interfaces. The process can of course be expected to continue. After many more collisions we would expect this interface region to expand to fill the whole volume, and the original slow- and fast-moving regions to disappear and be replaced by a single region where all molecules are moving at the same typical velocity.

As long as sufficient time is allowed to elapse, we can expect that any irregularities in our system will tend to average out as a result of collisions. Individual molecules moving with large momentum relative to their neighbors will slow down. Individual molecules moving with small momentum relative to their neighbors will

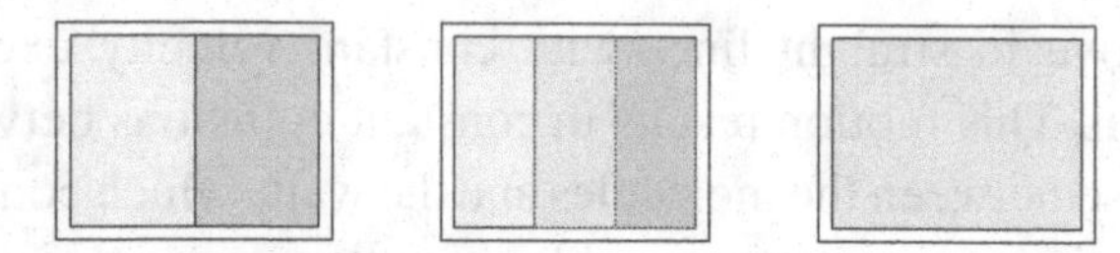

Figure 1.1 Sketch of averaging out of the density of molecules as a function of time once the separation between the two volumes is removed.

tend to speed up. Similarly, blocks of molecules with large average momentum relative to neighboring blocks will slow, losing average momentum, and blocks moving with small average momentum relative to neighboring blocks will speed up, gaining average momentum. After this sufficient time has elapsed, therefore, we can expect that the system will appear homogeneous. Molecular momenta at any one location in the system will be, on average, the same as molecular momenta at any other location.

Momentum is only one of the properties which we can expect to average out over a long timescale. Other properties such as molecular densities should average also. If the system starts with a high density of molecules in one region and a low density in a second region, then after some time we can expect that molecules will migrate out of the high-density region into the low-density region until density also averages out.

After a sufficient length of time, we can expect therefore that our system will achieve a uniform state where the only important quantities of physical interest are just those quantities which describe the typical average values in the system. For a gas of molecules, these quantities should include the density of molecules at a location, and the average momenta of those molecules.

All of the arguments just presented presuppose that the system we are describing is left undisturbed for sufficient time so that this averaging process can complete. We have implicitly presumed that there is no external source of disturbance which, for example, continuously causes molecules at some points to speed up or slow down relative to their neighbors. One of the basic assumptions of classical thermodynamics is that it is possible to arrange the environment of a physical system is such a way that no such disturbances act on it. The system is then said to be *thermally isolated*. Our thought experiments imply that, as long as sufficient time is allowed to elapse, a thermally isolated system will rearrange itself in such a way as to remove all atypical dynamics. At this time the system is said to be in *thermal equilibrium*. Thermal equilibrium is the first fundamental concept in classical thermodynamics and can be defined as follows:

Definition 1.1 A system is in *thermal equilibrium* if the state of the system does not change under time-independent external conditions.

A consequence of our picture of thermal equilibrium is that systems in thermal equilibrium exhibit a uniform dynamics throughout. Since the dynamics is uniform, the important parameters needed to describe that dynamics can be encapsulated in a small number of variables, which directly or perhaps indirectly define things like the average velocity and density of particles in the system. Classical thermodynamics proposes that the different thermal equilibrium configurations of a system are distinct *thermodynamic states*. Each such state is then describable in terms of the values of a small number of *state variables*. These state variables (as we shall see later) are related to quantities such as average molecular momenta and densities. A *change of state* or *thermodynamic transformation* of a system then refers to a change in the system from one state of thermal equilibrium to a different state of thermal equilibrium. A change of state will always be signaled by a change in the value of one or more of the state variables which define the different equilibrium states.

The classical state variables which have been found experimentally to be sufficient for describing the equilibrium thermal properties of a gas of molecules are:

(1) The number N of molecules.
(2) The volume V in which the molecules are contained.
(3) The pressure P of the gas.

Pressure measures the force per unit area that the gas exerts on the surface of any object placed in the gas. It is determined by the rate at which molecules cross uniform area, and is related to the mass and average velocity of the gas molecules. In Système International (SI) units, the unit of volume is m^3. The unit of pressure is $\mathrm{N/m}^2$ or equivalently $\mathrm{kg/(m\,s^2)}$. Atmospheric pressure is of the order 10^5 $\mathrm{kg/(m\,s^2)}$ at sea level. Atmospheric pressure is usually quoted in bars where 1 bar equals 10^5 $\mathrm{kg/(m\,s^2)}$.

The state variables defining an equilibrium state are often called *macroscopic* variables of the system, since they give general information about the very large number of molecules which make a complete system. In contrast, variables which describe details of an individual molecule's dynamics, are called *microscopic* variables. Classical thermodynamics deals only in macroscopic variables, and almost exclusively with systems which are in thermal equilibrium, since state variables are only defined for such systems. A state in thermal equilibrium can be uniquely defined by specifying values for all of its state variables. What makes thermodynamics such a powerful tool is that the state variables do not depend on the history of how the system reached a given equilibrium state. Consequently one can choose a path ingeniously and in this way come to conclusions which would have been very difficult or even impossible to reach by following another path.

1.3 The zeroth law of thermodynamics

Consider now what can happen when two thermally isolated systems which are separately in equilibrium are brought into contact with each other. At the point or surface of contact, we expect that the molecules from both systems will interact with each other. These molecules may be completely free moving, or they may be bound into separate solid walls for the materials contained in each separate system. In either case, before contact, the systems are each separately in equilibrium, and the molecules at the boundaries of each system will be moving with momenta which are typical of their systems of origin. Once contact is established between the two systems, molecules from one system will start to collide with molecules from the second system. The averaging process which we have described as occurring within a system will now begin to occur between systems. The two systems start behaving like a single larger system, and so long as no further outside disturbances are added we can expect that the larger combined system will eventually smooth out its irregularities as before, and a new different thermal equilibrium will be reached for the combined system. *Thermal contact* is defined to be contact of systems, which allows this re-averaging process to occur.

In general, our intuition suggests that when two thermally isolated systems are brought into thermal contact as just described, an averaging process will result, and quantities such as average molecular momenta in each system will change as our averaging process proceeds. There is however a special case to consider, that occurs when two systems are brought into thermal contact, and no changes in the individual systems are observed. For this to happen, the average molecular velocities in the separate systems must already be correctly matched so that no change occurs when they are brought into thermal contact. This view enables us to generalize the concept of thermal equilibrium to include systems which are not originally in thermal contact with each other. Two systems which are thermally isolated and separately in thermal equilibrium are also in thermal equilibrium with each other if, when the systems are brought into thermal contact, no changes occur in the state variables describing the separate systems.

Having considered the possibilities which occur with two thermally isolated systems, the next step is to look at what happens with three or more thermally isolated systems. This leads to the zeroth law of classical thermodynamics:

The zeroth law Thermal equilibrium is transitive, i.e. if X, Y, and Z are three thermal systems, and if X is in thermal equilibrium with Y, and X is in equilibrium with Z, then Y is in thermal equilibrium with Z.

Within our molecular model, this law describes exactly what we would expect. Thermal equilibrium implies matching of quantities such as average molecular momenta between systems. If these are matched between X and Y, and between X

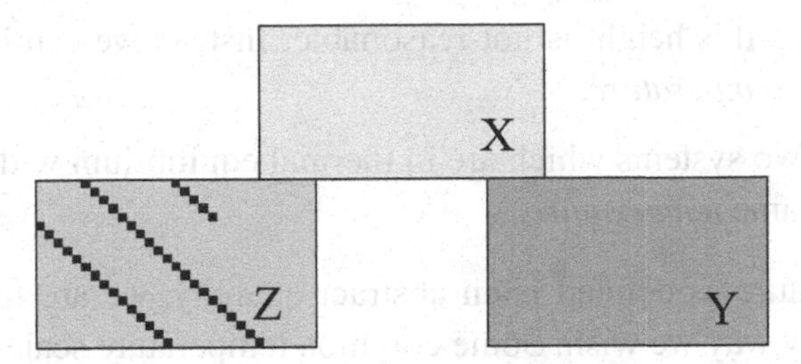

Figure 1.2 Transitivity of thermal equilibrium: If X and Y are in thermal equilibrium and X and Z are in thermal equilibrium, then Y and Z are in equilibrium.

and Z, then they must be matched between Y and Z also. As we shall immediately see, however, this law has a very profound consequence. In particular, it allows us to define a thermometer.

Systems in equilibrium, as we have seen, are described by just a few state variables. In our example these are pressure, volume, and number of particles. Different possible equilibrium configurations have different values for these state variables. So long as we only look at a single system these state variables are sufficient. If, however, we start to bring different systems into thermal contact, things get complicated. Thermal contact can change the state of a system, changing therefore the state variables describing that state. One of the fundamental pieces of information which we shall need to know is how state variables will change when systems are brought into contact. In principle, we can imagine doing a large number of experiments for all possible pairs of systems and studying what actually happens, for example, when a gas with one set of state variables is brought into thermal contact with a second gas with a different set of state variables. In practice, this is impossible. There are simply too many different possible systems, and too many ways for them to interact. The zeroth law comes to the rescue. Whenever we need to find how state variables change in a system in thermal contact, it is enough to consider the special case of interactions of the system with a special reference system. This reference system is the system X of the zeroth law and is normally called a *thermometer*.

One simple thermometer is a system where a column of mercury is confined in a long thin transparent tube. When this thermometer is brought into thermal contact with any system, the significant effect in the thermometer is that the height of the column of mercury changes until thermal equilibrium is established. This allows us to label different equilibrium states of a system by specifying the particular value of the height of the column of mercury for which equilibrium is established between the thermometer and the system. The mercury thermometer is only one of the possible thermometers that can be used. So directly indicating the particular equilibrium state by specifying this height is not reasonable. Instead we can introduce a new state variable, the *temperature*.

Definition 1.2 Two systems which are in thermal equilibrium with each other are said to have the same *temperature*.

Since temperature as defined is an abstract quantity, we are free to define its numeric value any way we wish. Some common temperature scales are defined by specifying a temperature value for water at its freezing point, and a temperature value for water at its boiling point. The Fahrenheit scale specifies the freezing point to be 32 °F, and the boiling point to be 212 °F. The Celsius scale specifies the freezing point to be 0 °C, and the boiling point to be 100 °C. Notice that the units used are completely arbitrary. We make an arbitrary decision that a particular system has a particular temperature when in one state, and a different temperature when in a different state. Other possible temperature values can then be defined by interpolating or extrapolating between these two reference values. For example a temperature of 50 °C can be defined as the temperature for which the column of mercury in a mercury thermometer is exactly halfway between its height at 0 °C and 100 °C. A thermodynamic system will have temperature 50 °C if a mercury thermometer in thermal contact with it has a column exactly this height. Clearly there is some ambiguity in these definitions. Happily, we will resolve this later in this chapter. The SI standard temperature scale is the Kelvin scale. We will define this scale explicitly later in this chapter. For the moment, we note that the Kelvin scale corresponds exactly to the Celsius scale except for an offset which is given by $0\,\text{K} = -273.15\,^\circ\text{C}$.

The zeroth law has allowed us to introduce a new state variable, the temperature. This variable is normally labeled T. Equilibrium states of our molecular gas are now described by four variables: N, V, P, and T. As we saw in the last section, N, V, and P are sufficient to describe the equilibrium state of a gas in isolation. We have now introduced a new state variable T which also specifies the equilibrium state. We thus expect T to be determined by the other state variables. The relationship which so determines T is called an *equation of state* and can be expressed in the generic form

$$f(T, V, P, N) = 0.$$

In particular, any substance with an equation of state of this type defines a thermometer.

For a sufficiently dilute gas at high temperature, the equation of state takes the explicit form

$$PV = NkT$$

where k is a constant of proportionality known as the Boltzmann constant. This form of the equation of state is known as the *perfect gas law*. For macroscopic systems, N is a very large number. The standard macroscopic unit of measure for molecule count is Avogadro's Number N_{A} given by

$$N_{\mathrm{A}} = 6.023 \times 10^{23}.$$

A mole of gas is defined to be that quantity of gas which contains N_{A} molecules. This quantity of gas will have mass of w grams, where w is the weight of an individual molecule expressed in atomic mass units. Using N_{A} we can rewrite the equation of state for a gas as

$$PV = nRT\,,$$

where $R = N_{\mathrm{A}}k$ is known as the gas constant, and $n = N/N_{\mathrm{A}}$ is the number of gram moles of the gas. The numerical values of k and R are

$$k = 1.38 \times 10^{-23}\ \mathrm{J/K} \qquad R = 8.314\ \mathrm{J/K\ per\ mol}$$

1.4 The first law, heat

Let us return now to consider what happens when two systems are brought into thermal contact. If they are in thermal equilibrium, that is, if they have the same temperature, then we know that no change of state will occur. If they are not initially in equilibrium then we expect both to change state. Since different equilibrium states are labeled by different temperatures, this change of state is usually (but not always!) indicated by a change in temperature. In this latter case, when thermal contact is established, the individual molecules will find that there is a momentum mismatch between them. Molecules in one system typically are moving slower than molecules in the second system. As the thermal contact continues, this mismatch begins to even out. Molecules in one system speed up on average, while molecules in the second system slow down on average. The immediate consequence, of course, is that a transfer of energy occurs between the two systems. Energy for a molecule in our simple example is just its momentum squared divided by twice its mass. If a molecule speeds up it gains energy, if it slows down it loses energy. Equally if all the molecules on average speed up, the energy of the system containing them increases, while if all the molecules on average slow down, the energy of the system containing them decreases.

The energy transfer which proceeds by molecular interactions of this kind is called *heat*. The symbol most often used to indicate heat in classical thermodynamics is Q. Adding heat to a system will change its internal energy, and will also change the equilibrium state of the system. As a result, when heat is added, at least some of the state variables describing the system will change. The generic case is at least that the temperature will change, and again generically this change in temperature

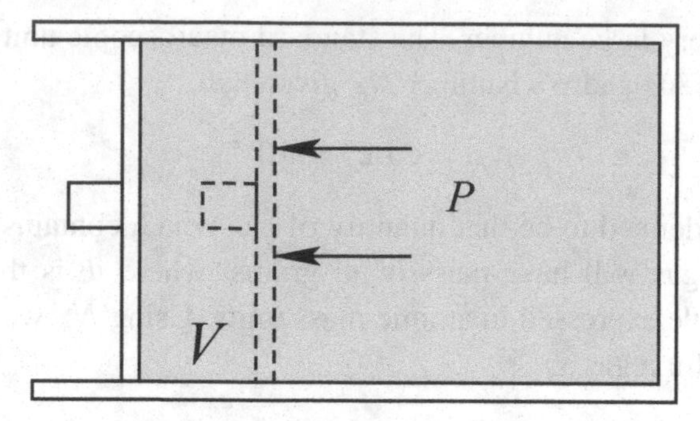

Figure 1.3 Executing work on a gas of pressure P by reducing its volume by ΔV.

is found to be proportional to the amount of heat. If ΔT is the change in temperature caused by the addition of an amount ΔQ of heat to a system, then we have

$$\Delta Q = c\Delta T.$$

The positive constant c here is the constant of proportionality and is called the *specific heat capacity* or simply *specific heat*. In general the value of c depends on the conditions under which the amount of heat ΔQ is added to the system. For gaseous systems, there are two special cases which are encountered. If ΔQ is added while keeping the volume fixed, then the specific heat is called *specific heat at constant volume*, and labeled c_V. If ΔQ is added while keeping the pressure fixed, then the the specific heat is called *specific heat at constant pressure*, and labeled c_P. One point to note is that these relations between temperature change and heat represent the generic case only. There are special cases where adding heat does not change the temperature or, even more startling, causes a negative change in temperature.

The classical unit used to measure heat is the *calorie*. This unit is defined to be the quantity of heat which must be added to one gram of water at 3.5 °C to raise its temperature by 1 °C. Heat is however a form of energy, so the modern SI unit to use for heat is the joule (J $\equiv$ kg m^2/s^2). One calorie is equal to 4.1855 J.

Heat is only one way to add energy to a thermodynamic system. An alternative way is to do work on the system. For a gaseous system one simple way to do work is to force it to contract in volume. This is illustrated in Figure 1.3. Since the gas is at pressure P, it exerts a force PA perpendicular to any boundary surface of area A which contains it. If the area of the wall which is moved in this figure is A, then the distance the wall moves is $\Delta V/A$. The work done, W, by the gas is therefore force times distance moved

$$\Delta W = (PA)(\Delta V/A) = P\Delta V\,.$$

Adding heat or doing work on a thermodynamic system changes its energy, and since energy is directly related to average molecular momenta this also will change. As a result the state variables describing a thermodynamic system will change if heat is added or work is done. There are two basic possibilities which we now need to distinguish. State variables suffice to describe a thermodynamic system only if that system is in equilibrium. During a transformation brought about by applying heat or doing work, the system may change so rapidly that it is unable to maintain equilibrium during the change. If, however, a thermodynamic transformation of a system proceeds at a rate that allows the system to maintain equilibrium at all times during the transformation, then the transformation is said to be *quasi-static*. Any thermodynamic transformation can be made quasi-static if it is carried out sufficiently slowly, since a system will always reach equilibrium provided it is given sufficient time. In a quasi-static transformation all state variables are defined throughout the transformation process. The alternative transformation possibility is non-quasi-static. In this case, the system changes too rapidly to allow it to maintain equilibrium during the transformation. Different parts of the system will exhibit different average properties during such a transformation, and state variables such as temperature and pressure will not be well defined during such a process.

Thermodynamic transformations can be further classified by the following definitions:

Definition 1.3 A thermodynamic transformation is

(1) *adiabatic* if no heat is added to the system during the transformation, i.e. $\Delta Q = 0$,
(2) *isothermal* if the temperature T remains constant during the transformation. Since T is only defined for systems in equilibrium, isothermal transformations must be quasi-static,
(3) *reversible* if it is quasi-static, and if the transformation can be reversed in time when the external conditions which drive the transformation are reversed in time,
(4) *cyclic* if the state variables describing the final state are the same as those which described the initial state.

The law of the conservation of energy valid in mechanics continues to hold for thermodynamic transformations. As we have seen on the basis of an atomic model of matter there are two basic ways to change the energy of a thermodynamic system, adding heat or doing work. There are also a multiplicity of different ways in which a transformation can proceed: quasi-statically, adiabatically, etc. We expect from our model however that energy should be conserved in all of these possible methods and mechanisms of energy transfer. This is embodied in the first law of classical thermodynamics:

The first law The internal energy U of any thermodynamic system is a state variable, which means that for any infinitesimal transformation δQ, δW

$$dU = \delta Q - \delta W$$

is an exact differential.

In words, this equation states that the change in energy of a system is the heat added minus the work done by the system. In any thermodynamic transformation, the total energy of the system plus environment is conserved and the change in the internal energy $\Delta U = U(\mathrm{B}) - U(\mathrm{A})$ is independent of the transformation that connects the two equilibrium states A and B.

An immediate corollary of the first law is the impossibility of a *perpetuum mobile*, i.e. a transformation for which $\Delta U = \Delta Q = 0$ but $\Delta W \neq 0$, where ΔQ is the heat added to the system, and ΔW is the work done by the system.

1.5 The second law, entropy

As far as the first law is concerned there is no restriction on the possibility of transferring energy from one form into another provided the total amount of heat energy is equal to the total amount of work energy in the process. It is experimentally observed that work can be totally converted into heat and that heat can flow, with no restrictions, from a hot body to a cold one. However, the reverse process of converting heat into work or transfer of heat from a cold object to a hot object does not occur. We have already described how interactions proceed for two systems in thermal contact. We expect that the system with the relatively slowly moving molecules will find that its molecules will speed up, and it will gain energy. The system with the relatively fast-moving molecules on the other hand, will find that its molecules tend to slow down and it will tend to lose energy. In other words, the cold system, with slow-moving molecules, heats up. The hot system, with fast-moving molecules, cools down. What we do not expect, however, is the opposite case with the hot system heating further and gaining energy at the expense of the cold system cooling further and losing energy.

It is a remarkable achievement in the development of classical thermodynamics that this qualitative picture describing our expectation of how complex systems interact can be formulated precisely into the second law which we will now describe.

The second law deals with fundamental properties of a system in thermal contact with other systems. In order to abstract the details of these other systems we introduce the concept of a *heat reservoir* or *heat bath* at temperature T. This is defined to be a system from which heat can be removed or added without changing

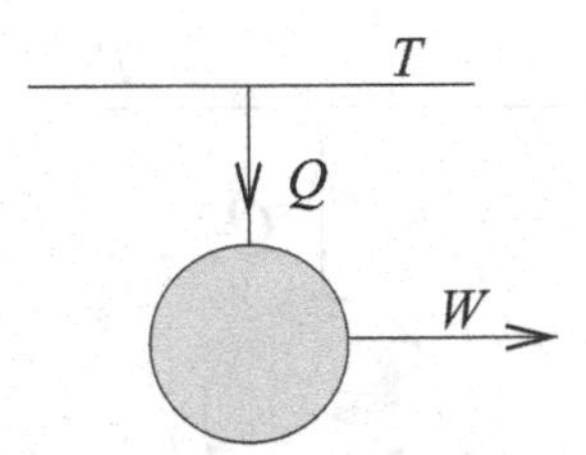

Figure 1.4 An engine which transforms heat entirely into work.

its temperature. The basic idea here is that we wish to concentrate on a given system which interacts with its environment by exchanging heat. In principle, this interaction can change the environment as well as the given system. We are, however, not interested in those cases where the environment changes significantly since we would then have to be specific about the environment. Thus we define an idealized environment which has a constant temperature, and an infinite capacity to absorb or provide heat for our system without appreciable change. Physically, we imagine that a reservoir is a very large thermodynamic system relative to the system in which we are interested. Amounts of heat transfer which have significant effects on our system are then too small to significantly alter the state of the reservoir.

A system interacting with its surroundings will, according to the first law, change energy as heat is added or work is done. We wish now to consider the special case where a system begins in some state, interacts in diverse ways with its surroundings and eventually returns to a new state which has the same energy with which the system began. This allows us to focus on the kinds of change that a system can undergo which are not simply due to change in energy. The second law addresses what is possible. There are different formulations of the second law; we state two of them.

The second law **Kelvin–Planck statement:** There exists no thermodynamic transformation whose sole effect is to extract a quantity of heat from a given heat reservoir and convert it entirely into work, Figure 1.4.
Clausius statement: There exists no thermodynamic transformation whose sole effect is to extract a quantity of heat from a given cold reservoir and transfer it to a hotter reservoir, Figure 1.5.

The key word in both statements is the word "sole." If we consider a gas whose internal energy is a function of temperature only then a reversible isothermal expansion of this gas converts heat totally into work. For such a transformation $\Delta U = 0$, which immediately gives $\Delta Q = \Delta W$, i.e. in this expansion, heat ΔQ has been totally converted to work ΔW. At the end of the process, however, the gas has

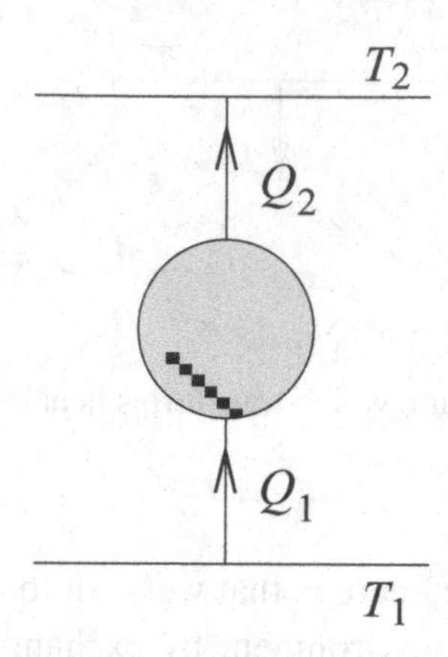

Figure 1.5 An engine whose sole effect is to transport heat from T_1 to T_2, $T_2 > T_1$.

expanded. Thus the transformation does not solely result in the conversion of heat into work. A gas expansion also occurs.

The second law has some very remarkable consequences. These include the existence of a universal temperature scale which is completely independent of any substance, and the existence of a completely new state variable for thermodynamic systems called *entropy*. We will establish these consequences in a series of problems and theorems.

Note the directionality contained in the second law. It does not allow heat to be converted to work in a cyclic manner but it does allow work to be converted to heat. Similarly it does not allow heat to be transferred from a cold to a hot reservoir in a cyclic manner but it does allow heat to be transferred from a hot to a cold reservoir.

An idealized arrangement in which to discuss the consequences of the second law is provided by a simple heat engine. This is a device which undergoes a cyclic thermodynamic transformation during which it extracts heat of amount Q_2 from a reservoir at temperature T_2, performs work of amount W, and ejects heat of amount Q_1 to a reservoir at temperature T_1. The temperatures involved are ordered so that $T_2 > T_1$. Since the transformation is by definition cyclic, the initial and final thermodynamic states of the engine are the same, and we have $\Delta U = 0$. The first law then gives

$$Q_2 - Q_1 = W.$$

Thus the engine converts heat $Q_2 - Q_1$ into work W. A simple heat engine is represented pictorially in Figure 1.6. A simple heat engine in which the cyclic thermodynamic transformation involved is reversible is called a *Carnot engine*.

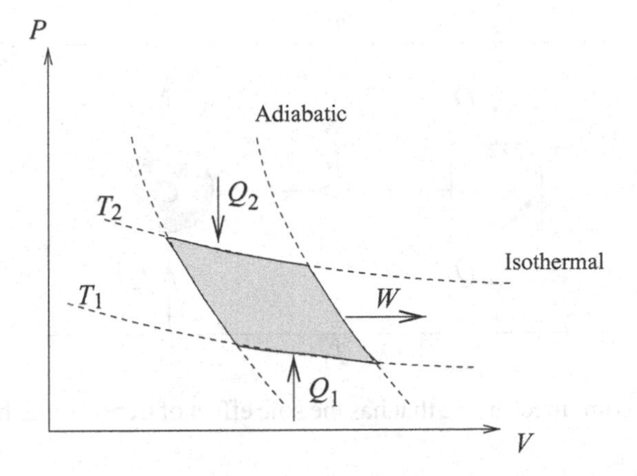

Figure 1.6 Carnot cycle.

The *efficiency* of a heat engine is the measure of how well it converts heat into work. It is defined as the ratio, η, of work performed to heat extracted from the reservoir

$$\eta = \frac{W}{Q_2} = \frac{Q_2 - Q_1}{Q_2} = 1 - \frac{Q_1}{Q_2} .$$

Note that the first law requires that $W \leq Q_2$. The maximum efficiency possible occurs when the engine ejects no heat, and converts all extracted heat into work. In this case $W = Q_2$, $Q_1 = 0$, and $\eta = 1$. The Kelvin–Planck statement of the second law expressly forbids this possibility, however, and therefore requires that all heat engines must have $\eta < 1$.

We leave it as a problem to the reader (see Problem 1.4) to show that the Kelvin–Planck statement and the Clausius statement of the second law are equivalent and continue by describing a fundamental result that follows directly from the second law and which states that

Theorem 1.1 No simple heat engine is more efficient than a Carnot engine.

Proof. Consider a simple engine E which absorbs heat Q_2 from a reservoir at temperature T_2 and rejects heat Q_1 to a reservoir at temperature T_1 with $T_2 > T_1$. This engine performs work $W_E = Q_2 - Q_1$. Its efficiency is, by definition,

$$\eta_E = \frac{Q_2 - Q_1}{Q_2} = \frac{W_E}{Q_2} .$$

Now consider a Carnot engine C operating between the same two reservoirs which does the same amount of work $W_C = W_E \equiv W$ by absorbing heat Q'_2 at T_2 and

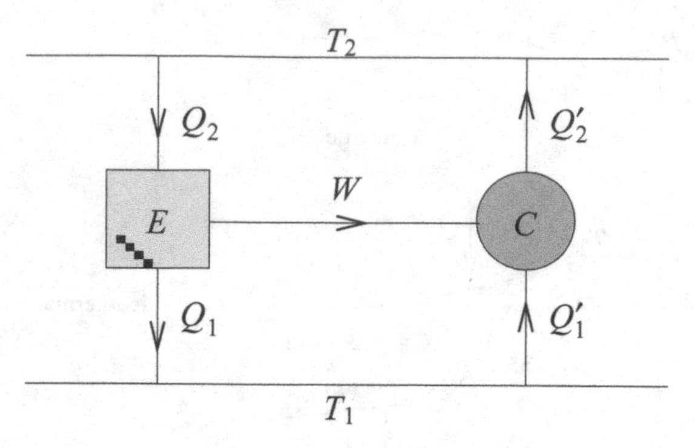

Figure 1.7 A combined engine that has the sole effect of transporting heat between T_2 and T_1.

ejecting heat Q_1' at T_1. The efficiency of the Carnot engine is

$$\eta = \frac{Q_2' - Q_1'}{Q_2'} = \frac{W}{Q_2'}.$$

Suppose now that the Carnot engine is the less efficient, so that $\eta_E > \eta$. This implies

$$\frac{W}{Q_2} > \frac{W}{Q_2'} \Longrightarrow Q_2 < Q_2'.$$

We now show that this is in contradiction to the second law. To establish this result we run the Carnot engine in reverse so that it now absorbs Q_1' at T_1 and rejects Q_2' at T_2. The combined effect of one cycle of the engine E and the Carnot engine C running in reverse is to absorb net heat $Q_1' - Q_1$ from the reservoir at temperature T_1, eject net heat $Q_2' - Q_2 > 0$ to the reservoir at T_2, and perform net work $W_E - W_C = 0$. The sole effect, therefore, is to transfer $Q_2' - Q_2 > 0$ to the reservoir at temperature T_2 (see Figure 1.7). This heat must come from the reservoir at temperature T_1, since no net work is performed by the combined system. We have arrived therefore at a violation of the Clausius statement of the second law. We must therefore have $\eta_E < \eta$. Note that no violation occurs in this latter case, when the Carnot engine is the more efficient. We find then that $Q_2' - Q_2 < 0$, and heat is taken from the reservoir at the higher temperature rather than ejected to it. □

Corollary 1.2 All Carnot engines working between the same reservoirs, have the same efficiency.

Proof. Consider two Carnot engines C_1 and C_2 working between the same reservoirs. Since C_2 is a heat engine, we have according to the theorem $\eta_{C_1} \geq \eta_{C_2}$.

Similarly we find $\eta_{C_2} \geq \eta_{C_1}$. Thus

$$\eta_{C_1} = \eta_{C_2}. \qquad \square$$

When we introduced temperature in Section 1.3, we were careful to point out that the scales we defined for temperature were arbitrary, and that they depended on the particular physical system which was used as a thermometer. A Carnot engine represents a physical system, and therefore can be used as a thermometer. The nice feature of Carnot engines which distinguishes them as thermometers, however, is the universality of efficiencies which we have just proved. This allows us to give a universal definition of a *thermodynamic temperature* which is independent of the substance used to measure it.

Definition 1.4 A Carnot engine defines an absolute temperature scale. In this scale, the temperatures of two systems are related by connecting them with a Carnot engine. We have

$$\frac{T_1}{T_2} \equiv \frac{Q_1}{Q_2} = 1 - \eta.$$

where Q_2 is the heat absorbed by the Carnot engine from the system at temperature T_2, and Q_1 is the heat ejected from the Carnot engine at temperature T_1.

With this definition temperature has taken on a quantitative interpretation which goes beyond just distinguishing between hotter and cooler reservoirs. Of course, the numerical value of temperature in the absolute scale is not defined by the preceding considerations. In fact, this value is seen to be completely arbitrary, since our definitions have involved only ratios. In SI units one takes 273.15 K as the numerical value of the absolute temperature of the triple point of water. This choice fixes the remaining arbitrariness and defines the Kelvin absolute scale of temperature.

An immediate corollary that follows from this definition together with the second law is that

Corollary 1.3 No thermodynamic state with absolute temperature $T \leq 0$ is possible.

Proof. Observe that, by the Kelvin–Planck statement of the second law, we must have $\eta = 1 - Q_1/Q_2 < 1$ since this statement requires that at least some heat Q_1 has to be ejected by a cyclic engine. For temperatures measured with the absolute temperature scale, we therefore have

$$0 < \frac{T_1}{T_2} = \frac{Q_1}{Q_2} < 1.$$

Thus all systems which can be connected by Carnot engines must have absolute temperatures with the same sign, and since the triple point of water has a positive temperature in the absolute scale, all other systems must also have positive temperatures. A negative T_1 would lead to $\eta > 1$ and thus violate the second law of thermodynamics.

Note further that the inequality $T_1/T_2 > 0$ is a strict inequality. The possibility that $T_1 = 0$ is excluded. No thermodynamic system, therefore, can have a zero absolute temperature. All systems must have a strictly positive absolute temperature (see Chapter 12 for further discussion on this point). □

From our definition of the thermodynamic temperature it is clear that

$$\frac{Q_1}{T_1} = \frac{Q_2}{T_2}$$

holds for any Carnot process. Thus $\Delta Q/T$ is independent of which reversible path is chosen to transform two states into each other. This suggests that we introduce a new variable of state – the *entropy*.

Definition 1.5 The difference of the entropy $\Delta S \equiv S(\mathrm{B}) - S(\mathrm{A})$ is defined as

$$\Delta S \equiv \int_{\mathrm{A}}^{\mathrm{B}} \frac{\delta Q}{T},$$

where the integral is taken along the path of a reversible process.

Theorem 1.4 The entropy function is a state function, that is, it depends only on the state of a system and not on the way the state was constructed.

Proof. Consider a system undergoing a reversible thermodynamic transformation which starts in state A, then passes through a sequence of different states, $\mathrm{B}_1, \ldots, \mathrm{B}_N$, and finally returns to state A. We imagine that an engine E is driving this transformation, and that each of the intermediate states B_i have well-defined temperatures T_i, and are occupied for short but finite time intervals. The system will smoothly move between a series of different states during the transformation process and the states B_i approximately represent the state a system occupies during a short time interval. For each infinitesimal transformation we then have

$$\Delta S = \frac{Q_i}{T_i}.$$

Now, let us take the limit when the discrete set becomes a continuum. We find then that

$$\sum_i \frac{\Delta Q_i}{T_i} \to \oint \frac{\delta Q}{T}$$

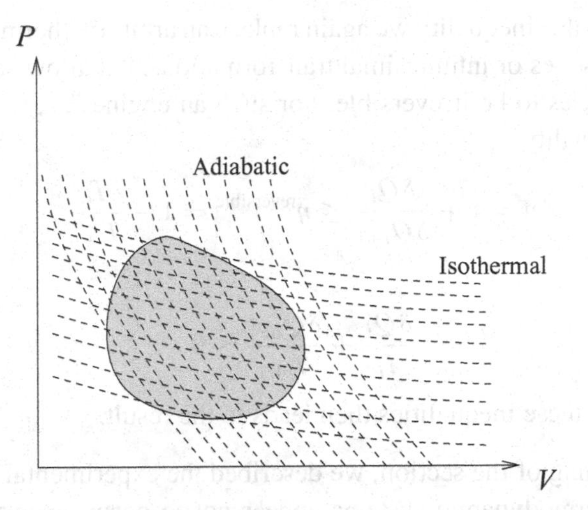

Figure 1.8 Any reversible transformation can be approximated by a series of Carnot cycles.

where $\oint$ denotes integration over a closed path. Recall that the transformation which the system undergoes starts and ends at A. If the closed path describes a reversible transformation we can approximate this contour to arbitrary accuracy by a composition of Carnot cycles as in Figure 1.8. But since ΔQ/T sums to zero for each Carnot cycle we have

$$\oint \frac{\delta Q}{T} = 0,$$

so that the integral,

$$S(\mathrm{B}) - S(\mathrm{A}) = \int_{\mathrm{A,reversible}}^{\mathrm{B}} \frac{\delta Q}{T}$$

is found to be independent of the particular reversible path joining states A and B, and so is a function of the path end points only. □

There is, in fact, a simple generalization of the theorem we just proved, known as

The clausius inequality For any arbitrary engine E which operates between a sequence of reservoirs at temperatures $T_1, T_2, \ldots, T_n$ absorbing (or ejecting) heat Q_1 at T_1, Q_2 at T_2, ..., Q_n at T_n the following inequality holds,

$$\sum_i \frac{Q_i}{T_i} \leq 0.$$

Proof. To prove this inequality we again replace an arbitrary thermodynamic transformation by a series of infinitesimal transformations, but allow some or all of the infinitesimal cycles to be irreversible. For such an engine $E_{i\to i+1}$ the second law implies the inequality

$$\eta = 1 + \frac{\delta Q_i}{\delta Q_{i+1}} \leq \eta^{\text{reversible}} = 1 - \frac{T_i}{T_{i+1}}\,.$$

Thus

$$\frac{\delta Q_i}{T_i} + \frac{\delta Q_{i+1}}{T_{i+1}} \leq 0\,.$$

Summing up all these inequalities then leads to the result. □

At the beginning of the section, we described the experimental fact that certain processes in thermodynamic systems appear not to occur, even though they are allowed energetically. For example, a hot system cools rather than heats further when it comes into contact with a cold system. The second law places a fundamental restriction on what is actually possible. As stated, the second law applies only to a very special case: that of an engine connecting two heat reservoirs. However, the introduction of the idea of entropy allows us to generalize the applications of the second law, and we can now state a fundamental theorem which applies to all systems and all thermodynamic transformations.

Theorem 1.5

(1) The entropy for an isolated system undergoing an irreversible change always increases. The entropy is thus maximal at equilibrium.
(2) S is a concave function of (U, V, N) (Figure 1.9).

Proof. Consider an arbitrary irreversible thermodynamic transformation, t_{I}, which transforms a system from a state A into a state B. Find a reversible thermodynamic transformation, t_{R}, which transforms the system back from B to A. Note that we assume that such a reversible transformation is always possible to find. By the Clausius inequality, we have

$$\int_{\text{A},t_{\text{I}}}^{\text{B}} \frac{\delta Q}{T} + \int_{\text{B},t_{\text{R}}}^{\text{A}} \frac{\delta Q}{T} \leq 0.$$

The contribution to this inequality coming from the integration over the reversible path t_{R} gives just the change in entropy between A and B, and we find therefore,

$$S(\text{B}) - S(\text{A}) \geq \int_{\text{A},t_{\text{I}}}^{\text{B}} \frac{\delta Q}{T}$$

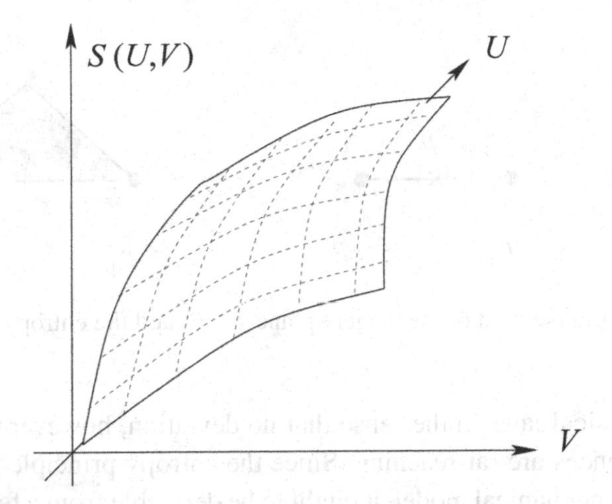

Figure 1.9 Sketch of the entropy function for fixed N.

If the system is thermally isolated as it undergoes the irreversible transformation t_{I}, then we have $\Delta Q = 0$ during the transformation, and so

$$S(\mathrm{B}) - S(\mathrm{A}) \geq \int_{\mathrm{A},t_{\mathrm{I}}}^{\mathrm{B}} \frac{\delta Q}{T} = 0$$

which gives

$$S(\mathrm{B}) \geq S(\mathrm{A}).$$

To prove the second part of the theorem we consider a pair of systems which are separately in thermodynamic equilibrium. If we now take a fraction λ of one system $(0 \leq \lambda \leq 1)$ and bring it in thermal contact with a portion $(1-\lambda)$ of the second system such that they are thermally isolated from the rest of the world the state of maximum entropy will occur when the two subsystems are in equilibrium with each other. Thus according to our above result

$$\begin{aligned} S(\lambda U_1 + (1-\lambda)U_2, \lambda V_1 + (1-\lambda)V_2, \lambda N_1 + (1-\lambda)N_2) \\ \geq \lambda S(U_1, V_1, N_1) + (1-\lambda)S(U_2, V_2, N_2), \end{aligned}$$

which is equivalent to the statement that S is a concave function in variables (U, V, N). □

This theorem represents the full consequence of the second law of thermodynamics. The essence of the second law is the statement that all adiabatic processes (slow or violent, reversible or not) can be quantified by a unique entropy function, S, on the equilibrium states of all macroscopic systems, whose increase is a necessary and sufficient condition for such a process to occur. It is one of the few really

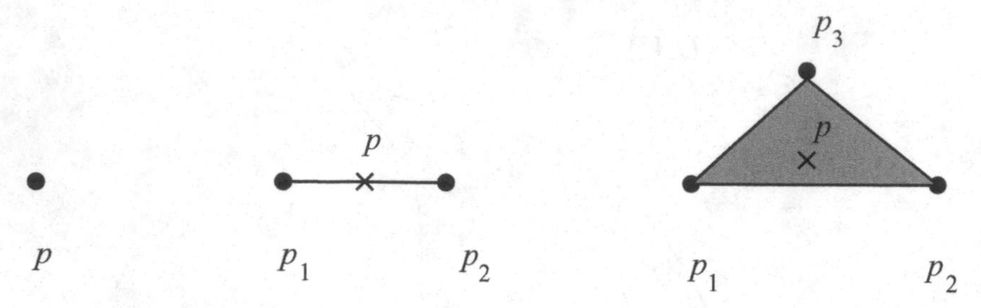

Figure 1.10 Intersection of the tangent plane $T_\Sigma(p)$ and the entropy surface Σ.

fundamental physical laws in the sense that no deviation, however tiny, is permitted. Its consequences are far-reaching. Since the entropy principle is independent of any statistical mechanical model, it ought to be derivable from a few logical principles without recourse to Carnot cycles, ideal gases and other assumptions about such things as "heat", "hot" and "cold", "temperature", "reversible processes", etc. Indeed, temperature is a consequence of entropy rather than the other way around, as we will see in the next section.

1.6 Gibbs phase rule

An important observation is that the entropy is not, in general, strictly concave. This means that equality is indeed possible in the last line of the proof of Theorem 1.5. As we will now see this property is related to the possibility of coexistence of more than one phase, such as ice and water. In order to elaborate on this claim in full generality we need to make use of two basic properties. The first is that each equilibrium state corresponds to a point on the entropy surface, Figure 1.9, which we will denote by Σ. The second property is that each point represents either a pure phase or a *unique* mixture of pure phases. If this second property were not satisfied then there would be no deterministic description of a thermodynamic state. We will therefore assume that this is a property of all physical systems.

To decide if a given point, p, on Σ corresponds to a pure phase we can consider the tangent plane $T_\Sigma(p)$, at the point p. Note that since $(\partial S/\partial U) = 1/T$ and $(\partial S/\partial V) = P/T$, the intersection $T_\Sigma(p) \cap \Sigma$ corresponds to constant temperature and pressure which are just the quantities which are typically held fixed during a phase transition. If the intersection $T_\Sigma(p) \cap \Sigma$ is a point, then p is a pure phase. More generally, if Σ is not strictly concave, then this intersection can be a higher dimensional *simplex*, i.e. a segment of a straight line or a triangle (see Figure 1.10).

In the case of a line we write $p = \lambda p_1 + (1-\lambda)p_2$, $0 \le \lambda \le 1$, where p_1 and p_2 are pure phases. In the case of a triangle we have $p = \lambda_1 p_1 + \lambda_2 p_2 + \lambda_3 p_3$, $\lambda_1 + \lambda_2 + \lambda_3 = 1$. That the intersection is necessarily of this form follows from the

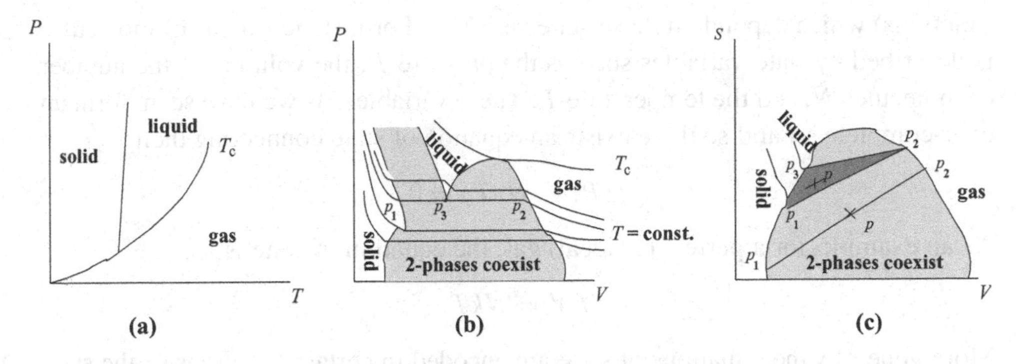

Figure 1.11 Phase diagram of a one-component fluid: (a) in the P, T plane; (b) in the P, V plane; (c) in the S, V plane projected along the U-axis. The shaded triangle in (c) parametrizes the mixture of three phases.

property that each equilibrium state which is not a pure phase is a unique mixture of pure phases. In particular, this would not be the case if the boundary of $T_\Sigma(p) \cap \Sigma$ were a circle.

As an illustration we consider the *phase diagram* of a one-component fluid (e.g. water) in the P, T plane, the P, V plane, and in the S, V plane in Figure 1.11. The line $\overline{p_1 p_2}$ in Figure 1.10c corresponds to the mixture of two phases, while the triangle p_1, p_2, p_3 describes a mixture of three phases. Note that in the P, V plane there is no unique decomposition of p into pure phases. In order to decompose p uniquely we have to work with the variables S and V.

Let us now consider a fluid consisting of n types of molecules, e.g. H_2O, $CO_2, \ldots$ In this case Σ is a higher-dimensional surface. We then consider the entropy per mole, $s(u, v, c_1, \ldots, c_{n-1})$, where $u = U/\sum_i N_i$, $v = V/\sum_i N_i$ and $c_i = N_i/\sum_i N_i$ are the concentrations. Thus Σ is a $n+1$ dimensional hyper surface. The intersection $T_\Sigma(p) \cap \Sigma$ is then a simplex of dimension $d \leq n+1$, representing $d+1$ coexisting phases. In particular, we conclude that at most $n+2$ phases can coexist simultaneously. Furthermore, assuming that k phases coexist, we define the *degeneracy* by the number, f, of variables that can be varied without changing the composition of phases. Since the dimension of Σ is $n+2$ we have

$$f \leq n + 2 - k.$$

This is the Gibbs phase rule.

1.7 Thermodynamic potentials

In the development we have followed to describe the zeroth, first, and second laws of thermodynamics we have defined thermodynamic state variables (and state functions) which depend on those state variables. For instance, a gas of molecules is described by state variables such as the pressure P, the volume V, the number of molecules N, and the temperature T. These variables, as we have seen, form an over-complete set and so there exists an equation of state connecting them,

$$f(P, V, N, T) = 0.$$

As an example, for a perfect (or ideal) gas, the equation of state is

$$PV = NkT.$$

More generally the equations of state are encoded in certain functions of the state variables such as the internal energy U and the entropy S. The internal energy is defined on any state of the system, whether in equilibrium or not. The entropy is defined only on equilibrium states.

For the following we will confine our attention only to systems whose equilibrium states are defined by the thermodynamic variables P, V, T and N. These comprise a wide variety of solid, liquid, and gaseous systems, and so are not particularly restrictive. Since any equilibrium state can be defined by specifying values for P, V, T, and N, the internal energy and entropy must be functions of these variables,

$$U \equiv U(P, V, T, N) \quad \text{and} \quad S \equiv S(P, V, T, N).$$

Not all these variables are independent of course, since they are connected by the equation of state. Thus we are at liberty to choose any set of variables which uniquely define the different equilibrium states of the system. All other thermodynamic functions can then be considered as functions of the independent set we have chosen. For example, we can choose P, V, and N as the independent set, and we would find that T, U, and S would all be functions of these three independent variables. However the variables P, V, N, and T are not in any way special. The variables U and S are also thermodynamic, and we are at liberty to choose these also as members of our basic independent set. For example, we could choose T, S, and N as the independent variables, and we would find U, P, and V would be functions of these. There are clearly many different choices of variables available. As we will now describe, the choice of an appropriate set of variables depends on the external conditions imposed on the thermodynamical system we want to study.

Consider first the thermodynamic laws defining internal energy and entropy. The first law, when applied to an infinitesimal transformation, states

$$\mathrm{d}U = \delta Q - \delta W = \delta Q - P\mathrm{d}V,$$

where we have used the fact that the work done by a system undergoing a volume change is $\delta W = P\mathrm{d}V$. The form of the second law which is most useful to us now is the statement that entropy increases in any irreversible transformation. In differential form this is given by

$$\mathrm{d}S \geq \frac{1}{T}\delta Q.$$

For a reversible transformation, this becomes

$$\mathrm{d}S = \frac{1}{T}\delta Q.$$

If we combine these two equations, we have for a reversible thermodynamic transformation

$$\mathrm{d}U = T\mathrm{d}S - P\mathrm{d}V.$$

Since we have already assumed, in defining entropy, that all equilibrium states of a thermodynamic system can be connected by a reversible thermodynamic transformation, we now find that we can calculate the internal energy difference between any states by integrating this last equation along a reversible path. As we move along this reversible path, we find that U changes only when S or V changes. This implies that the natural viewpoint to adopt for the internal energy is that it is a function of S and V. So far we have not allowed the possibility that the number of particles in our system can change, and we expect that adding particles would also add energy, and that U must also change as N changes. The natural set of independent variables to choose when describing the internal energy is therefore

$$U \equiv U(S, V, N).$$

In addition we have from the equation for $\mathrm{d}U$ that,

$$\left(\frac{\partial U}{\partial S}\right)_{V,N} = T, \quad \text{and} \quad \left(\frac{\partial U}{\partial V}\right)_{S,N} = -P.$$

The subscripts V, N and S, N indicate the fact that the corresponding partial derivatives must be evaluated while keeping V, N fixed in the first case, and S, N fixed in the second case. We will denote the magnitude of the variation of U as the number of particles in the system varies by the chemical potential μ,

$$\left(\frac{\partial U}{\partial N}\right)_{S,V} = \mu,$$

leading to the final formula for the differential of U allowing for a variable particle number

$$\mathrm{d}U(S, V, N) = T\mathrm{d}S - P\mathrm{d}V + \mu\mathrm{d}N\,.$$

We summarize properties of the internal energy U as a function of S, V, and N by formulating the following theorem.

Theorem 1.6

(1) $U(S, V, N)$ is homogeneous, $U(\lambda S, \lambda V, \lambda N) = \lambda U(S, V, N)$.
(2) U is convex and is minimal at equilibrium.
(3) The derivatives of U with respect to S, V and N provide the complete set of equations of state.

Thus $U(S, V, N)$ encodes the complete information about the system at thermodynamic equilibrium. Such a function is called a *thermodynamic potential.*

Proof. The first property follows simply from the observation that U is an extensive state variable depending only on variables which are themselves extensive quantities. The second property follows from the concavity of S (see Problem 1.5). Finally the third property is clear from the form of the differential dU given above. □

Depending on the external conditions applied to the system, the variables (S, V, N) may not be best suited to describe the thermodynamics of the system. For instance instead of a thermally isolated system we may want to consider a system in contact with a thermal heat bath. Thus we seek a thermodynamic potential, $F(T, V, N)$ say, such that

$$\mathrm{d}F = \mathrm{d}F(\mathrm{d}T, \mathrm{d}V, \mathrm{d}N).$$

Recalling that $T = (\partial U/\partial S)_{V,N}$ we find that

$$\mathrm{d}(U - TS) = -S\mathrm{d}T - P\mathrm{d}V + \mu\mathrm{d}N,$$

so that

$$F(T, V, N) = U(S, V, N) - TS + \text{const.}$$

Setting the irrelevant constant to zero, F is just the Legendre transform of the internal energy U with respect to S. The function $F(T, V, N)$ is called the *Helmholtz free energy* and defines a thermodynamic potential equivalent to the internal energy U, but expressed in the set of variables (T, V, N). It is clear that $F(T, V, N)$ is a homogeneous function of the extensive variables (V, N). Furthermore, we have:

Lemma 1.7 For a system undergoing an infinitesimal transformation at constant temperature during which no work is done, the change in the Helmholtz free energy F satisfies

$$\Delta F \leq 0.$$

Proof. To see this we start from the differential form of Clausius' inequality

$$\Delta S \geq \frac{\Delta Q}{T}.$$

On the other hand we recall that from the first law we have

$$\Delta U = \Delta Q - \Delta W.$$

Combining these two expressions we have

$$\Delta Q = \Delta U + \Delta W \leq T \Delta S.$$

For changes at fixed temperature T

$$\Delta(U - TS) + \Delta W \leq 0.$$

For changes during which no work is done $\Delta W = 0$ and we have

$$\Delta F = \Delta(U - TS) \leq 0$$

as claimed. □

The equations of state, in turn, follow from the following identities. For a gas undergoing a reversible change, we have

$$S = -\left(\frac{\partial F}{\partial T}\right)_{V,N}, \quad P = -\left(\frac{\partial F}{\partial V}\right)_{T,N}, \quad \text{and} \quad \mu = \left(\frac{\partial F}{\partial N}\right)_{T,V}.$$

This follows simply by comparing

$$\mathrm{d}F = \mathrm{d}U - T\mathrm{d}S - S\mathrm{d}T = -S\mathrm{d}T - P\mathrm{d}V + \mu \mathrm{d}N,$$

and

$$\mathrm{d}F = \left(\frac{\partial F}{\partial T}\right)_{V,N} \mathrm{d}T + \left(\frac{\partial F}{\partial V}\right)_{T,N} \mathrm{d}V + \left(\frac{\partial F}{\partial N}\right)_{T,V} \mathrm{d}N.$$

We may also use the latter identity to express U in terms of S, that is

$$U = \frac{\partial}{\partial(1/T)}\left(\frac{F}{T}\right)_{V,N}.$$

Consequently, recalling the definition of c_V (see also Problem 1.2),

$$c_V = -T\left(\frac{\partial^2 F}{\partial T^2}\right)_{V,N}.$$

If, in addition to the temperature, the pressure is kept fixed then the appropriate thermodynamic potential is the *Gibbs free energy* G:

$$G(T, P, N) = U - TS + PV, \quad \mathrm{d}G = -S\mathrm{d}T + V\mathrm{d}P + \mu \mathrm{d}N.$$

For a system undergoing a transformation at constant temperature and pressure the change in the Gibbs free energy G is a monotonously decreasing function

$$\Delta G \leq 0.$$

Indeed, we have already seen that

$$\Delta(U - TS) + \Delta W \leq 0.$$

At constant pressure we then have $\Delta W = P\Delta V$ so that

$$\Delta G = \Delta(U - TS + PV) \leq 0,$$

as claimed. The Gibbs free energy plays an important role for phase transitions and chemical reactions. At fixed temperature and pressure the equilibrium condition of a system at the interface of two different phases is $\mathrm{d}G = 0$. Thus

$$\mu_1 \mathrm{d}N_1 + \mu_2 \mathrm{d}N_2 = 0$$

where μ_i and N_i are the chemical potential and number of molecules in each phase. Because of the conservation of the total number of particles $\mathrm{d}N_1 + \mathrm{d}N_2 = 0$, we conclude that equilibrium between two phases requires the two chemical potentials to be equal,

$$\mu_1 = \mu_2.$$

Similarly, if several phases are present, such as, for example, at the triple point for water, then equilibrium between any two phases then requires

$$\mu_1 = \mu_2 = \cdots = \mu_\nu$$

where ν is the number of phases. The number and types of phase that can coexist simultaneously in different systems is determined by the Gibbs phase rule derived in Section 1.6.

Finally, we may introduce a fourth thermodynamic potential, the *enthalpy* H,

$$H(S, P, N) = U + PV, \quad \mathrm{d}H = T\mathrm{d}S + V\mathrm{d}P + \mu\mathrm{d}N.$$

We leave it as an exercise to the reader to show that the enthalpy is a monotonously decreasing function in an adiabatic expansion with no work done.

To summarize, the functions S, F, G and H determine which transformations are possible for thermodynamic systems subject to various external conditions. In particular, for transformations in a thermally isolated system, S increases. This means the equilibrium configuration of a thermally isolated system maximizes S. For transformations where the temperature is kept fixed and no work is done, F decreases. This means the equilibrium state for such changes minimizes F. Similarly for changes which take place under conditions of constant temperature

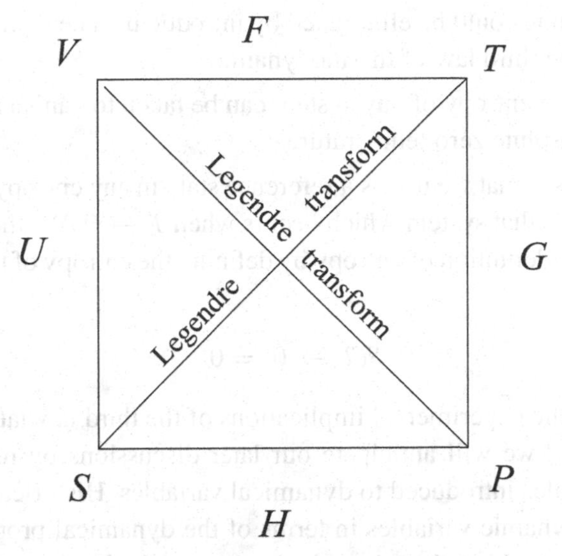

Figure 1.12 The corners of the square are labeled with the natural variables, V, T, S, P in which the thermodynamic potentials "between" the two variables, F, G, U, H are expressed. The diagonal lines indicate the Legendre transforms which relate the various natural variables.

and pressure, the equilibrium state minimizes G. We will use these fundamental properties throughout this book. A convenient way to remember how the different thermodynamic potentials are related is to present the potential as in Figure 1.12.

1.8 The third law

In this chapter we have so far reviewed classical thermodynamics conveniently described in terms of the zeroth, first, and second laws. The motivation for each of these laws was our view of the random nature of the motion of atoms and molecules in macroscopic matter. The arguments we presented were phenomenological. The formal quantitative development of thermodynamics is the subject of statistical mechanics which we begin to present in the next chapter.

We then introduced the important thermodynamic state functions, entropy S, the free energy F, and the Gibbs free energy G. We also noted that each of the laws of thermodynamics led to a state function, i.e. a function which depended only on the equilibrium properties of the macroscopic system and not on its previous history. The zeroth law led to the concept of temperature T. The first law introduced the internal energy function U. The second law introduced the entropy function S. In our development of the concept of entropy we were very careful to note that the evaluation of entropy of a state gave only an entropy difference between that state and some reference state. It was proposed by Nernst that the arbitrariness inherent in this reference state could be eliminated by introducing a new phenomenological law. This law is the third law of thermodynamics.

The third law The entropy of any system can be taken to vanish as the system approaches the absolute zero temperature.

This law proposes that we use as a reference state in any entropy calculation the equilibrium state of that system which occurs when $T \to 0$. We then eliminate the arbitrariness in the definition of entropy by defining the entropy of this $T \to 0$ state to be

$$S(T \to 0) = 0.$$

We will consider the experimental implications of the third law later in Chapter 7.

For the moment we will anticipate our later discussions by relating the thermodynamic variables introduced to dynamical variables. Historically the interpretation of thermodynamic variables in terms of the dynamical properties of atoms and molecules was achieved by introducing a model of a gas as a collection of molecules called the kinetic theory of gases. It was then shown that the temperature of a gas was related to the average kinetic energy of the molecules of the gas while the internal energy of the gas was related to the total energy of the system. What was not immediately clear was the dynamical interpretation of the entropy function. This was provided by Boltzmann. Boltzmann's interpretation of entropy also led to the conclusion that the third law had a dynamical interpretation only in terms of a quantum mechanical description.

We started this chapter by emphasizing the difference between the dynamical state of a system and its thermodynamic state. For a gas of molecules the dynamical state involved specifying $6N$ variables for $N > 10^{20}$ molecules. The thermodynamic state was described by a small number of variables such as the gas temperature, the gas pressure, and the gas volume. From our remarks regarding the interpretation of temperature and internal energy it follows that many dynamical states correspond to the same thermodynamic state since clearly a gas of molecules of fixed energy can be in many different dynamical configurations. For instance, a molecule moving along the x axis with momentum p has the same kinetic energy as a molecule moving with momentum p in the y direction. For a collection of non-interacting molecules these are two different configurations having the same energy. In studying the kinetic theory of gases, Boltzmann introduced a count, π, of the number of dynamical states that correspond to a given thermodynamic state and showed that the entropy S was related to π by the formula

$$S = k \ln \pi$$

where k is the Boltzmann constant.

At first sight application of Boltzmann's formula seems to impose a serious practical difficulty. In a system of molecules described by continuous coordinates and momenta, the dynamical states are infinite in number. Indeed for a system with $3N$ degrees of freedom there are ∞^{6N} number of states since each position and momentum components can take any real value. To determine S unambiguously using Boltzmann's relation seems hopeless.

A practical way out of this difficulty is to divide the phase space, that is the $6N$ dimensional space of coordinates and momenta, of the system into cells each of volume τ. If we suppose that the dynamical states corresponding to molecules in a given cell are equivalent, then the number of dynamical configurations corresponding to a given thermodynamic state can be calculated in units of τ, and S can be determined using Boltzmann's formula. However, if the size of τ is changed π changes and this introduces an ambiguity in S. There is no way of resolving this ambiguity using ideas of classical mechanics. The situation is, however, saved by quantum mechanics. The reason is that, in quantum theory, there is a canonical volume element $\tau_q = \hbar^3$ for phase space. When states are counted using this canonical volume, then we find

$$S(T \to 0) = 0 \quad \Longrightarrow \quad \pi = 1.$$

Thus, the third law of thermodynamics is simply a statement that the equilibrium thermodynamic state of a system at absolute zero corresponds to a unique dynamical state.

Problems

Problem 1.1 Calculate the work done by 10 grams of oxygen expanding from 2 liters to 10 liters isothermally at 300 K.

Problem 1.2 Show that for any thermodynamic system $c_V = (\partial U/\partial T)_{V,N}$ and $c_P = (\partial H/\partial T)_{P,N}$, where $H = U + PV$. Show furthermore that for one mole of an ideal gas $c_P - c_V = R$.

Problem 1.3 Calculate the dependence of the temperature of the atmosphere on the height above sea level. Hint: since air is a poor conductor of heat you may assume that air rises adiabatically in the atmosphere without exchange of heat.

Problem 1.4 Show that the Kelvin–Planck statement and the Clausius statement of the second law are equivalent.

Problem 1.5 Using the result that the entropy is a concave function of (U, V, N) show that the inverse function $U(S, V, N)$ is convex. Hint: show first that U is a monotonously increasing function of S.

Problem 1.6 Making use of the properties of the thermodynamic potentials show that $(\partial S/\partial V)_T = (\partial P/\partial T)_V$ and $(\partial S/\partial P)_T = -(\partial V/\partial T)_P$. These relations are two examples of the so-called *Maxwell relations* which are the set of identities obtained using the symmetry of the second derivatives of thermodynamic potentials.

Problem 1.7 Stefan–Boltzmann law: in classical electrodynamics one shows that the energy density and pressure of an electromagnetic field in a cavity are given by

$$\frac{E}{V} = \frac{1}{8\pi}(\mathbf{E}^2 + \mathbf{B}^2) \quad \text{and} \quad P = \frac{1}{24\pi}(\mathbf{E}^2 + \mathbf{B}^2) = \frac{1}{3}\frac{E}{V},$$

where **E** and **B** are the electric and magnetic fields respectively. The chemical potential vanishes since the particle number for electromagnetic radiation can vary. Show that the density of internal energy $u = U/V$ of the cavity at finite temperature is proportional to T^4.

Problem 1.8 Show that the internal energy of a substance which obeys the perfect gas law, $PV = RT$, is a function of temperature only and is independent of the volume.

Problem 1.9 Show that the enthalpy is a monotonically decreasing function during the process of an adiabatic expansion with no work done by the system.

Problem 1.10 Show that the equilibrium condition for the chemical reaction, $2H_2 + O_2 \rightleftharpoons 2H_2O$, at fixed temperature and pressure is given by $2\mu_{H_2} + \mu_{O_2} - 2\mu_{H_2O} = 0$.

Problem 1.11 Determine the temperature dependence of the saturated vapor pressure, that is of a system consisting of a liquid and its vapor in equilibrium.

Problem 1.12 Show that the third law of thermodynamics requires the thermal capacity c_P of a solid body at constant pressure to vanish at absolute zero.

Problem 1.13 Give an algebraic derivation of the Gibbs phase rule using the fact that the Gibbs free energy in the phase r can be written as

$$G_r = \sum_{i=1}^{n} \mu_{ir} N_{ir},$$

where μ_{ir}, the chemical potential of the substance i in the phase r, is a homogeneous function of T, P and N_{ir}, $i = 1, \ldots, n$. Generalize the phase rule to the case when chemical reactions between the different molecules are included.

Historical notes

The fact that heat is a form of energy was established by Mayer, Joule, and Helmholtz in the 1840s. Before that, heat was regarded as a fluid whose total amount was conserved. Indeed, using such a heat fluid theory, Carnot in 1824 came to a reasonably clear understanding of the limitations involved in the transformations of heat into work. Carnot's ideas form the basis of what is now the second law of thermodynamics. The modern formulation of the second law was presented by Clausius in 1850. Clausius succeeded in converting the qualitative work of Carnot into the form with which we are now familiar.

Boltzmann (1844–1906) decided to explore the consequences of an atomic theory of matter and understand thermodynamics in atomic terms. In this he was spectacularly successful. Boltzmann defined entropy as $\log \pi$ where π is the number of microstates accessible to the system. From this definition all of thermodynamics and statistical physics flows. Boltzmann also explained the second law of thermodynamics in statistical terms by introducing irreversibility as a postulate that characterized randomness. The specific assumption he made was criticized as violating the inherent reversibility of the laws of mechanics used by Boltzmann to describe atomic motion. Boltzmann strongly defended atomism against such criticisms from those who felt that the hypothesis of an atomic structure for matter was uncalled for and did not have any experimental basis. Those opposed to the atomic hypothesis were the "Energists". Boltzmann took his own life in 1906. It is said that he was upset by the continuing criticism of the Energists and felt that his ideas were not accepted by the community and that the battle against Energism was lost.

The third law, as it eventually became known, was proposed by Nernst in 1906. Nernst was involved in work on the measurement of specific heats at the time. The quantum nature of the third law was immediately realized.

Further reading

A self-contained account of classical thermodynamics can be found in E. Fermi, *Notes on Thermodynamics and Statistics*, Phoenix (1966). Classic texts well worth consulting are M. Planck, *Thermodynamics*, Dover (1945) and A. Sommerfeld, *Thermodynamics and Statistical Physics*, Academic Press (1956). For a more recent text with a detailed discussion of the principles of thermodynamics see J. R. Waldram, *The Theory of Thermodynamics*, Cambridge University Press (1985) and H. B. Callan, *Thermodynamics and an Introduction to Thermostatistics*, John Wiley (1985). We have chosen in this book to present thermodynamics using the idea of heat engines to get at the elusive concept of entropy. A more mathematical approach based on the notion of "adiabatically inaccessible states in the neighborhood of a given state" was formulated by the mathematician Caratheodory. An account of the second law of thermodynamics following Caratheodory can be found in W. Pauli's book on *Thermodynamics and Kinetic Theory of Gases*, MIT Press (1973) and also R. Giles, *Mathematical Foundations of Thermodynamics*, Pergamon (1964). A recent, mathematical account of the mathematical structure of the second law can be found in E. H. Lieb and J. Yngvason, *Notices of the Amer. Math. Soc.* **45** (1998). A personal but scholarly historical account of thermodynamics can be found in C. Treusdell, *Tragicomical History of Thermodynamics 1822-54*, Springer (1980) and, for the third law in particular, in J. Wilks, *The Third Law of Thermodynamics*, Oxford University Press (1961). The history of theories of the properties of matter starting with Boyle and ending with Landau and Onsager is given in S. Brush, *A Kind of Motion We Call Heat*, North-Holland, Volumes I and II (1983).

2

Statistical mechanics

2.1 Fundamental ideas

In this chapter we introduce the basic principles of statistical mechanics. As we saw in Chapter 1 the basic problem is to find a relation between functions of the microscopic variables of a macroscopic system and its macroscopic equilibrium thermodynamic variables. It is clear that, for instance, the position vector or the momentum vector are not appropriate observables in thermodynamics as they change with time and hence cannot themselves be related to equilibrium, time-independent, thermodynamic variables. The simplest possibility is to try and relate thermodynamic state variables to the conserved quantities of a microscopic system. To be specific we take as our system a gas of N non-interacting identical molecules of mass m. A conserved quantity for this system is the kinetic energy

$$E_N = \sum_{i=1}^{N} \frac{|\mathbf{p}_i|^2}{2m},$$

where $\mathbf{p}_i$ is the momentum of the ith molecule. One possibility could thus be to identify E_N with the internal energy function U of the system. Such an identification is indeed possible and leads to a formulation of statistical mechanics known as the micro canonical ensemble. We will return to examine this possibility later. For the moment we proceed somewhat differently. Although we know that E_N is defined we do not know the precise value which E_N takes. Under these circumstances it would be useful to know the probability, $P(E_N)$, that the system has energy E_N. Once $P(E_N)$ is known the average $\sum E_N P(E_N)$ could be identified with the internal energy function U and thus a bridge between the microscopic variables present in E_N and thermodynamics would be established. Transferring attention from the mechanical variable E_N to the probability function $P(E)$ for the system to have energy E is a key idea of statistical mechanics. The case where the energy of the system is fixed at some value E_N^0, for instance, corresponds to the special choice $P(E_N) = C^{-1}(E)\delta_{E_N,E_N^0}$ where

$$\delta_{E_N,E_N^0} = \begin{cases} 1 & \text{if} \quad E_N = E_N^0 \\ 0 & \text{if} \quad E_N \neq E_N^0 \end{cases}$$

and $C(E)$ is a phase space factor counting the number of states with energy E. Different formulations of statistical mechanics correspond to different choices for the probability. Justifying the choice made for the probability function associated with a macroscopic system, in essence, corresponds to trying to understand how thermal behavior can be obtained from a microscopic system governed by the laws of mechanics or quantum mechanics. We will consider this aspect of statistical mechanics briefly in Chapter 12. For the moment, we regard a choice for $P(E)$ as the basic postulate of statistical mechanics.

2.2 The canonical ensemble

We proceed to determine the form that $P(E)$ must take on general grounds. Let $\rho(E)\,\mathrm{d}^D q\,\mathrm{d}^D p$ be the probability to find the system in a small volume of phase space with energy E. The fact that ρ depends only on E but not on $\mathbf{q}$ and $\mathbf{p}$ is a consequence of the *postulate of equal a priori probability* which expresses the fact that all configurations with the same energy are equally probable. We furthermore require that the probability function $P(E)$ has the following property: if E_1, E_2 are two energy values which the system can have then, since energy is only defined to an additive constant E, we insist that

$$\frac{\rho\,(E_1)}{\rho\,(E_2)} = \frac{\rho\,(E_1 + E)}{\rho\,(E_2 + E)}.$$

The only function ρ, with this property is

$$\rho(E) \propto \mathrm{e}^{-\beta E}.$$

Thus

$$P\,(E) = \frac{1}{Z_N} C(E)\,\mathrm{e}^{-\beta E},$$

where

$$Z_N\,(\beta) = \sum_{\mathbf{q},\mathbf{p}} \mathrm{e}^{-\beta E} = \sum_E C(E)\mathrm{e}^{-\beta E}.$$

Here $\sum_E$ represents a sum over all the allowed energy values of the system of N particles. $Z_N\,(\beta)$ is known as the *partition function*. The form for $P(E)$ we have just obtained defines what is known as the *canonical ensemble* of statistical mechanics.

It was first introduced by Gibbs. Note that E is a microscopic variable while the constant β is a property of the thermal system. It determines the probability for a system to be in a state with given energy E. By introducing a physical requirement for the probability function $P(E)$ we have been able to determine its form, but $P(E)$ itself does not have a simple interpretation in thermodynamics. However, we can close this gap by making two identifications. The first is the natural identification of the internal energy U with the average energy $\langle E \rangle$ of the system, defined as

$$\begin{aligned}\langle E \rangle &\equiv \sum_E E \, P_\beta(E) \\ &= -\frac{\partial}{\partial \beta} \ln Z_N(\beta).\end{aligned}$$

It is clear that $\langle E \rangle$ depends on the parameter β. In order to establish a quantitative correspondence between mechanical and thermodynamic variables we need a thermodynamic interpretation for β. This is provided by the second identification. To justify this identification we recall the following identity derived in Section 1.7 of the last chapter

$$U = \left[\frac{\partial}{\partial\,(1/T)}\left(\frac{F}{T}\right)\right]_V,$$

where U is the internal energy. Comparing the two formulas above and using the identification $U = \langle E \rangle$ suggests that we identify β with $1/kT$ and relate the free energy F to $Z_N(\beta)$ by the formula

$$F = -\frac{1}{\beta} \ln Z_N(\beta)$$

with $\beta = 1/kT$. Note that the dimensional parameter k is required for both sides of the equation to have the same dimensions. Clearly, βE must be dimensionless, that is kT must have the dimensions of energy. We have thus established a quantitative correspondence between the microscopic definition of the energy E and the thermodynamic potentials U and F. This link is provided by the partition function which is the fundamental object in statistical mechanics. Note that the temperature, introduced via the identification $\beta = 1/kT$ is a global non-dynamical property of the system.

We have set up the link between the microscopic dynamical variables and thermodynamics as quickly as possible by introducing the notion of probability into the system. The validity of the prescription proposed above is justified by the many successful applications of statistical mechanics. Establishing this correspondence axiomatically based on the laws of classical (or quantum) mechanics is one of the major problems in statistical mechanics. We will briefly discuss different views on this problem in Chapter 12.

We should also emphasize that in relating the thermodynamic potential to the microscopic definition of the energy we have not made any use of the specific expression for the energy in terms of the phase space variables. The prescription described therefore applies to any system for which an expression for energy E is available. The form for $\rho(E)$ obtained on plausible, intuitive grounds is now elevated to be a postulate of statistical mechanics valid for any system for which a conserved energy E can be defined. The details of the microscopic system in question are encoded in the phase space factor $C(E)$. This will be illustrated for the perfect gas in the next section.

2.3 The perfect gas

In this section we illustrate the basic concept of statistical mechanics for the case of a perfect gas for which the thermodynamical properties are already known. In the process we will determine the last remaining unknown, that is the dimensional constant k that enters in the definition of the partition function. We consider a system of N non-interacting molecules in a cubic box of side L and volume $V = L^3$ with

$$E = \sum_{i=1}^{N} \frac{|\mathbf{p}_i|^2}{2m}.$$

We calculate the partition function $Z_N(\beta)$ and construct the free energy F for the system. To calculate $Z_N(\beta)$ we are instructed to sum over all possible energy values the system can have, multiplied by the canonical probability function, $P(E)$, which includes the phase space factor $C(E)$. In our example E is independent of the location of the molecules. However, in general we expect the molecules to interact and hence E will depend on $\{\mathbf{p}_i\}$, the set of momenta the molecules have, as well as the set $\{\mathbf{x}_i\}$ of their locations. Then the sum over all possible energy values for the system, multiplied by the phase space factor, must correspond to an integration over all possible momenta and position variables of the N molecules contained in a cubic box of side L and volume $V = L^3$. Thus

$$\sum_E C(E) = \frac{1}{h^{3N}} \int \mathrm{d}^3 p_1 \int \mathrm{d}^3 p_2 \ldots \int \mathrm{d}^3 p_N \int \mathrm{d}^3 x_1 \ldots \int \mathrm{d}^3 x_N.$$

The momentum integrals are over all of $\mathbf{R}^3$ whereas the integral over the positions of the molecules is, of course, restricted to the volume $V = L^3$. The factor h^{3N} is introduced to make the right-hand side dimensionless. In the quantum mechanical treatment this constant will be identified with Planck's constant, but for now it is just some constant with the dimension of an action. This $6N$ dimensional space of position and momenta coordinates is the *phase space* of the system.

Thus

$$Z_N = \sum_E C(E)\,\mathrm{e}^{-\beta E}$$
$$= \frac{1}{h^{3N}} \int \mathrm{d}^3 p_1 \ldots \int \mathrm{d}^3 p_N \int \mathrm{d}^3 x_1 \ldots \int \mathrm{d}^3 x_N\, \mathrm{e}^{-\frac{\beta}{2m}(\mathbf{p}_1{}^2+\cdots+\mathbf{p}_N{}^2)}\,.$$

As explained above, in the absence of interactions, the integral over the position of each molecule just produces a factor V^N so that

$$Z_N = V^N \left(\frac{1}{h^3} \int \mathrm{d}^3 p\, \mathrm{e}^{-\frac{\beta}{2m}\mathbf{p}^2} \right)^N .$$

Each p-integration is thus a simple Gauss integral. Recalling the text book formula

$$\int_{-\infty}^{\infty} \mathrm{e}^{-\frac{1}{2}ax^2} \mathrm{d}x = \sqrt{\frac{2\pi}{a}}\,,$$

we end up with the simple result

$$Z_N = V^N \left(\frac{2\pi m}{\beta h^2} \right)^{\frac{3N}{2}}, \quad \beta = \frac{1}{kT}$$

for the partition function of a gas of non-interacting particles contained in a volume V.

To continue we construct the free energy F for this system following the prescription given in the previous subsection, that is

$$F = -\frac{1}{\beta} \ln Z_N = -NkT \ln \left(\frac{V}{\lambda^3} \right),$$

where $\lambda = h/\sqrt{2\pi mkT}$. We thus have an explicit expression for the free energy as a function of the volume V and the temperature T. To recover the equation of state we recall that

$$P = -\left(\frac{\partial F}{\partial V} \right)_T = \frac{NkT}{V}\,.$$

Comparing this with the ideal gas equation of state from kinetic theory, $PV = NkT$, we then conclude that the constant of proportionality k introduced in the definition of β is the same as the Boltzmann constant and that a perfect gas can be regarded as a collection of N non-interacting molecules.

The quantity λ introduced above is known as the *thermal wavelength* of the particle. The name suggests that this scale is quantum mechanical in nature. To see how this comes about recall that the de Broglie wavelength of a particle of mass m and kinetic energy E_{kin} is given by

$$\lambda = \frac{h}{|\mathbf{p}|} = \frac{h}{\sqrt{2mE}}\,.$$

Replacing the kinetic energy E_{kin} by the typical thermal energy $E_{\text{th}} = 3kT$, we get

$$\lambda = \frac{h}{\sqrt{6mkT}} \simeq \frac{h}{\sqrt{2\pi mkT}} .$$

To give an example, the thermal wavelength of an electron at ambient temperature is $\lambda \simeq 10^{-9}$ m. It is amusing to note that, although we are presently working strictly with classical statistical mechanics, two intrinsically quantum mechanical constants have made their appearance, namely Planck's constant and the de Broglie wavelength. In the next two sections we will encounter more hints that statistical mechanics and quantum mechanics are intrinsically linked together.

2.4 The Gibbs paradox

It turns out that there is, in fact, a problem with our identification in the last section which we must resolve. This is known as the Gibbs paradox. Consider a gas consisting of N identical molecules at temperature T. Suppose we introduce a fictitious partition into the system so that there are N_1 molecules in volume V_1 on one side of the partition and N_2 molecules on the other side in volume V_2 where $N = N_1 + N_2$ and $V = V_1 + V_2$. From our expression for Z_N we have, for the free energies of N_1, N_2 molecules, the expressions:

$$F_1 = -\frac{1}{\beta} N_1 \ln\left(\frac{V_1}{\lambda^3}\right)$$
$$F_2 = -\frac{1}{\beta} N_2 \ln\left(\frac{V_2}{\lambda^3}\right).$$

We may regard this as the initial configuration of the system. For the final configuration we disregard the partition and consider the system as a collection of $N = N_1 + N_2$ molecules in volume $V = V_1 + V_2$. The free energy is then given by

$$F = -\frac{1}{\beta}(N_1 + N_2) \ln\left(\frac{V_1 + V_2}{\lambda^3}\right).$$

The change considered in this problem is reversible at fixed temperatures. We should have no change in the free energy (Figure 2.1). However, using the expressions given

$$\begin{aligned}\Delta F &= F - (F_1 + F_2) \\ &= -\frac{1}{\beta} N_1 \ln\left(\frac{V_1 + V_2}{V_1}\right) - \frac{1}{\beta} N_2 \ln\left(\frac{V_1 + V_2}{V_2}\right) < 0,\end{aligned}$$

that is the free energy has decreased. This is a disastrous result. The process considered was fictitious. We could have introduced further fictitious subdivisions $F_1 \to F_{11} + F_{22}$, $F_2 \to F_{21} + F_{22}$ which would lead to further reductions in F. This indicates that the value of F depends on the path followed in its construction

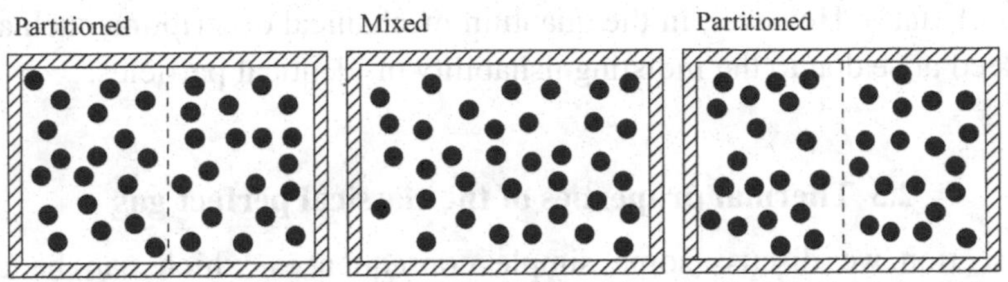

If entropy is a measure of disorder, we expect no increase in entropy for truly identical particles.

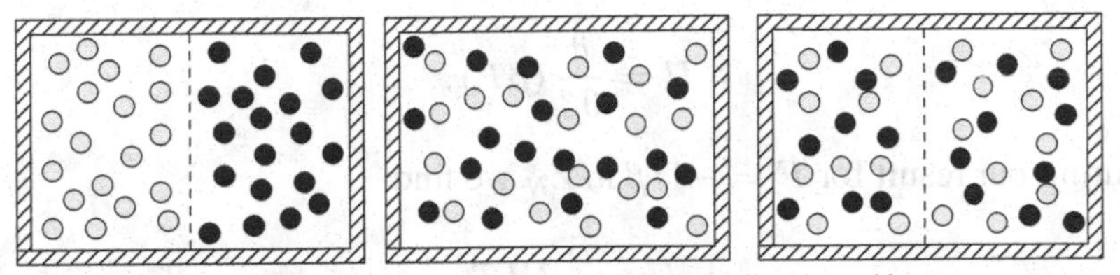

For non-identical particles there is an obvious increase in disorder and hence entropy.

Figure 2.1 Mixture of identical and non-identical gases. Only for non-identical gases the entropy increases.

which contradicts the basic definition of a state function. Remember a state function depends only on the equilibrium state of the system and not on its past history.

Gibbs resolved this paradox by noting that if, in the expression for F_1, F_2, and F the volume term could be replaced by specific volume terms, then $\Delta F = 0$, i.e. if V_1 is replaced by V_1/N_1, V_2 by V_2/N_2, $V_1 + V_2$ by $(V_1 + V_2)/(N_1 + N_2)$, where $V_1/N_1 = V_2/N_2 = (V_1 + V_2)/(N_1 + N_2)$, then $\Delta F = 0$. Such a replacement occurs if the expression for Z_N is slightly modified. Namely if we take

$$Z_N = \frac{1}{N!h^{3N}} \int \mathrm{d}^{3N} p \int \mathrm{d}^{3N} q \, \mathrm{e}^{-\beta \sum_i E_i}$$

and use Stirling's approximation

$$N! \simeq \sqrt{2\pi}\, N^N \,\mathrm{e}^{-N},$$

we indeed end up with

$$Z_N = \left(\frac{V}{\lambda^3 N}\right)^N \frac{\mathrm{e}^N}{\sqrt{2\pi}}.$$

At this stage this modification is purely ad hoc, and is called proper *Boltzmann counting* of states. However, in the quantum mechanical description such a factor does indeed arise due to the indistinguishability of identical particles.

2.5 Thermal properties of the classical perfect gas

In this section we discuss some simple consequences which can be derived straightforwardly from the partition function of the perfect gas. We begin by evaluating the internal energy and the entropy of the non-relativistic, perfect gas. As explained at the beginning of this chapter we have

$$U = \frac{\partial}{\partial \beta}(\beta F).$$

Substituting our result for $F = -1/\beta \ln Z_N$ we find

$$U = \frac{3}{2} NkT.$$

This is an example of what is known as the *equipartition of energy*. In thermodynamic equilibrium each degree of freedom contributes the same amount of energy, $1/2\,(kT)$, to the internal energy U. An immediate consequence of this result is that the specific heat at constant volume is constant. Indeed

$$c_V = \left(\frac{\partial U}{\partial T}\right)_V = \frac{3}{2} Nk.$$

Experimentally this result is reasonably well confirmed for noble gases at ambient temperature.

Similarly we obtain for the entropy

$$\begin{aligned} S &= -\left(\frac{\partial F}{\partial T}\right)_V \\ &= Nk \ln V + \frac{3}{2} Nk \ln\left(\frac{2\pi mkT}{h^2}\right) + \frac{3}{2} Nk - kN(\ln N - 1). \end{aligned}$$

Note that the expression for the entropy S has the property that as T vanishes, S goes to $-\infty$. This violates the experimentally valid requirement of the third law of thermodynamics which states that S vanishes as T goes to zero. Thus it is clear that a major revision of this formula is necessary where T is small (i.e. at low temperatures). As we shall see in Chapter 7, a proper resolution of this problem requires ideas from quantum theory.

To close this section, let us discuss how the equation of state is modified in a situation where a non-relativistic treatment is no longer justified. For this we note that as long as the Hamilton function $H(x, p)$ is independent of x, i.e $H(x, p) = E(p)$, where $E(p)$ is any arbitrary function of $(\mathbf{p}_1 \ldots \mathbf{p}_N)$, then the volume dependence of the free energy F is not modified. Indeed, the only modification in the partition function is that the thermal wave length λ will be replaced by an unknown expression $\lambda \to \lambda(m, \beta, N)$ where

$$(\lambda(m, \beta, N))^{-3N} = \frac{1}{N!h^{3N}} \int \mathrm{d}^{3N} p\, \mathrm{e}^{-\beta E(p)}.$$

which is independent of V. Thus the pressure

$$P = -\left(\frac{\partial F}{\partial V}\right)_T = \frac{NkT}{V}$$

is not modified. In particular, for a relativistic, perfect gas we have have

$$E(\mathbf{p}) = c \sum_i \sqrt{(\mathbf{p}_i)^2 + m^2 c^2},$$

so that the equation of state will be the same as for the non-relativistic gas.

2.6 Paramagnetic systems

A simple example of a system with an energy dispersion relation different from that of a perfect gas is that of a paramagnetic system of magnetic moments. Such a system can be regarded as a collection of N fixed magnetic dipoles each of magnetic moment μ. In the presence of an external magnetic field $\mathbf{B}$, these dipoles experience a torque tending to align them in the direction of the field. The energy of such an assembly of dipoles, neglecting their mutual interactions, can be taken to be

$$E = -\sum_{i=1}^{N} \mu_i \cdot \mathbf{B} = -\mu B \sum_{i=1}^{N} \cos\theta_i$$

where we take $\mathbf{B}$ to point along the z-axis and μ is equal to the absolute value of μ. The partition function for such a system is then given by

$$Z_N(\beta) = \sum_E C(E)\, \mathrm{e}^{-\beta E}$$

with E as above.

The sum-over-energy configurations would correspond to integrating over all angles that the dipoles can point to. This is the classical model. For a solid a quantum treatment might be more appropriate as quantum effects become important at short distances. Let us however for the moment consider the classical model.

Then

$$Z_N(\beta) = \prod_{i=1}^{N} \left[\int \mathrm{d}\Omega_i \mathrm{e}^{\beta\mu B \cos\theta_i} \right]$$

with $\int \mathrm{d}\Omega_i = \int_0^{\pi} \sin\theta_i \mathrm{d}\theta_i \int_0^{2\pi} \mathrm{d}\phi_i$ where (θ_i, ϕ_i) represent the direction of the ith dipole. This gives

$$Z_N(\beta) = \left[4\pi \frac{\sinh(\beta\mu B)}{\beta\mu B} \right]^N .$$

We now observe that the mean magnetization of the system in the direction of the external field is given by

$$M_z = < \sum_{i=1}^{N} \mu \cos\theta_i > = \frac{1}{\beta} \frac{\partial}{\partial B} \ln Z_N(\beta) = N\mu L(\beta\mu B)$$

where $L(x)$ is known as the Langevin function and is given by

$$L(x) = \coth x - \frac{1}{x}.$$

Let us consider the classical model in the high-temperature region, that is, as x tends to zero ($x = \beta\mu B$). Using

$$L(x) \to \frac{x}{3} - \frac{x^3}{45} + \cdots$$

for small x, we have $M_z = N\mu^2/3kTB$. A relevant quantity to compare with experiment is the magnetic susceptibility

$$\chi = \left(\frac{\partial M}{\partial B} \right) |_{B=0} = \frac{N\mu^2}{3kT}.$$

This result is in qualitative agreement with the experimentally observed magnetic susceptibility at high temperatures. At low temperature, however, quantum effects become important.

In order to get a taste of the predictions modified in a quantum mechanical treatment we write

$$\boldsymbol{\mu} = \mu \mathbf{L}$$

where $\mathbf{L}$ represents the angular momentum vector of a dipole in the external magnetic field $\mathbf{B}$ pointing in the z-direction. In quantum mechanics the angular momentum operator has the property that L_z has eigenvalues $m = -j\,, -j+1\,, \ldots\,, j-1\,, j$.

Writing $\boldsymbol{\mu} \cdot \mathbf{B} = \mu B m$ and introducing the parameter $x = \beta \mu B j$, the partition function is given by

$$Z_N(\beta) = \left[\sum_{m=-j}^{j} \mathrm{e}^{\frac{mx}{j}}\right]^N .$$

That is the integration over angles in the classical case is replaced by the discrete sum over the permitted values of m. Carrying out the sum we get

$$Z_N(\beta) = \left(\frac{\sinh\left(x + \frac{x}{2j}\right)}{\sinh\left(\frac{x}{2j}\right)}\right)^N .$$

This gives for the mean magnetic moment of the system

$$M_z = N\mu j\left[\left(1+\frac{1}{2j}\right)\coth\left(\left(1+\frac{1}{2j}\right)x\right) - \frac{1}{2j}\coth\left(\frac{x}{2j}\right)\right].$$

This simple example illustrates how the ideas of statistical mechanics can be used to analyze the thermal properties of different systems.

In the model we have just examined the quantum treatment involved simply taking into account the fact that the energy configurations are to be determined from the eigenvalues of a quantum operator. The significant difference between the quantum and the classical model was that, in the classical system, the range of possible energy configurations was infinite while, in the quantum case, the range was discrete and finite.

2.7 The one-dimensional Ising model

We can refine the simple model for paramagnetism considered in the last section by including an interaction between dipoles. The simplest model which includes such interactions is the model introduced by Ising (1924). In this model, the N elementary dipoles are represented by N spin variables S_i, $i = 1, \ldots, N$. Each of these spin variables can take one of two values $S_i = \pm 1$. The Hamiltonian for this system is

$$H = -g\sum_{i=1}^{N} S_i S_{i+1} - B\sum_{i=1}^{N} S_i .$$

The spins S_i interact with an external magnetic field B but now they also interact with each other. We imagine that the spins are arranged in a circle (Figure 2.2), and that any given spin interacts only with its nearest neighbors. The circle is closed by specifying that the neighbors of S_N are S_{N-1} and S_1. If we want this model

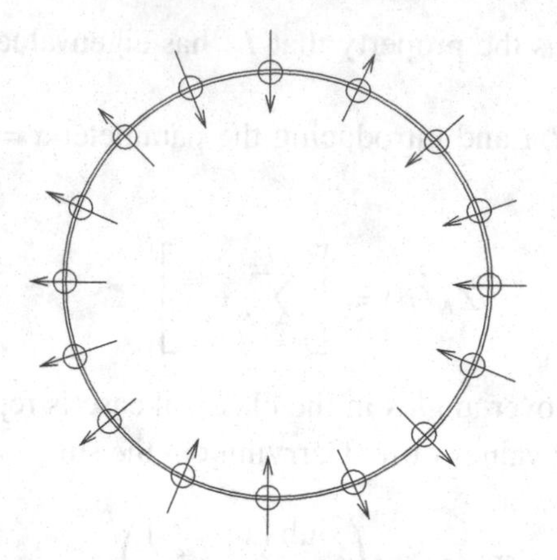

Figure 2.2 One-dimensional spin chain with periodic boundary conditions.

to represent a ferromagnet then we should choose $g > 0$ so that the energy of the system is lowered when spins point in the same direction.

Our basic problem is to evaluate the partition function as a sum over the set $\{c\}$ of all possible configurations

$$\begin{aligned} Z_N &= \sum_{\{c\}} \mathrm{e}^{-\beta H[c]} \\ &= \sum_{\{c\}} \prod_{i=1}^{N} \mathrm{e}^{\beta[gS_iS_{i+1}+\frac{B}{2}(S_i+S_{i+1})]} \end{aligned}$$

where we have replaced $\sum_i S_i$ by $\sum_i \left(\frac{S_i+S_{i+1}}{2}\right)$ which is valid in view of $S_{N+1} = S_1$. To continue, let us introduce a convenient notation for the exponential of the energy. We write

$$\mathrm{e}^{\beta[gS_iS_{i+1}+\frac{B}{2}(S_i+S_{i+1})]} \equiv \langle S_i|\mathsf{P}|S_{i+1}\rangle.$$

The matrix P introduced is an example of what is known as a transfer matrix. Let us write down P explicitly as a 2×2 matrix. We have

$$\begin{bmatrix} \langle +1|\mathsf{P}|+1\rangle & \langle +1|\mathsf{P}|-1\rangle \\ \langle -1|\mathsf{P}|+1\rangle & \langle -1|\mathsf{P}|-1\rangle \end{bmatrix} = \begin{bmatrix} \mathrm{e}^{\beta[g+B]} & \mathrm{e}^{-\beta g} \\ \mathrm{e}^{-\beta g} & \mathrm{e}^{\beta[g-B]} \end{bmatrix}.$$

The partition sum is then expressed as a product of the transfer matrices

$$\begin{aligned} Z_N &= \sum_{\{c\}} \langle S_1|\mathsf{P}|S_2\rangle\langle S_2|\mathsf{P}|S_3\rangle \ldots \langle S_N|\mathsf{P}|S_1\rangle \\ &= \sum_{S_1,S_2,\ldots,S_N} \langle S_1|\mathsf{P}|S_2\rangle\langle S_2|\mathsf{P}|S_3\rangle \ldots \langle S_N|\mathsf{P}|S_1\rangle \end{aligned}$$

where we note that the sum over configurations corresponds to a sum over all the values which the variables $S_1, S_2, \ldots, S_N$ can take. We have also used $S_{N+1} = S_1$. We now observe that $\langle S_i|\mathsf{P}|S_{i+1}\rangle$ can be regarded as the $\mathsf{P}_{S_i,S_{i+1}}$ matrix element of a matrix P. From the rules of matrix multiplication it follows that

$$Z_N = \sum_{S_1}\langle S_1|\mathsf{P}^N|S_1\rangle = \mathrm{Tr}\,\mathsf{P}^N .$$

Thus determining Z_N has been reduced to evaluating the trace of the Nth power of the transfer matrix P.

Now the trace of a matrix is an invariant under similarity transforms, i.e. $\mathsf{P} \to \mathsf{SPS}^{-1}$ and the symmetric matrix P can be diagonalized by a suitable similarity transform so that

$$\begin{aligned} Z_N &= \mathrm{Tr}(\mathsf{SPS}^{-1})^N \\ &= \lambda_+^N + \lambda_-^N \\ &= \lambda_+^N\left[1 + \left(\frac{\lambda_-}{\lambda_+}\right)^N\right]. \end{aligned}$$

The eigenvalues of this matrix are easy to determine. They are

$$\lambda_\pm = \mathrm{e}^{\beta g}\left[\cosh\beta B \pm \sqrt{\cosh^2(\beta B) - 2\mathrm{e}^{-2\beta g}\sinh(2\beta g)}\right].$$

Since $\lambda_-/\lambda_+ < 1$ we have $Z_N \to \lambda_+^N$, as $N \to \infty$. The free energy in this limit is thus given by

$$F = -\frac{N}{\beta}\ln\lambda_+ .$$

A simple calculation gives for the equilibrium magnetization M per spin

$$\begin{aligned} M &= -\frac{1}{N}\left(\frac{\partial F}{\partial B}\right) \\ &= \frac{\sinh(\beta B)}{\sqrt{\cosh^2(\beta B) - 2\mathrm{e}^{-2\beta g}\sinh(2\beta g)}}. \end{aligned}$$

This function is plotted in Figure 2.3. The relevant quantity to compare with experiment is the magnetic susceptibility

$$\chi = \frac{\partial M}{\partial B}\Big|_{B=0} = \beta\mathrm{e}^{2\beta g} .$$

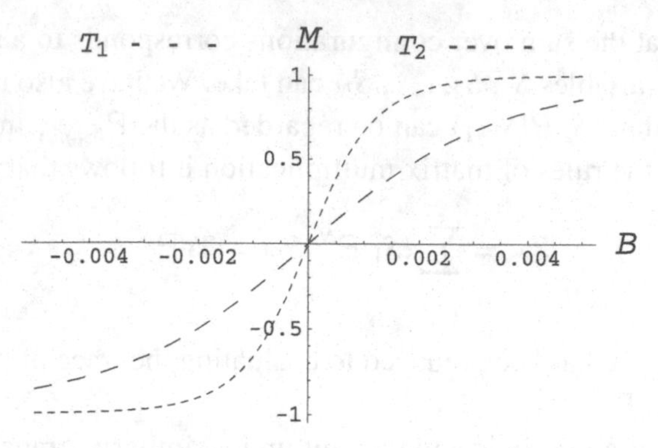

Figure 2.3 Magnetization as a function of B at two different temperatures, $T_1 > T_2$.

Thus the magnetic susceptibility diverges at zero temperature. This is to be expected: in the absence of thermal fluctuation all spins will align with the external magnetic fields however small this field is. On the other hand our result for the magnetization shows that there is no spontaneous magnetization, in the absence of a magnetic field, for any positive temperature. This is due to the fact that the increase in entropy, by flipping a single spin, wins over the cost in energy to do this flip. We will come back to this point in Chapter 13.

2.8 Applications in biology

The Ising model as well as other statistical models have found a variety of applications in molecular biology. To give a taste of this fascinating area we will model two such systems using the Ising model: hemoglobin which carries the oxygen in blood cells, and deoxyribonucleic acid (DNA) which is a very long molecule composed of millions of atoms and which contains the genetic code of living organisms. The role played by the Ising model in these systems is to model the *cooperativity* which appears to be responsible for some observed properties in these molecules. For instance it is observed experimentally that if a molecule of oxygen is already bound to a hemoglobin molecule then this will increase the probability of binding a second molecule. This is observed by measuring the percentage of oxygenated hemoglobin as a function of the partial pressure sketched in Figure 2.4. This property of hemoglobin is important to ensure that the molecule binds oxygen where it is abundant, i.e. the lungs, and releases it where it is rare.

To model this phenomenon we consider a single molecule with N binding sites for oxygen. In a realistic molecule, $N = 4$. To each site we associate an occupation number $n_i \in \{0, 1\}$, counting the number of oxygen molecules bound to this site. We can relate this number to the spin variable S_i in the Ising model by writing

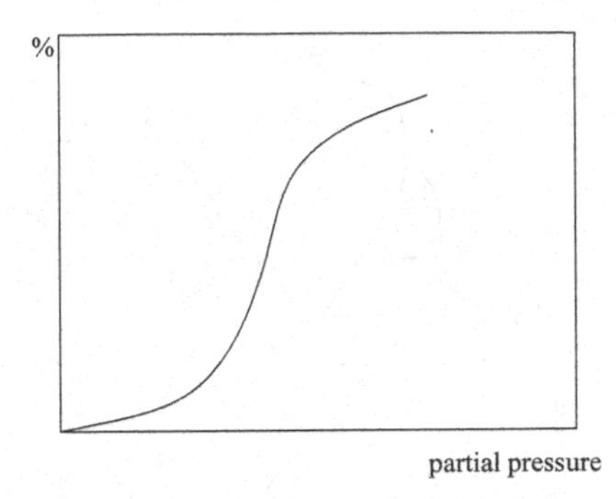

Figure 2.4 Proportion of oxygenated hemoglobin as a function of the partial pressure of oxygen.

$n_i = 1/2(1 + S_i)$. The state of the molecule is then completely described by the "spins" $S_1, \ldots, S_N$, and the number of oxygen molecules bound to the hemoglobin is given by $M = \sum_{i=1}^{N} 1/2(1 + S_i)$.

Let us first consider the case when there is no interaction between the different sites. We parametrize the probability for a site to be occupied or not by $p(S_i) = Ce^{BS_i}$, where B is a parameter to be determined and $C = (2\cosh B)^{-1}$ to ensure that the sum of the two probabilities add up to one. Up to the shift by $N/2$, the expression for the average number of oxygen molecules bound to the hemoglobin is thus identical with that for the total magnetization M, in the Ising model for $\beta = 1$ and in the absence of nearest neighbor interactions ($g = 0$), that is

$$M = \frac{1}{2C^N} \sum_{\{S_i\}} \sum_{i=1}^{N} (1 + S_i) \mathrm{e}^{B \sum_{r=1}^{N} S_r}$$
$$= \frac{N}{2}(1 + \tanh B)\,.$$

But what is the interpretation of B in this model? For this we notice that $z \equiv \mathrm{e}^{2B}$ is the ratio of the probabilities of a site being occupied and empty respectively. This suggests that we interpret z as a measure for the concentration of oxygen.

Next we take the interactions between neighboring binding sites for oxygen into account. We shall assume that only nearest neighbors interact. This assumption can be justified by the geometry and the chemical composition of the hemoglobin molecule. In addition we will assume periodic boundary conditions. If we parametrize the strength of the interaction, as in the Ising model, by the coupling constant g we can read off the partition function for this model from the Ising model result in Section 2.7, that is

$$Z_N = \lambda_+^N + \lambda_-^N\,,$$

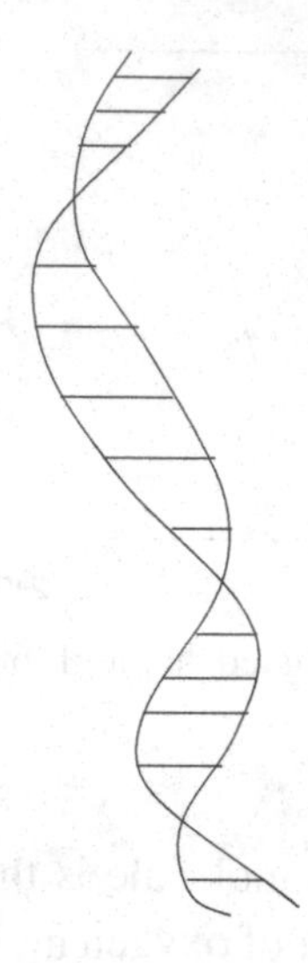

Figure 2.5 Sketch of a double-stranded DNA with hydrogen bonds between the two strands.

where λ_+ and λ_- are obtained from the corresponding expression in Section 2.7 after setting $\beta = 1$. The average occupation number M is thus given as before by

$$M = \frac{N}{2} + \frac{1}{2}\frac{\partial}{\partial B} \log Z_N \, .$$

The quantity of interest to compare with experiment is the filling fraction $f(z) = M/N$. In the limit of large g this function takes the simple form

$$f(z) = \frac{z^N}{1 + z^N} \, .$$

This result is known as the *Hill equation* and was originally introduced as a fit to the experimental curve for the filling fraction. The Ising model is thus in qualitative agreement with the experimentally measured oxygen saturation curve for hemoglobin.

Let us now turn to the deoxyribonucleic acid or DNA molecule. In the Watson–Crick model this molecule is represented as a double-stranded helix with the two strands attached to each other by hydrogen bonds as in Figure 2.5. The physical phenomena which we would like to describe is the "melting" of DNA, that is the breaking up of the hydrogen bonds as the temperature increases. Experimentally one observes a sharp transition between the two regimes as the temperature increases beyond $60 - 80\,^\circ$C and it is thought that cooperative phenomena play an important role in this process as well. Just like the occupation number in hemoglobin we can write the number of broken bonds as $M = \sum_{i=1}^{N} 1/2(1+S_i)$, where S_i is now the "spin" associated to each hydrogen bond. To continue we parametrize the probability for a given hydrogen bond to be broken or not by $p(S_i) = Ce^{BS_i}$, where B will now have to depend on the temperature T, since the probability of breaking a bond increases with T. If we furthermore assume that the interaction between different bonds is described by the Ising model interaction with strength g then the average number of broken bonds is a gain given by $M = (N/2) + (1/2)(\partial/\partial B)\log Z_N$ in complete analogy with the model for hemoglobin. However, here, since the number of bonds N is of the order of 10^6, we can take the large N result for the partition function in the Ising model, that is $\log Z_N = N\lambda_+$ where λ_+ is again obtained from the expression in the last section after setting $\beta = 1$. Thus

$$M = \frac{1}{2}\left[1 + \frac{\sinh B}{\sqrt{\cosh^2(B) - 2e^{-2g}\sinh^2(2g)}}\right].$$

Finally we need to make an ansatz for the temperature dependence of B. The simplest choice is $B = a(T - T^*)$ where a and T^* are fixed such as to fit the experimental result.

While this model for DNA goes some way to fit the experimental data it is nowadays generally discarded as a model for DNA melting. One serious shortcoming of the Ising model approximation is that it does not take into account the entropy associated with loops of broken bonds between segments of bound strands. There is a big phase space volume associated with the various geometric forms of such loops which is not taken into account in this approximation. A popular model which includes these configurations was proposed by Poland and Scheraga in 1966. This model improves considerably on the simple approximation presented here but still has some difficulties reproducing the first order transition (see Chapter 13) observed in the melting of DNA at the transition temperature.

Problems

Problem 2.1 Compute $c_V(T)$ and $U(T)$ for a relativistic gas of N non-interacting particles with rest mass m.

Problem 2.2 Determine the chemical potential $\mu(T, P, c)$, where $c = N/V$ for a perfect gas in the canonical ensemble. Hint: use Stirling's approximation $N! \simeq \sqrt{2\pi}N^N e^{-N}$, valid for large N.

Problem 2.3 Show that the equipartition law

$$\left\langle \frac{p_i^2}{2m} \right\rangle = \frac{1}{2} kT,$$

holds for all components p_i of a given particle in the classical canonical ensemble with Hamiltonian

$$H(\{\mathbf{q}_i\}, \{\mathbf{p}_i\}) = \sum_{i=1}^{N} \frac{|\mathbf{p}_i|^2}{2m} + V(\{\mathbf{q}_i\}).$$

Problem 2.4 Compute the partition sum for the one-dimensional Ising model with free boundary conditions in the absence of an external magnetic field, B. Calculate the correlation function

$$\langle s_i s_j \rangle = \frac{1}{Z_N} \sum_{\{s\}} s_i s_j \, \mathrm{e}^{-\beta H(\{s\})}.$$

Hint: introduce a position dependent coupling $g(i)$.

Problem 2.5 For a $d > 2$ dimensional lattice the exact solution for the Ising model is not known. One way to proceed is to make a *molecular field approximation*. For this we consider the Hamiltonian for a particular spin S_j, that is

$$H(S_j) = -BS_j - gS_j \sum_{i:\text{neighbors of } j} S_i .$$

We then introduce the *molecular field*

$$B_{MF} \equiv \frac{1}{N} \sum_{i=1}^{N} \langle S_i \rangle.$$

In the molecular field approximation one replaces the Ising model Hamiltonian by an effective Hamiltonian for S_j given by

$$H(S_j) = -(B + gnB_{MF})S_j$$

where n is the number of nearest neighbors. Find an implicit equation for B_{MF} and solve this equation for high temperatures $T \to \infty$. Calculate the magnetic susceptibility χ for high temperatures in the molecular field approximation. The result is known as *Curie's law*.

Problem 2.6 Critical temperature and surface tension
In order to estimate the critical temperature of the two-dimensional Ising model we approximate the Ising model as a *solid on solid model*. Concretely we compare configurations with boundary conditions as shown.

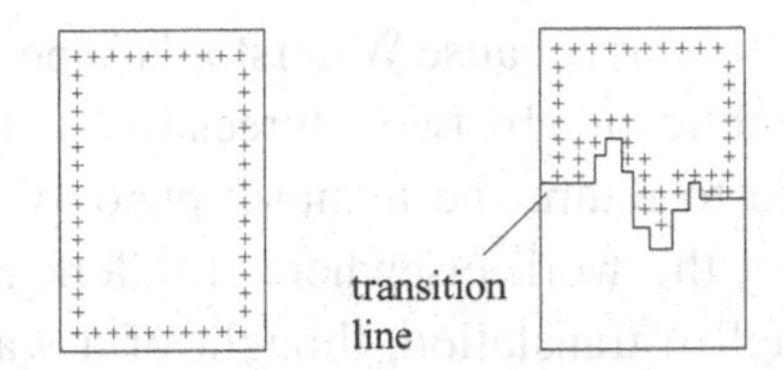

(1) Assuming only nearest neighbor interactions, what is the energy difference ΔE, between the two configurations for a generic transition line, parametrized by the height function h_i, measuring the elongation of the transition line from the straight line at the lattice site $i = 1, \ldots, N$?
(2) Compute the difference in the free energy of the two configurations by averaging over all possible transition lines weighted by $e^{-\beta \Delta E}$. Hint: for simplicity we assume that h_i can take all integer values between $-\infty$ and ∞.
(3) Find an implicit equation for the critical temperature T_c for which the *surface tension*,

$$\sigma(T) \equiv \frac{\Delta F}{N}$$

vanishes. Here N is assumed to be large.
(4) Compare the so-determined value for T_c with the critical temperature found in the *mean-field approximation* and with the *exact value*, $g\beta_c = g/k_B T_c = \frac{1}{2}\log(1+\sqrt{2})$.

Problem 2.7 Find the generalization of the Hill equation for finite nearest neighbor interaction g by evaluating the filling fraction for $N = 4$ and g arbitrary. The result reproduces a model suggested by Pauling in 1935.

Historical notes

The transition from the kinetic theory of gases, which tried to provide a mechanical basis for the thermal properties of the gas, to statistical mechanics represents a shift from mechanical models to a more abstract mathematical viewpoint. The problem of how a mechanical system with time-reversal symmetry properties by a process of averaging can become an irreversible system remains a major conceptual problem. Indeed some feel such an endeavor is pointless and impossible. A satisfactory resolution of the problem is still not available.

Historically, the kinetic theory of gases was only taken seriously after Clausius, in 1857, suggested a realistic model for a gas. With the help of the model and some ad hoc hypotheses, Clausius was able to make the model consistent with the then known experimental results. The previous work of J. J. Waterston (1845) although clear, written in an axiomatic style and full of insight, was generally unknown until a decade after his death. This was because Waterston's paper submitted to the Royal Society was rejected for publication by two referees who called it nonsense. Worse, the Royal Society refused to return the manuscript to Waterston, who was thus prevented from publishing the work elsewhere. J. Clerk Maxwell, after reading Clausius' paper in an English translation, thought of a way to refute the kinetic theory by deducing consequences which could be shown experimentally to be false. Maxwell thought he had such a result when he showed that a prediction of Clausius' model was that the viscosity of a gas should be independent of its density and should increase with temperature. This prediction of density independence was initially thought to be ruled out by experiments on the damping of pendulum swings in air. However, to convince himself, Maxwell, in 1865, decided to measure the effect and found to his surprise that the viscosity of air was indeed constant over a wide range of densities. This result was subsequently confirmed by others. The kinetic theory of gases had thus crossed a landmark. An unexpected prediction had been confirmed by experiment.

Before 1865 Maxwell had already made a significant contribution to the kinetic theory, converting it from a theory of a mechanical atomic nature to a statistical theory. In 1860 Maxwell had determined the statistical distribution of molecules of different velocities in a gas of temperature T, using abstract symmetry and probability arguments for the distribution function. Conceptually, this work of Maxwell was quite revolutionary. It showed the power of abstract symmetry arguments.

Further reading

There are numerous excellent text books which describe the basic concepts of statistical mechanics. A standard text with many applications is L. D. Landau and E. M. Lifshitz, *Statistical Physics*, Pergamon (1959). Another classic text, which includes a treatment of the Kramers–Wannier duality in two dimensions, is K. Huang, *Statistical Mechanics*, John Wiley (1987). For further reading about the interpretation of entropy and applications to electromagnetism in matter see R. Balian, *From Microphysics to Macrophysics; Methods and Applications of Statistical Physics*, I & II, Springer-Verlag (1991). L. E. Reichl, *A Modern Course in Statistical Physics*, Edward Arnold (1980), gives an encyclopedic treatment of statistical mechanics. A well-written graduate-level text with a comprehensive set of problems is R. K. Pathria, *Statistical Mechanics*, Butterworth-Heinemann (1996). A very readable text with many worked examples is C. Kittel and H. Kroemer, *Thermal Physics*, Freeman (1980). For another modern text see W. Greiner, L. Neise and H. Stocker, *Thermodynamics and Statistical Mechanics*, Pergamon Press (1995).

Further applications of the Ising model to biology can be found in C. J. Thompson, *Mathematical Statistical Mechanics*, Princeton University Press (1972). A good modern book describing the physics of soft matter, using tools of statistical mechanics is R. A. L. Jones, *Soft Condensed Matter*, Oxford Press (2002). For an advanced text on further exactly soluble models see R. J. Baxter, *Exactly Solved Models in Statistical Mechanics*, Academic Press (1982).

3
Variations of a theme

In our discussion so far we described the canonical ensemble of N identical particles or molecules. We found that from the canonical partition sum we can recover the free energy which is one of the thermodynamic potentials introduced in the first chapter. A natural question is whether there are other approaches to statistical mechanics which are in turn related to other state functions such as the entropy. In this chapter we will see that this is indeed the case. We will end up with the complete picture of how different probability measures in statistical mechanics are related to the various potentials in thermodynamics. In the process we will also uncover a simple statistical interpretation of the entropy function in thermodynamics.

3.1 The grand canonical ensemble

In the previous chapter we considered a statistical system with a fixed number N of identical molecules. We have argued that although the energy E of the system is a constant its precise value is not known. Hence we considered the probability $P(E)$ that the system had energy E and used it to relate the average value of the energy of the system (involving the microscopic properties of the system) to the macroscopic thermodynamic variable U, the internal energy. In this section we will generalize this approach to include a variable number of molecules, Figure 3.1. We note that the number of particles N in a volume, although a constant, is similarly not precisely known. In the spirit of what was done before we introduce a probability function

$$P(E_N, N) = \frac{1}{Z_\Omega} C(E_N, N)\, \mathrm{e}^{-\beta(E_N - \mu N)}$$

where

$$Z_\Omega = \sum_N \sum_{E_N} C(E_N, N)\, \mathrm{e}^{-\beta(E_N - \mu N)}$$

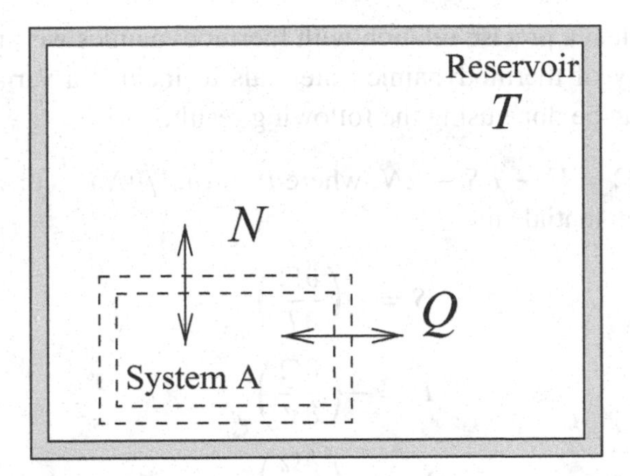

Figure 3.1 In the grand canonical ensemble exchange of energy and particles with a reservoir is assumed.

and $C(E_N, N)$ is the phase space factor for fixed energy and particle number. The function $P(E_N, N)$ is the probability that the system has N molecules and these N molecules have a total energy E_N. The parameters μ and β are undetermined at the moment but will be identified with the temperature and the chemical potential. Our procedure will be as before; that is, we identify the average value of the energy with the internal energy U, and define the average number of molecules as a new state function which we will also denote by N. We then have

$$\frac{1}{Z_\Omega}\sum_N\sum_{E_N} C(E_N, N)\, E_N\, \mathrm{e}^{-\beta(E_N-\mu N)} = \langle E\rangle = U$$

$$\frac{1}{Z_\Omega}\sum_N\sum_{E_N} C(E_N, N)\, N\, \mathrm{e}^{-\beta(E_N-\mu N)} = \langle N\rangle.$$

We now observe that

$$-\frac{\partial}{\partial\beta}\ln Z_\Omega = \langle E\rangle - \mu\langle N\rangle$$

while

$$\frac{\partial}{\partial\mu}\ln Z_\Omega = \beta\langle N\rangle.$$

Combining these two equations we find for the average energy $\langle E\rangle$

$$\langle E\rangle = -\frac{\partial}{\partial\beta}\ln Z_\Omega + \frac{\mu}{\beta}\frac{\partial}{\partial\mu}\ln Z_\Omega.$$

In order to establish a precise relation with thermodynamics we first need to generalize the family of thermodynamic potentials to include a variable number of particles. This can be done using the following result:

Theorem 3.1 If $\Omega = U - TS - \mu N$, where $\mu = (\partial U/\partial N)_{S,V}$ then Ω is a thermodynamic potential and

$$S = -\left(\frac{\partial \Omega}{\partial T}\right)_{\mu,V}$$

$$P = -\left(\frac{\partial \Omega}{\partial V}\right)_{T,\mu}$$

$$N = -\left(\frac{\partial \Omega}{\partial \mu}\right)_{T,V}$$

where μ is the chemical potential and N represents the number of molecules present in the system.

Proof. We recall that the first law of thermodynamics for a gas of molecules is

$$\mathrm{d}U = T\mathrm{d}S - P\mathrm{d}V + \mu \mathrm{d}N$$

where μ is the chemical potential. Thus

$$\Omega = U - TS - \mu N$$

is the Legendre transform of U with respect to S and N. Furthermore

$$\mathrm{d}\Omega = -S\mathrm{d}T - P\,\mathrm{d}V - N\mathrm{d}\mu$$

which implies

$$\left(\frac{\partial \Omega}{\partial T}\right)_{V,\mu} = -S\,, \quad \left(\frac{\partial \Omega}{\partial V}\right)_{T,\mu} = -P\,, \quad \left(\frac{\partial \Omega}{\partial \mu}\right)_{T,V} = -N$$

and thus completes the proof. □

Since taking the Legendre transform twice is the identity transformation we can express the internal energy U as the Legendre transform of Ω which we write as

$$\begin{aligned} U &= \frac{\partial}{\partial\left(\frac{1}{T}\right)}\left(\frac{\Omega}{T}\right) - \mu\frac{\partial}{\partial \mu}\Omega \\ &= \frac{\partial}{\partial \beta}(\beta\Omega) - \frac{\mu}{\beta}\frac{\partial}{\partial \mu}(\beta\Omega)\,. \end{aligned}$$

Comparing this last equation with our expression for $\langle E \rangle$ we get agreement provided we identify

$$Z_\Omega = \mathrm{e}^{-\beta\Omega}, \quad \beta = \frac{1}{kT},$$

and μ is identified with the chemical potential. The probability function $P(E, N)$ defines the grand canonical ensemble. By performing the partial sum over the energy levels at fixed N we can write the grand canonical partition sum as a sum over the canonical partition sums at fixed N

$$Z_\Omega = \sum_N z^N Z_N,$$

where $z = \mathrm{e}^{\beta\mu}$ is called the *fugacity*, and $Z_N = \sum_{E_N} C(E_N, N)\mathrm{e}^{-\beta E_N}$ is the partition function for the canonical ensemble. We should point out that in carrying out $\sum_{E_N}$, proper Boltzmann counting should be done, that is, for N molecules a factor of $1/N!$ has to be included in order to avoid the Gibbs paradox which we discussed earlier.

We will find the grand canonical very useful when we calculate the equation of state of a gas of interacting molecules in the next chapter. As a warm up we first reconsider the perfect gas in this approach.

To begin with we recall that the canonical partition sum for a perfect gas at fixed N is an integral over the $6N$-dimensional phase space (p, q). The integral over q just produced a volume factor V^N whereas the integral over the momenta was a $3N$-dimensional Gaussian integral producing a factor λ^{-3N}. Taking the Boltzmann counting factor $1/N!$ into account to avoid the Gibbs paradox we arrive at the simple expression for the grand canonical partition sum

$$Z_\Omega = \sum_{N=0}^{\infty} \frac{z^N}{N!} \frac{V^N}{\lambda^{3N}}.$$

This, however, is just the series expansion of the exponential function. We can thus evaluate the sum over N to arrive at

$$Z_\Omega = \exp\left(\frac{Vz}{\lambda^3}\right)$$

and consequently for the grand canonical potential Ω

$$\Omega = -\frac{1}{\beta} \ln Z_\Omega = -\frac{1}{\beta}\left(\frac{Vz}{\lambda^3}\right).$$

This allows us to express the pressure in terms of (T, μ) as

$$P = -\left(\frac{\partial \Omega}{\partial V}\right)_{\mu,T} = \frac{1}{\beta}\left(\frac{z}{\lambda^3}\right).$$

Thus we have, at this stage, related P to the temperature T and the fugacity z. In order to get an equation relating P, V, and T we need to replace the fugacity z by a function of V and T. This is done by using

$$N = -\left(\frac{\partial \Omega}{\partial \mu}\right)_{T,V} = \frac{Vz}{\lambda^3} = \frac{V}{\lambda^3}\mathrm{e}^{\beta\mu}.$$

Combining these two equations we then end up with the familiar identity

$$PV = NkT\,.$$

Thus we recover the equation of state for the perfect gas. Furthermore, we can solve the second equation for μ, that is

$$\mu = \frac{1}{\beta}\ln\left(\frac{N\lambda^3}{V}\right).$$

This result is relevant, for instance, to determine the equilibrium condition in chemical reactions described in the first chapter.

3.2 Density fluctuations

We have just seen that the same equations of state for the perfect gas can be derived either in the canonical, or in the grand canonical ensemble. This suggests that the two descriptions are equivalent. We now want to show in full generality that when the size of the system is large enough the two descriptions are indeed equivalent. We will do this by showing that in this limit the fluctuations of the particle number in the grand canonical ensemble is negligible.

Let us define the *average value* $\bar{f}$ of an extensive physical observable f in a statistical ensemble by

$$\bar{f} \equiv \sum_n f(n)\rho(n)$$

where the sum is over all states denoted by n, with a weight factor $\rho(n)$, i.e. the probability for the system to be in this state. We then decompose the system into R large statistically independent subsystems (R is supposed to be large as well). Let $f = \sum_a f_a$ and $\bar{f}_a$ denote the average of f in the subsystem a, $a = 1, \ldots, R$. We then have

$$\bar{f} = \overline{\sum_a f_a} = \sum_a \bar{f}_a \simeq R f^*$$

where we denote by f^* the average value of $\bar{f}_a$. Let us now consider the fluctuations $\Delta f \equiv f - \bar{f}$. We then have

$$\begin{aligned}
\overline{(\Delta f)^2} &= \overline{\left(\sum_{a=1}^{R} \Delta f_a\right)^2} \\
&= \sum_{a=1}^{R} \overline{(\Delta f_a)^2} + \sum_{a \neq b} \overline{\Delta f_a \Delta f_b} \\
&= R \overline{(\Delta f_a)^2}
\end{aligned}$$

where the last equality follows from the fact that the mixed term in the second line vanishes due to statistical independence of the different subsystems. Combining these two equations, we then end up with

$$\frac{\sqrt{\overline{(\Delta f)^2}}}{\bar{f}} \propto \frac{1}{\sqrt{R}} \to 0$$

as R tends to infinity. The ratio just introduced is called the *root mean square*, or *RMS*-fluctuation, and measures the statistical fluctuations compared to the expected value of the observable. Thus what we have just shown is that for any physical observable the *RMS*-fluctuations tend to zero as the size of the system tends to infinity. The key hypothesis we made in this derivation is that the subsystems are statistically independent. If this hypothesis is not satisfied then the proof fails. This situation can arise in thermodynamics in the presence of a phase transition with long-range correlations and will be discussed in Chapter 13.

To prove the equivalence between the canonical and grand canonical ensemble we then choose the particle number as the physical observable of interest, $f \equiv N$. In this case the above result implies

$$\frac{\sqrt{\overline{(\Delta N)^2}}}{N} \propto \frac{1}{\sqrt{R}} \to 0 .$$

Thus for large systems it is irrelevant whether we impose fixed particle number by hand or not since the contributions to the grand canonical partition sum from a system with particle number different from the expected value are negligible.

3.3 Entropy and the Boltzmann constant

We identified the average value of the energy in a system with the internal energy function U. We also learnt how to calculate thermodynamic state functions like the free energy F and the entropy S in terms of the microscopic model provided by statistical mechanics but we have not analyzed the microscopic interpretation of the thermodynamic function S except to state without discussion the interpretation of S due to Boltzmann.

We recall that S was defined rather abstractly. From a discussion of the second law, following a rather circuitous route, it was established that a thermodynamic system contained a state function called the entropy. Furthermore this function attained a maximum value for a closed system in its equilibrium state. Let us examine the expression for the entropy S we obtained for a gas of non-interacting molecules. From $Z_N = 1/N!(V/\lambda^3)^N = \mathrm{e}^{-\beta F}$, where F is the free energy, it follows that

$$S \cong k \ln\left(\frac{V}{\lambda^3}\right)^N + \text{ constant term.}$$

We recall that the temperature-dependent length scale, λ^3, is, roughly speaking, the volume of a single molecule. We can then interpret this expression as follows. For a collection of N non-interacting molecules, the energy of a molecule is independent of where in V the molecule is located. Since the molecules do not interact, the number of ways the state $\langle E_N \rangle$ can be realized for N molecules is thus proportional to $(V/\lambda^3)^N$. The expression for the entropy S obtained thus implies that S is proportional to the logarithm of the number of ways a state of given internal energy can be constructed. The proportionality constant is Boltzmann's constant k. Therefore the statistical mechanics approach in this case makes contact with the interpretation of entropy introduced by Boltzmann, which we stated in Chapter 1. This interpretation is useful in describing phase transitions in Chapter 13 and in describing the micro-canonical ensemble, to which we now turn.

3.4 Micro canonical ensemble

In going from the canonical to the grand canonical ensemble we replaced a system with a fixed value for the number of molecules by a system which had a probability $P(N)$ for containing N molecules. We then saw how such an approach could be related to thermodynamics through the thermodynamic function Ω. It is reasonable to ask if the canonical ensemble where the energy was treated in terms of probabilities could be related to another ensemble where the energy had a fixed value. This is indeed possible. The corresponding ensemble is known as the *micro canonical ensemble.*

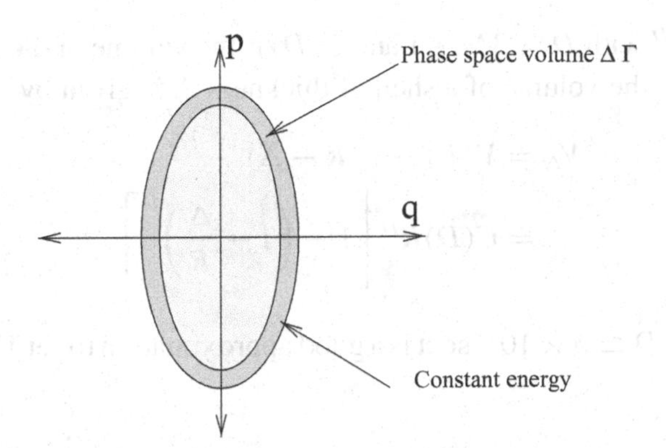

Figure 3.2 The volume in the phase space with energy $E_0 - \Delta < E < E_0$.

In the micro canonical ensemble the probability that the system has energy E is given by (see Fig. 3.2)

$$P(E) = \begin{cases} \frac{1}{\Delta\Gamma} & \text{for } E_0 - \Delta < E < E_0 \\ 0 & \text{otherwise} \end{cases}$$

where $\Delta\Gamma$ is the volume of the region in phase space with $E_0 - \Delta < E < E_0$, i.e.

$$\Delta\Gamma = \int_{E_0 - \Delta < E < E_0} \mathrm{d}^{3N} x \, \mathrm{d}^{3N} p.$$

In order to establish the correspondence with thermodynamics in the micro canonical ensemble we make the following identifications.

(1) The internal energy U is identified with E_0.
(2) The entropy S is identified with $k \ln \Delta\Gamma$.

The first identification does not need further justification. Concerning the second identification, we will now show that this identification is correct in the case of the perfect gas. We consider a system of N non-interacting particles of mass m, contained in volume V. In this case,

$$\Delta\Gamma = V^N \int_{E_0 - \Delta \le E(p) \le E_0} \mathrm{d}^{3N} p, \quad E(p) = \frac{1}{2m} \sum_{i=1}^{N} | \mathbf{p}_i |^2 .$$

In geometrical terms $\Delta\Gamma = V^N \times \Omega_D$, where Ω_D is the volume of the $D = 3N - 1$ dimensional sphere in momentum space with radius $R = \sqrt{2mE_0}$ and thickness proportional to Δ. Fortunately the value of this volume is insensitive to Δ. This is easy to show. Let us write the volume of the sphere in $3N$ dimensions as

$V(R) = C(D)R^D$ with $D = 3N - 1$ and $C(D)$ is the volume of the D dimensional unit sphere. Then the volume of a shell of thickness Δ is given by

$$V_\Delta = V(R) - V(R - \Delta)$$
$$= C(D)R^D \left[1 - \left(1 - \frac{\Delta}{R}\right)^D\right].$$

For a normal gas, $D \simeq 3 \times 10^{23}$ so it is a good approximation to set $V_\Delta \simeq C(D)R^D$, so that

$$\Delta\Gamma \simeq V^N C(D) R^D.$$

To determine $C(D)$ we prove the following lemma.

Lemma 3.2 The volume of the D-dimensional unit sphere $C(D)$ is given by

$$C(D) = \frac{2\pi^{\frac{D+1}{2}}}{\Gamma(\frac{D+1}{2})}.$$

Proof. Consider

$$I(D+1) = \int_{-\infty}^{+\infty} \mathrm{d}x_1 \ldots \int_{-\infty}^{+\infty} \mathrm{d}x_{D+1} \mathrm{e}^{-(x_1^2+\cdots+x_{D+1}^2)} = \pi^{\frac{D+1}{2}}$$

On the other hand, in polar coordinates we can write

$$I(D+1) = \int_0^\infty \mathrm{e}^{-r^2} r^D C(D) \mathrm{d}r$$
$$= \frac{1}{2}\Gamma\left(\frac{D+1}{2}\right) C(D).$$

Here $r^{D-1}C(D)$ is the surface area of a D-dimensional sphere. Comparing the two expressions for $I(D)$ we get

$$C(D) = \frac{2\pi^{\frac{D+1}{2}}}{\Gamma\left(\frac{D+1}{2}\right)}.$$

□

If we now substitute this result into the expression for $\Delta\Gamma$ we find

$$\Delta\Gamma \simeq \frac{\pi^{\frac{D+1}{2}}}{\Gamma\left(\frac{D+1}{2}\right)} R^D V^N$$

with equality when the number of particles tends to infinity. Consequently we have for N

$$\begin{aligned} S &\simeq k \ln \Delta\Gamma \\ &\simeq kN \ln\left(V\pi^{\frac{3}{2}}(2mU)^{\frac{3}{2}}\right) - \frac{3N}{2}\ln\left(\frac{3N}{2}\right) + \frac{3N}{2}, \end{aligned}$$

where we have used the fact that $R = \sqrt{2mU}$. If we now identify S with the entropy, then we infer from the second law that

$$\frac{1}{T} = \left(\frac{\partial S}{\partial U}\right)_V = \frac{3Nk}{2U}, \quad \text{i.e.} \quad U = \frac{3}{2}NkT$$

and

$$P = T\left(\frac{\partial S}{\partial V}\right)_U = T\frac{\partial}{\partial V}(Nk\ln V) = \frac{NkT}{V}.$$

We have thus verified that the identification with the entropy is correct for the case of the perfect gas.

It is possible to show on general grounds that the entropy function S defined in this way is a concave function of (U, V, N) which is maximal at equilibrium. Furthermore, at equilibrium, S is an extensive function of (U, V, N). Thus S has all the properties of the entropy function we derived in Chapter 1 and therefore defines a consistent thermodynamics.

We close this section by noting that it can again be shown, by estimating the energy fluctuations, that the micro canonical and the canonical ensembles are equivalent. We leave the proof of this assertion as an exercise to the reader.

3.5 The full picture

We have now described three different approaches to statistical mechanics, the micro canonical, the canonical, and the grand canonical ensembles. Each of these ensembles is related to certain thermodynamic potentials upon appropriate identification. The different approaches in statistical mechanics are summarized in Table 3.1.

We have seen that all three approaches to statistical mechanics are equivalent if the size of the system considered is very large compared to the microscopic length scales of these systems. From the point of view of thermodynamics this equivalence is reflected in the equivalence of the corresponding thermodynamic potentials S, F, and Ω. Therefore we can choose whichever of the ensembles provides the best

Table 3.1. *Summary of micro canonical, canonical, and grand canonical ensembles in statistical mechanics.*

Ensemble	Probability	Thermodynamics
micro canonical	$P(E) = \begin{cases} \frac{1}{\Delta\Gamma} & E_0 - \Delta \leq E \leq E_0 \\ 0 & \text{otherwise} \end{cases}$	$S = k \ln \Delta\Gamma$
E, N fixed	$\Delta\Gamma = \int d^{3N} p \int d^{3N} x$	$U = E_0$
	$E_0 - \Delta \leq E \leq E_0$	$T = \frac{\partial S}{\partial V}$
canonical	$P(E) = \frac{C(E)}{Z_N} e^{-\beta E}$	$F = -\frac{1}{\beta} \ln Z_N = U - TS$
T, N fixed	$Z_N = \sum_E C(E) e^{-\beta E}$	$\beta = \frac{1}{kT}$
		$U =< E >= -\frac{\partial \ln Z_N}{\partial \beta}$
grand canonical	$P(E_N, N) = \frac{C(E_N)}{Z_\Omega} e^{-\beta(E_N - \mu N)}$	$\Omega = -\frac{1}{\beta} \ln Z_\Omega$ $= U - TS - \mu N$
T, μ fixed	$Z_\Omega = \sum_{E_N, N} C(E_N, N) e^{-\beta(E_N - \mu N)}$	μ = chemical potential
		$N = \frac{1}{\beta} \frac{\partial \ln Z_\Omega}{\partial \mu}$

approach to the concrete problem under consideration. For instance we will find in the next chapter that the grand canonical ensemble is best suited to include interactions between molecules in a realistic gas.

For completeness we should mention that in addition to the ensembles described in this chapter there should also be an ensemble for which the corresponding thermodynamic potential is the Gibbs free energy $G(T, P, N)$. Such an ensemble exists and is treated as a problem for the reader. It is normally omitted because it is rarely used in practical applications.

Problems

Problem 3.1 Assuming that the chemical potential, μ, is independent of N for fixed P and T, show that $G = \mu N$.

Problem 3.2 Show that the perfect gas law, $PV = NkT$, is recovered even if the proper Gibbs counting factor of $N!$ is omitted.

Problem 3.3 Work done by magnetic moments: our definition of the energy for a magnetic substance in Sections 2.6 and 2.7 suggests that the internal energy for a magnetic substance is a function of S and B, i.e. $U = U(S, B)$. This is consistent with the observation that in the process of moving a magnetic substance through an inhomogeneous magnetic field the internal energy varies as

$$dU = -\int_{B_0}^{B_1} M\, dB\,.$$

With this interpretation the quantity $-kT \log Z$ is naturally interpreted as the free energy $F(T, B)$. This is the point of view taken in Section 2.7. However, there is also an alternative interpretation where one considers the work done to magnetize the substance. For this, consider a solenoid of length ℓ, cross section A and n windings per unit of length, through which a current j creates a vacuum magnetic field $B = (4\pi/c)nj$. This solenoid is then filled uniformly with the magnetic substance. Using Faraday's law for the induced voltage by a varying magnetic flux, $V = -(4\pi n\ell A/c)\, dM/dt$, compute the voltage induced as the magnetic moment of the substance is changed by the amount ΔM. Show that the resulting work done by the battery driving the current in the solenoid equals $\Delta W = B\Delta M$. Thus

$$dU = \int_{M_0}^{M_1} B\, dM\,.$$

Convince yourself that from this point of view it is natural to interpret the partition sum in term of the Gibbs free energy as $-kT \log Z = G(T, B)$, with $M = -\partial G/\partial B$.

Problem 3.4 Information theoretic interpretation of the entropy (*Jaynes' Principle*): consider a family of registers with each register consisting of bits taking the values 0 and 1. We then define an entropy as $S \equiv c \ln \Omega$, where c is a positive constant and Ω counts the number of possible states of registers. Clearly, S defined in this way is a measure for the lack of information about the system. For two systems A and B we have

(1) $S_A > S_B \Leftrightarrow \Omega_A > \Omega_B$
(2) $S = 0 \Leftrightarrow \Omega = 1$
(3) $S_A + S_B = S_{A\cup B}$

We can similarly consider M equivalent physical systems ($M \rightarrow \infty$), with each system taking one of its possible states E_i, $i = 1, \cdots, \mu$, with probability w_i, i.e. $n_i = w_i M$ systems are in the state E_i.

(1) Compute Ω for this ensemble.
(2) Using Stirling's formula, express S for one system in terms of the w_i's only. The function of w_i obtained in this way is called the *Shannon function*.
(3) Determine w_i by maximizing S, subject to the condition $\sum_i w_i = 1$, using the method of Lagrange multipliers.
(4) Show that $0 \leq S(w_i) \leq \mathrm{c} \ln \mu$.
(5) Determine w_i by maximizing S, subject to the condition $\sum_i E_i w_i = U$, where U is the internal energy.

Problem 3.5 Show that the canonical and the micro canonical ensembles are equivalent by calculating the energy fluctuations in the canonical ensemble.

Problem 3.6 Construct the partition sum of the "isothermal-isobar" ensemble and discuss its relation to the Gibbs free energy.

Problem 3.7 Consider an ideal gas in the presence of a surface with N_0 distinguishable sites capable of absorbing a single molecule of the ideal gas due to a sink of potential energy, $\Delta E = -\epsilon$. Determine the fraction of occupied sites $f(T, P) \equiv n/N_0$. Hint: describe the surface as grand canonical ensemble. The isothermal curve $(P, f(P))$ for fixed T is the so-called *absorption isothermal* of Langmuir.

Historical notes

The problem of understanding the molecular basis of Maxwell's distribution law was taken up by Boltzmann who was able to extend Maxwell's law to the case where external fields such as gravity are present. Such considerations suggested to Boltzmann that the basic principle of statistical mechanics was that the relative probability of a molecular state with total energy E was $\mathrm{e}^{-E/kT}$. Boltzmann did not succeed in proving this principle but one of his attempts to justify it involved the extremely interesting idea of equal a priori probabilities: for a gas consisting of M non-interacting molecules of total energy E, Boltzmann postulated that every micro state of the system, defined by assigning N_1 units of energy to particle 1, N_2 units of energy to particle 2, etc. such that the system has fixed energy E, has equal probability. In modern day terminology, Boltzmann had introduced the micro canonical ensemble.

The next major step in statistical mechanics came with Gibbs. In his book *Elementary Principles of Statistical Mechanics* (1902), Gibbs introduced the canonical and the grand canonical ensembles and showed how they could be used to discuss the thermal properties of systems thus establishing the framework of modern statistical mechanics. A remark made regarding Gibbs' book is worth repeating: "A little book, little read because it is a little difficult."

J. W. Gibbs was born in New Haven in 1839. He studied mechanical engineering and went on to get the first Ph.D. in engineering in the United States in 1863. For three years (1866–69) Gibbs toured Europe and attended lectures and seminars in Paris, Berlin, and Heidelberg. Two years after returning from his European tour Gibbs was appointed the first Professor of Mathematical Physics at Yale where he spent the rest of his life. Gibbs formulated statistical mechanics clearly and his way of understanding phase space, phase transitions and thermodynamic surfaces using elegant geometrical constructions revolutionized the subject. It was only after his work had been translated to German and French that his ideas received wide recognition in Europe. Gibbs introduced the concept of chemical potential, the concept of the ensemble, and he presented his work as providing a rational foundation for thermodynamics. Besides his great work on thermodynamics and statistical mechanics, Gibbs pioneered the use of vector analysis which he built on the bases of the work of Grassmann and Hamilton.

A quotation from Gibbs may be of interest: "...The usual point of view in the study of mechanics is that attention is mainly directed to the changes which take place in the course of time in a given system... for some purposes, however, it is desirable to take a broader view of the subject. We may imagine a great number of systems of the same nature, but differing in the configurations and velocities which they have ... here we may set the problem, not to follow a particular system through its succession of configurations, but to determine how the whole number of systems will be distributed among the various conceivable configurations..."

Further reading

A thorough discussion of the various ensembles in classical statistical mechanics can be found in K. Huang, *Statistical Mechanics*, John Wiley (1987). For a clear mathematical treatment of statistical mechanics see C. J. Thompson, *Mathematical Statistical Mechanics*, Princeton (1972). A careful discussion of magnetic systems is found in C. Kittel, *Thermal Physics*, John Wiley (1969). A standard text on ergodic theory in statistical mechanics is E. Farquhar, *Ergodic Theory in Statistical Mechanics*, John Wiley (1965). The use of probability theory to establish the basic mathematical results for statistical mechanics can be found in A. I. Khinchin, *Mathematical Foundations of Statistical Mechanics*, Dover (1949) and also N. S. Krylov, *Works on the Foundations of Statistical Physics*, Princeton (1979).

4

Handling interactions

4.1 Statement of the problem

Up to now we have applied the formalism of statistical mechanics to molecules and particles without interactions. However, the formalism developed in the last two chapters is not restricted to these cases. We will now study the effects of intermolecular interactions. Unfortunately, once interactions are included, it is no longer possible to completely determine the canonical or grand canonical partition functions analytically. In view of this we develop in this chapter an approximate method for determining the grand canonical partition function known as the *Ursell–Mayer cluster expansion*. In this approach the short range nature of the intermolecular interactions is exploited. We start with the Hamiltonian $H(x, p)$ for a system of N identical molecules of mass m.

$$H(x, p) = \sum_{i=1}^{N} \frac{|\mathbf{p}_i|^2}{2m} + \sum_{i<j} V(|\mathbf{x}_i - \mathbf{x}_j|).$$

Here $|\mathbf{p}_i|^2/2m$ represents the kinetic energy of the ith molecule while $V(|\mathbf{x}_i - \mathbf{x}_j|)$ is the interaction potential energy between molecules located at $\mathbf{x}_i$ and $\mathbf{x}_j$ respectively. The interaction potential can in principle be determined in scattering experiments and is assumed to be known. We show how this can be done in Chapter 9. Let us list some general features of $V(|\mathbf{x}|)$, where $|\mathbf{x}|$ denotes the separation distance between two molecules.

(1) $V(|\mathbf{x}|)$ leads to a large repulsive force at short distance. This means that a large amount of energy is needed to push molecules very close together.
(2) $V(|\mathbf{x}|)$ leads to a weakly attractive force between molecules when they are close but not too close together. These forces are responsible for binding molecules together to form larger objects.
(3) $V(|\mathbf{x}|)$ rapidly approaches zero as the distance between the molecules increases beyond some critical distance.

A simple model which incorporates these features of $V(|\mathbf{x}|)$ is:

$$V(|\mathbf{x}|) = \begin{cases} \infty & \text{if } |\mathbf{x}| \leq a, \\ -V_0 & \text{if } a < |\mathbf{x}| \leq a + L, \\ 0 & \text{if } |\mathbf{x}| > a + L. \end{cases}$$

In principle $V(|\mathbf{x}|)$ can be determined from scattering experiments, as we have already stated, or it can be theoretically determined from the structure of the molecules involved. The important point is that $V(|\mathbf{x}|)$ is not a small quantity. This means that any method of calculating the effect of interactions on the equation of state which regards $V(|\mathbf{x}|)$ as small is unreliable. We would like to introduce a calculational approach in which the interaction effects can be introduced in a step-by-step way, which exploits the short range nature of $V(|\mathbf{x}|)$ and the fact that the attractive force between molecules is weak, i.e. V_0 is small. This is done by observing that it is not $V(|\mathbf{x}|)$ but $\mathrm{e}^{-\beta V(|\mathbf{x}|)}$ which is relevant in the partition function. For our model $V(|\mathbf{x}|)$ let us consider the variation of $\mathrm{e}^{-\beta V(|\mathbf{x}|)} - 1 \equiv f(|\mathbf{x}|)$, as a function of $|\mathbf{x}|$. We have

$$f(|\mathbf{x}|) = \begin{cases} -1 & \text{if } |\mathbf{x}| \leq a, \\ \beta V_0 & \text{if } a < |\mathbf{x}| \leq a + L, \quad \text{assuming } \beta V_0 \text{ is small} \\ 0 & \text{if } |\mathbf{x}| > a + L. \end{cases}$$

4.2 Example: van der Waals equation of state

Before we develop the general theory let us consider the effect of such two-body interactions on the equation of state, assuming that V_0 and a are small in a sense to be clarified below. Consider now the canonical partition sum

$$Z_N = \frac{1}{N!h^{3N}} \int \mathrm{d}^{3N}p \int \mathrm{d}^{3N}x \, \mathrm{e}^{-\beta \frac{\sum |\mathbf{p}_i|^2}{2m}} \mathrm{e}^{-\beta \sum_{i<j} V(|\mathbf{x}_i - \mathbf{x}_j|)}.$$

We begin by carrying out the momentum integrals which are the same as for the perfect gas, i.e.

$$Z_N = \frac{1}{N!} \lambda^{-3N} \int \mathrm{d}^{3N}x \, \mathrm{e}^{-\beta \sum_{i<j} V(|\mathbf{x}_i - \mathbf{x}_j|)}.$$

In the spirit of the cluster expansion, we then introduce $f_{ij} = \mathrm{e}^{-\beta V_{ij}} - 1$, where $V_{ij} = V(|\mathbf{x}_i - \mathbf{x}_j|)$. The canonical partition sum is then written as

$$Z_N = \frac{1}{N!} \lambda^{-3N} \int \mathrm{d}^{3N}x \prod_{i<j} (1 + f_{ij}).$$

Note that f_{ij} is a bounded object. We then expand the above product to the first order in f_{ij}, i.e.

$$\prod_{i<j}(1+f_{ij}) \simeq 1+\sum_{i<j}(f_{ij}).$$

This is reasonable if V_0 and the size a of the molecule are both small. Substituting this expression in Z_N we get

$$\begin{aligned} Z_N &\simeq \frac{1}{N!}\lambda^{-3N}\int\left(1+\sum_{i<j} f_{ij}\right)\mathrm{d}^{3N}x \\ &= \frac{1}{N!}\lambda^{-3N}\left[V^N+\sum_{i<j}\int \mathrm{d}^3x_1\cdots\int \mathrm{d}^3x_i\cdots\int \mathrm{d}^3x_j\cdots\int \mathrm{d}^3x_N\, f_{ij}\right]. \end{aligned}$$

We observe that the labels i and j are dummy labels which are summed over. Furthermore

$$\int \mathrm{d}^3x_1\cdots \mathrm{d}^3x_i\cdots \mathrm{d}^3x_j\cdots \mathrm{d}^3x_N\, f_{ij}$$

gives the same result no matter which singular pair of labels i and j is chosen for f_{ij}. Then choosing $i=1$, $j=2$ we get

$$Z_N \simeq \frac{1}{N!}\lambda^{-3N}\left\{V^N+\frac{N(N-1)}{2}V^{N-2}\cdot\int \mathrm{d}^3x_1\mathrm{d}^3x_2 f_{12}\right\}$$

where $N(N-1)/2 \simeq N^2/2$ represents the number of terms in $\sum_{i<j}$ and V^{N-2} is the result of integrating over the positions of the molecules that do not participate in the interaction. To continue we note that $f_{12}=f(|\mathbf{x}_1-\mathbf{x}_2|)$ only depends on the separation distance between $\mathbf{x}_1$ and $\mathbf{x}_2$. This suggests to change to center of mass and relative coordinates, $\mathrm{x}=(\mathrm{x}_1+\mathrm{x}_2)/2$ and $\mathbf{r}=\mathbf{x}_1-\mathbf{x}_2$. Then

$$\begin{aligned} Z_N &\simeq \frac{1}{N!}\lambda^{-3N}\left\{V^N+\frac{N^2V^{N-1}}{2}\int \mathrm{d}^3r\, f(|\mathbf{r}|)\right\} \\ &= \frac{1}{N!}\lambda^{-3N}V^N\left[1+\frac{N^2}{2V}\kappa(\beta,a,L)\right] \end{aligned}$$

where

$$\kappa(\beta,a,L)=4\pi\int r^2\mathrm{d}r f(|\mathbf{r}|)=4\pi\int r^2\mathrm{d}r\left(\mathrm{e}^{-\beta V(|\mathbf{r}|)}-1\right).$$

This integral is easily evaluated to give $\kappa(\beta, a, L) = -A + B\beta$, where

$$A = \frac{4\pi}{3}a^3, \quad \text{and} \quad B = \frac{4\pi}{3}((a+L)^3 - a^3)V_0.$$

Thus

$$\begin{aligned} F &= \frac{-1}{\beta} \ln Z_N \\ &= \frac{-1}{\beta} \ln \left\{ Z_N^0 \left(1 + \frac{N^2}{2V}\kappa(\beta) \right) \right\} \end{aligned}$$

where $Z_N^0 = \lambda^{-3N} V^N / N!$ is the partition sum for the perfect gas. Assume now that V_0 and a are small in the sense that $N^2\kappa(\beta)/2V \ll 1$. Then

$$F \simeq -\frac{1}{\beta} \ln Z_N^0 - \frac{1}{\beta} \cdot \frac{N^2\kappa(\beta)}{2V} = -\frac{1}{\beta} \ln Z_N^0 - \frac{1}{\beta}\frac{N^2}{2V}(-A + B\beta).$$

The equation of state of a gas with interactions of the type just described is thus given by

$$P = -\left(\frac{\partial F}{\partial V}\right)_T = \frac{NkT}{V} - \frac{N^2}{2\beta V^2}(-A + B\beta)$$

which upon substitution of our results for A and B leads to

$$P = \frac{NkT}{V} + \left(\frac{N^2 A}{2V^2}\right) \cdot kT - \frac{N^2 B}{2V^2}$$

or

$$\left(P + \frac{N^2 B}{2V^2}\right)\left(V - \frac{NA}{2}\right) = NkT.$$

This equation of state is known as the van der Waals equation of state which describes a realistic gas at low density and with a shallow attractive potential between the molecules. The condition $N^2\kappa(\beta)/2V \ll 1$ implies that the particle density times the volume taken by the hard core of the molecules is much smaller than one. Note that the van der Waals equation of state can be written in the form of the ideal gas law

$$P_{\text{eff}} V_{\text{red}} = NkT,$$

where V_{red} is "reduced" due to the finite size of the molecules and P_{eff} takes the attraction between the molecules into account. In particular, the measured pressure is reduced by the attractive force.

4.3 General theory: the cluster expansion

In the example above we have seen how interactions modify the equation of state in the case when the interactions can be treated as a small perturbation of the perfect gas. In this section we now develop a formalism for systematically summing up all contributions coming from the expansion of the partition sum in terms of the perturbation f_{ij}. This will lead us to a formalism called cluster expansion. We will find that the grand canonical ensemble is best suited to perform the summation explicitly. However, for the moment it will be convenient to concentrate on the canonical partition sum at fixed N. As shown in the previous section we then have

$$Z_N = \frac{1}{N!}\lambda^{-3N} Q_N$$

where

$$Q_N = \int d^{3N}x \prod_{i<j}(1 + f_{ij}).$$

We now develop a method to compute Q_N. In order to understand the structure of Q_N it is useful to consider the first non-trivial case, $N = 3$, explicitly. We have

$$Q_3 = \int d^3x_1 \int d^3x_2 \int d^3x_3 \,(1 + f_{12})(1 + f_{13})(1 + f_{23})\,.$$

Let us rewrite Q_3 as

$$Q_3 = Q_3^{(1)} + Q_3^{(2)} + Q_3^{(3)}$$

where $Q_3^{(1)}$ refers to the terms in Q_3 in which no interactions are present, i.e.

$$Q_3^{(1)} = \int d^3x_1 \int d^3x_2 \int d^3x_3 = V^3,$$

and $Q_3^{(2)}$ refers to the terms in Q_3 in which two molecules interact, i.e.

$$\begin{aligned} Q_3^{(2)} &= \int d^3x_1 \int d^3x_2 \int d^3x_3 \,[f_{12} + f_{13} + f_{23}] \\ &= 3V \int d^3x_1 \int d^3x_2 \, f_{12} \\ &= 3V^2 \int d^3x \, f(\mathbf{x}), \qquad f(\mathbf{x}) = e^{-\beta V(|\mathbf{x}|)} - 1 \end{aligned}$$

where we have changed to the center of mass and relative coordinates, $\mathbf{R} = 1/2\,(\mathbf{x}_1 + \mathbf{x}_2)$ and $\mathbf{x} = \mathbf{x}_1 - \mathbf{x}_2$ respectively. Finally $Q_3^{(3)}$ refers to the term in Q_3 in which three molecules interact i.e.

$$Q_3^{(3)} = \int d^3x_1 \int d^3x_2 \int d^3x_3 \, [f_{12}f_{13} + f_{12}f_{23} + f_{13}f_{23} + f_{12}f_{13}f_{23}].$$

The decomposition of Q_3 can be understood in terms of clusters. In this example three molecules are involved. The first term $Q_3^{(1)}$ can be described as consisting of three 1-clusters, i.e. three non-interacting molecules. The second term $Q_3^{(2)}$ can be described as involving one 1-cluster and one 2-cluster. A 2-cluster represents two molecules that interact. Each of these clusters contribute the same to $Q_3^{(2)}$ which explains the factor of 3 in the expression for $Q_3^{(2)}$. Finally for three molecules there is only one 3-cluster possible, i.e. only one term in which all three molecules interact, this is $Q_3^{(3)}$. This example suggests that we write Q_3 as

$$\begin{aligned} Q_3 &= S(3,0,0) + S(1,1,0) + S(0,0,1) \\ &= \sum_{m_1,m_2,m_3} S(m_1,m_2,m_3), \quad \sum_{l=1}^{3} l m_l = 3 \end{aligned}$$

where $S(m_1, m_2, m_3)$ refers to terms in Q_3 in which there are m_1 1-clusters, m_2 2-clusters, and m_3 3-clusters. The numbers m_1, m_2, m_3 must satisfy the constraint $m_1 + 2m_2 + 3m_3 = 3$, since the total number of molecules in this example was three.

After having set up the notation we are now ready to give the general expression for Q_N, that is

$$Q_N = \sum_{m_1,m_2,\ldots} S(m_1, m_2, \ldots, m_l, \ldots, m_N)$$

subject to the constraint $\sum_{l=1}^{N} l m_l = N$. Here $S(m_1, m_2, \ldots, m_l, \ldots)$ represents a term in Q_N which consists of m_1 1-clusters, m_2 2-clusters, …, m_l l-clusters. This is the cluster expansion representation for Q_N. It is convenient to represent clusters pictorially as follows. We replace $\int d^3x_1$ by

and $\int d^3x_1 \int d^3x_2 \, f_{12}$ by

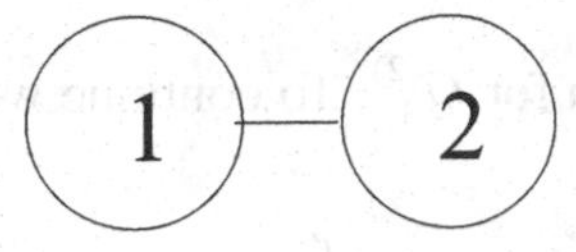

Note this is a 2-cluster. Then the terms in Q_3 can be graphically represented as

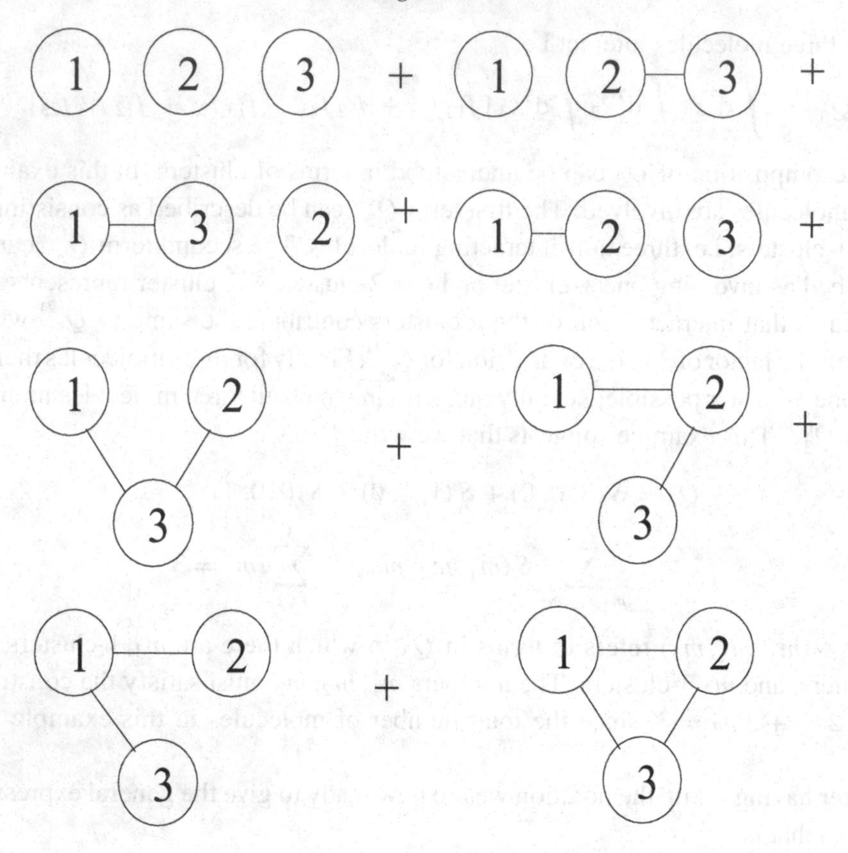

The last four terms together form a 3-cluster. Observe that the value of

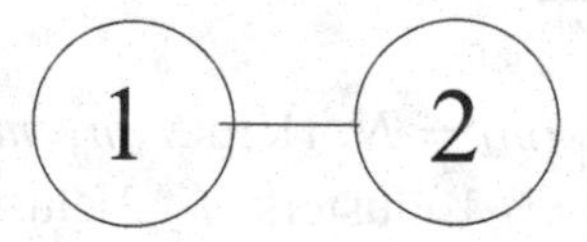

does not depend on the labels 1 and 2 since the labels 1, 2 are integration variables. We used this feature to note that

(1) (2)—(3) = (1)—(2) (3)

= (1)—(3) (2)

This simplifies the expression for $Q_3^{(2)}$. To continue we define a cluster integral c_l as follows:

$$c_l = \int d^3x_1 \cdots \int d^3x_l \; \mathcal{C}_l(\mathbf{x}_1 \cdots \mathbf{x}_l)\,,$$

where $C_l(\mathbf{x}_1 \cdots \mathbf{x}_l)$ represents an l-cluster, that is the sum of all possible contractions between the l molecules. Note that $c_l \propto V$ as a result of the integral over the center of mass coordinate. The value of c_l is independent of the particular set of l molecules picked from N-molecules to form the l-cluster. We have already seen this for 2-clusters in Q_3.

For the sake of being explicit we look at one term in Q_4 involving different 3-clusters, namely $S(1, 0, 1, 0)$, and check that each one of these 3-clusters makes the same contribution to $S(1, 0, 1, 0)$. Recall $S(1, 0, 1, 0)$ represents one 1-cluster and one 3-cluster, involving altogether four molecules. We can get the contribution to Q_4 from $S(1, 0, 1, 0)$ by using the definition of Q_4 and picking out terms from Q_4 which involve interaction between groups of three molecules. We then find

$$Q_4 = \int \mathrm{d}^3x_1 \int \mathrm{d}^3x_2 \int \mathrm{d}^3x_3 \int \mathrm{d}^3x_4 \,(1 + f_{12})(1 + f_{13})(1 + f_{14}) \\ (1 + f_{23})(1 + f_{24})(1 + f_{34})\,.$$

The contribution to Q_4 from $S(1, 0, 1, 0)$ is explicitly given by

$$S(1, 0, 1, 0) = \int \mathrm{d}^3x_1 \int \mathrm{d}^3x_2 \int \mathrm{d}^3x_3 \int \mathrm{d}^3x_4 \\ [f_{12}f_{13} + f_{12}f_{14} + f_{12}f_{23} + f_{12}f_{24} + f_{13}f_{14} + f_{13}f_{23} \\ + f_{13}f_{34} + f_{14}f_{24} + f_{14}f_{34} + f_{14}f_{34} + f_{23}f_{24} + f_{23}f_{34} \\ + f_{24}f_{34} + f_{12}f_{23}f_{13} + f_{12}f_{14}f_{24} + f_{23}f_{34}f_{24} + f_{13}f_{34}f_{14}].$$

All other terms in Q_4 involve either interaction between more or less than three molecules. We now observe that this expression can be written as the sum of four 3-clusters. Namely the clusters

(1) $f_{12}f_{24} + f_{12}f_{14} + f_{14}f_{24} + f_{12}f_{14}f_{24}$
(2) $f_{12}f_{23} + f_{13}f_{12} + f_{13}f_{23} + f_{12}f_{23}f_{13}$
(3) $f_{23}f_{34} + f_{24}f_{34} + f_{24}f_{23} + f_{23}f_{34}f_{24}$
(4) $f_{13}f_{34} + f_{14}f_{13} + f_{14}f_{34} + f_{13}f_{34}f_{14}$

Each one of these combinations gives the same contribution to $S(1, 0, 1, 0)$ since the labels distinguishing the contributions represent variables of integration only. There are thus four 3-clusters each contributing the same to S.

We are now ready to calculate Q_N in terms of cluster integrals. A general term $S(m_1, m_2, m_3, \ldots)$ has the pictorial structure

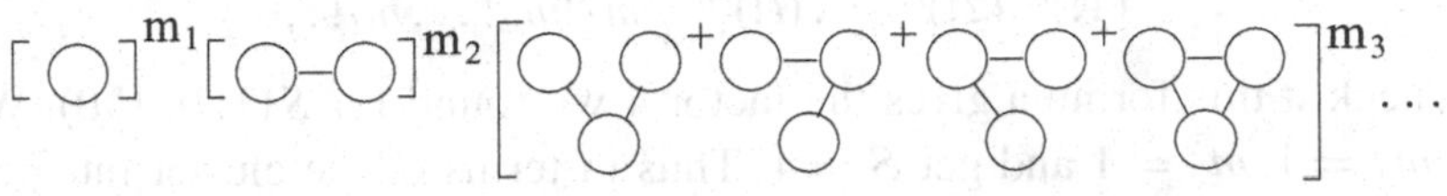

$$[\bigcirc]^{m_1}\,[\bigcirc\!-\!\bigcirc]^{m_2}\,[\,\cdots + \cdots + \cdots + \cdots\,]^{m_3} \ldots$$

with $\sum lm_l = N$. The contribution such a graph makes to $S(m_1, m_2, m_3, \ldots)$ depends on the number of ways a configuration $(m_1, m_2, \ldots)$ can be realized multiplied by the value of the corresponding cluster integrals for any specific labeling. For instance in the case of Q_4 we found that for $S(1, 0, 1, 0)$ there were four terms each contributing the same to $S(1, 0, 1, 0)$. We can thus write

$$S(m_1, m_2, \ldots, m_l, \ldots) = \left[c_1^{m_1} c_2^{m_2} \ldots c_l^{m_l} \ldots\right] \times \mathcal{S}.$$

The first factor gives the value of any configuration containing m_1 1-clusters, m_2 2-clusters and so on. $\mathcal{S}$ is the symmetry factor that counts the number of distinct ways a configuration labeled by $m_1, m_2, m_3, \ldots, m_l, \ldots$ can be realized. This number $\mathcal{S}$ can be calculated by noting that there are N integrals $\int \mathrm{d}^3x_i$ in Q_N. Each one of these pictorially corresponds to a box. A particular contribution corresponds to labeling these boxes with integers between 1 and N. This can be done in $N!$ ways. However, many labelings lead to the same configuration, for instance

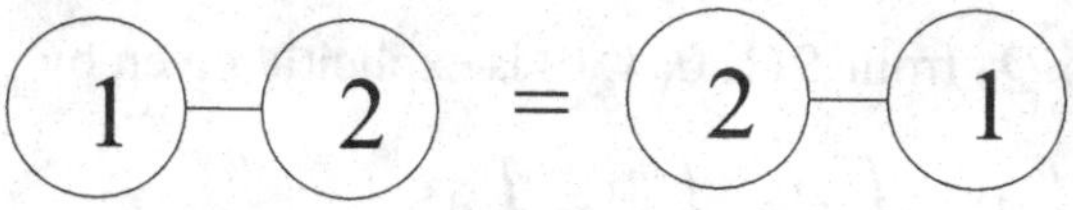

(1)—(2) = (2)—(1)

and

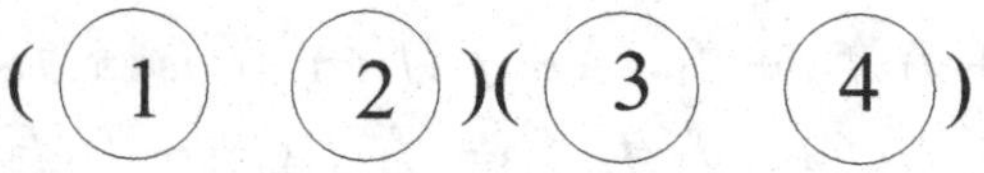

((1) (2))((3) (4))

= ((3) (4))((1) (2))

We have to weed these out. This can be done by dividing $N!$ by configurations which do not lead to distinct configuration of clusters. There are, as our simple example demonstrates, two ways in which identical cluster decompositions can be obtained. The first is if the labels within a given l-cluster are permuted among themselves. This can be done in $l!$ ways for each of the m_l l-clusters present. The second is if the m_l clusters are permuted among themselves. This can be done in $m_l!$ ways. Thus the number of distinct configurations $\mathcal{S}$ each of which contributes the same to $S(m_1, m_2, \ldots)$ is

$$\mathcal{S} = \frac{N!}{(1!)^{m_1}(2!)^{m_2}\ldots(l!)^{m_l}} \cdot \frac{1}{m_1!m_2!\ldots m_l!\ldots}.$$

Let us check if this formula gives the factor 4 we found for $S(1, 0, 1, 0)$. We set $N = 4$, $m_1 = 1$, $m_3 = 1$ and get $\mathcal{S} = 4$. Thus in terms of the cluster integrals c_l, defined earlier we have

$$Q_N = \sum_{m_1, m_2, \ldots} \prod_l \left(\frac{c_l}{l!}\right)^{m_l} \frac{N!}{m_l!}$$

subject to the constraint $\sum l m_l = N$. We have thus reduced the calculation of the partition sum $Z_N = Q_N/N!\lambda^{3N}$ to a sum over cluster integrals c_l which can be evaluated on a computer. Note that the sum over the clusters is a constraint sum to ensure that the total number of particles in the system equals N.

We can remove this complication if we consider the grand canonical partition sum instead of the canonical one. The grand canonical ensemble partition function is given by

$$Z_\Omega = \sum_{N=0}^{\infty} \frac{z^N}{N!} Q_N \frac{1}{\lambda^{3N}}.$$

Since in this expression N can range to ∞ we can write

$$Z_\Omega = \sum_{m_1, m_2, \ldots} \prod_l \left(\frac{c_l z^l}{l!\lambda^{3l}}\right)^{m_l} \frac{1}{m_l!},$$

where we have used $\prod_l \lambda^{3m_l l} = \lambda^{3N}$ and now allow both the product over l as well as the sum over m_l's to be unrestricted between $l = 1$ to ∞. Thus

$$Z_\Omega = \exp\left[\sum_{l=1}^{\infty} \frac{z^l c_l}{l!\lambda^{3l}}\right],$$

which is clearly simpler than the expression for Z_N above. The grand canonical potential is then simply

$$\Omega = -PV = -\frac{1}{\beta} \ln Z_\Omega = -kT \sum_{l=1}^{\infty} \frac{z^l c_l}{l!\lambda^{3l}}.$$

This last expression makes the reduction of the problem of including two-body interactions in a perfect gas to the computations of the cluster integrals most transparent. We note in passing that the leading correction c_2 can be computed exactly, as an infinite power series for the so-called *Lennard-Jones potential* to be introduced in Chapter 6.

4.4 Relation to experiment: the virial expansion

In our first look at the problem we found that including interactions modifies the perfect gas law to the van der Waals equation of state. Now that we have presented the general formalism to treat interactions it is of interest to see how the corresponding equation of state is modified. Experimentally, deviations from perfect gas law behavior have been studied carefully. Generically, such deviations, for a gas of molecules of mass m, are expressed as

$$\frac{Pv}{kT} = \sum_{l=1}^{\infty} a_l(T) \left(\frac{\lambda^3}{v}\right)^{l-1}$$

where $v = V/N$, $\lambda = h/\sqrt{2\pi MkT}$. This expansion is known as the *virial expansion*. Clearly, the first coefficient $a_1 = 1$ since this gives the equation of state for an ideal gas. The first correction to the ideal gas law is determined by a_2. For example, the experimental value for nitrogen at 100 K is $a_2\lambda^3 = 160\,\text{cm}^3/\text{mol}$. Our theoretical task is to relate the experimentally determined virial coefficients $a_l(T)$ to some function of the interaction potential between molecules.

In order to get the equation of state we need to express z in terms of (P, V, T). For this we use $N = -(\partial\Omega/\partial\mu)_{V,T}$ to get

$$\frac{1}{v} = \frac{1}{\lambda^3} \sum_{l=1}^{\infty} l\, b_l z^l\,, \quad \text{where} \quad b_l = \frac{\lambda^{3-3l}}{Vl!} c_l\,.$$

Combining this with our expression for PV in the last section we get

$$\frac{Pv}{kT} = \frac{\sum b_l z^l}{\sum l\, b_l z^l}.$$

We can now relate the virial coefficients $a_l\,(T)$ to the cluster integrals $c_l(T)$ or $b_l\,(T)$ by comparing this last expression with the virial expansion at the beginning of this section. This leads to

$$\frac{\sum b_l z^l}{\sum b_l l z^l} = \sum a_l \left(\sum_n b_n n z^n\right)^{l-1}.$$

Evaluating both sides order by order in z and remembering that $b_1 = 1$, the coefficients a_l are determined recursively. After some straightforward though a little tedious algebra one finds

$$\begin{aligned} a_1 &= 1 \\ a_2 &= -b_2 \\ a_3 &= 4b_2^2 - 2b_3 \\ a_4 &= -20b_2^3 + 18b_2b_3 - 3b_4 \\ &\text{etc.} \end{aligned}$$

Thus the goal to relate the experimentally measured virial coefficients a_l to the theoretical cluster integrals b_l has been achieved. In fact, it can be shown that the equations for the coefficients a_l simplify if they are expressed in terms of *irreducible clusters*. Concretely one finds

$$a_l = -(l-1)b_l^{\text{irr}} \quad l \geq 2,$$

where b_l^{irr} is obtained from b_l by retaining only those graphs which remain connected if one line between any two circles is removed. For $l = 3$, for example, there is only one such graph.

Problems

Problem 4.1 Calculate b_2 and b_3 for the potential

$$V(|\mathbf{x}|) = \begin{cases} \infty, & |\mathbf{x}| < a \\ 0, & |\mathbf{x}| > a. \end{cases}$$

Problem 4.2 For a mixture of two gases with atomic fractions x_1 and x_2 such that $x_1 + x_2 = 1$, show that the second virial coefficient is of the form

$$B = B_{11}x_1^2 + 2B_{12}x_1x_2 + B_{22}x_2^2.$$

Problem 4.3 Compute $c_V - c_P$ for a gas with a van der Waals equation of state.

Problem 4.4 Show by explicit verification that $a_3 = -2b_3^{irr}$. Hint: use center of mass and relative coordinates to evaluate the integral expression.

Problem 4.5 For an ideal gas we have $U = U(T)$ which implies that during a free, adiabatic expansion the temperature does not change. Show that the leading order correction to the Joule coefficient $J = (\partial T/\partial V)_U$ is given by

$$J = -\frac{RT^2}{c_V v^2}\left(\frac{\mathrm{d}(a_2\lambda^3)}{\mathrm{d}T}\right)_U.$$

Further reading

A detailed description of classical and quantum cluster expansion can be found in R. K. Pathria, *Statistical Mechanics*, Butterworth-Heinemann (1996) which also includes a thorough discussion of calculations of higher-order terms in the cluster expansion for both classical and quantum systems. Another reference for the quantum cluster expansion is Huang, *Statistical Mechanics*, John Wiley (1987). For further details on equations of state, virial expansion and the Joule–Thomson expansion see e.g. W. Castellan, *Physical Chemistry*, Addison-Wesley (1971).

5
Monte Carlo integration

In the canonical ensemble one defines expectation values of observables, $A(p, q)$ by expressions of the form

$$\langle A \rangle = \frac{\int \mathrm{d}\Gamma \exp(-\beta H) A}{\int \mathrm{d}\Gamma \exp(-\beta H)},$$

where the integral is over the phase space of the system. We have discussed a number of analytical methods to evaluate these integrals in previous chapters. Although these techniques can provide us with considerable information about statistical mechanics systems they are in general not sufficient for all purposes. In particular, this is the case when the interactions between molecules are strong such that they can not be treated as a perturbation. In this chapter we will describe techniques to evaluate canonical expectation values numerically on a computer.

5.1 Numerical integration techniques

Recall that the evaluation of the expectation value of A in the canonical ensemble is essentially just the evaluation of an integral in a very high number of dimensions. We will first consider two basic numerical techniques which are commonly used to evaluate integrals, and discuss why these techniques fail in the present case. However, the reasons for failure of the techniques are quite informative, and will suggest a workable integration technique which we will then describe. This workable technique is called *Monte Carlo integration*, and it is one of the two cornerstones of numerical simulation of statistical mechanics systems. In the next chapter we will develop the other cornerstone, *molecular dynamics*.

The most straightforward numerical technique to evaluate an integral of the form

$$I = \int_a^b \mathrm{d}x f(x),$$

is to split the interval from a to b into n equal segments each of length h where $h = (b - a)/n$. If we define $x_k = a + kh$, then the ith segment is bounded by $x_{i-1} < x < x_i$, and provided the function f is sufficiently smooth, we can approximate the integral I by the sum

$$I \approx \frac{h}{2} \sum_{i=0}^{n-1} (f(x_i) + f(x_{i+1})).$$

The limit as $h \to 0$ of this formula is a possible definition of the Riemann integral of $f(x)$. For h finite, this formula can be evaluated on a computer, and a simple numerical algorithm to approximate I is to evaluate this formula for a decreasing sequence of values of h or equivalently an increasing sequence of values of n. When h is sufficiently small, the results of these evaluations will normally converge to a fixed value which becomes the numerical estimate of the integral I.

This integration technique works quite well for a function of a single variable x. It can be extended to multidimensional integrals involving functions of more than one dimension, but at a cost. Each separate dimension must be split into n segments of size h as in the one-dimensional case, and the integral approximation becomes a multidimensional sum over all the dimensions in the problem. For a one-dimensional problem, $n + 1$ function evaluations are required since n segments will share $n + 1$ endpoints. For a two-dimensional problem, $(n + 1)^2$ function evaluations will be required, and for an N-dimensional problem, $(n + 1)^N$ function evaluations will be required. The problem is further complicated because the actual integral required is obtained by taking the limit of the resulting multidimensional sum as n gets large.

For a statistical system with N particles the integrals needed will be $6N$-dimensional integrals (3 coordinates, and 3 momenta per particle). Even for a relatively small system with only 100 particles, the number of function evaluations needed to evaluate a statistical mechanics average would be $(n + 1)^{600}$. This number of evaluations would take too long to complete on any imaginable computer, and so the simplest numerical integration is not applicable to statistical systems.

A second technique which is often used to evaluate integrals numerically is based on *random numbers*. Many different algorithms exist by which a computer can generate a sequence of numbers which behave as if they are random. Typically one or more of these algorithms are encoded in a function `rand()` which returns real values x uniformly distributed in the interval $0 \leq x < 1$. Executing a sequence of calls of the form

```
preceding          x = rand()
```

in a computer program will produce a sequence of different values for the variable `x`. These values will be uniformly distributed in the interval $0 \leq x < 1$. With sufficient ingenuity, this basic `rand()` function can be used to generate random numbers with any arbitrary probability distribution.

Consider now the problem of evaluating an N-dimensional integral over a region R of the form

$$\int_R \mathrm{d}^n x\, w(\mathbf{x}) f(\mathbf{x})$$

where $\mathbf{x} = (x_1, \ldots, x_n)$ is an n-component vector, $w(\mathbf{x})$ is a weight function which is positive over the region R, and $f(\mathbf{x})$ is some function to be integrated. If we define

$$Z = \int_R \mathrm{d}^n x\, w(\mathbf{x})$$

then

$$\phi(\mathbf{x}) = w(\mathbf{x})/Z$$

is a probability density function for points x over the region R. Modulo the constant Z, the evaluation of the integral of $f(\mathbf{x})$ is then just the evaluation of the expectation value $\langle f \rangle$ of f for this probability density

$$\langle f \rangle = \int_R f(\mathbf{x}) \phi(\mathbf{x}) \mathrm{d}^n x.$$

Computer random number generation techniques can now be used to generate a sequence of independent vectors $\mathbf{x}_i$, $i = 1, \ldots, N$ distributed with this probability density. The quantities $f_i \equiv f(\mathbf{x}_i)$, and $\bar{f} \equiv \sum_{i=1}^{N} f_i/N$ are then *unbiased estimators* of the expectation value $\langle f \rangle$. To show this we simply need to evaluate the expectation values of these estimators. The expectation value $\langle f_i \rangle$ of f_i is given by the probability that $\mathbf{x}_i$ takes the value $\mathbf{x}$ times the value $f(\mathbf{x})$, summed over all possible values of $\mathbf{x}$

$$\langle f_i \rangle = \int_R f(\mathbf{x}) \phi(\mathbf{x}) \mathrm{d}^n x = \langle f \rangle.$$

The expectation value of $\bar{f}$ then follows trivially since the different $\mathbf{x}_i$ are independent

$$\langle \bar{f} \rangle = \frac{1}{N} \sum_{i=1}^{N} \langle f_i \rangle = \langle f \rangle.$$

Next we calculate the variance of these estimators,

$$\langle (f_i - \langle f \rangle)^2 \rangle = \left\langle f_i^2 \right\rangle - \langle f \rangle^2 = \langle f^2 \rangle - \langle f \rangle^2$$

and

$$
\begin{aligned}
\langle(\bar{f}-\langle f\rangle)^2\rangle &= \langle\bar{f}^2\rangle-\langle f\rangle^2 \\
&= \frac{1}{N^2}\sum_{i=1}^{N}\sum_{j=1}^{N}\langle f_i f_j\rangle-\langle f\rangle^2 \\
&= \frac{1}{N^2}\sum_{i=1}^{N}\langle f_i^2\rangle+\frac{1}{N^2}\sum_{i\neq j}\langle f_i f_j\rangle-\langle f\rangle^2 \\
&= \frac{1}{N}\langle f^2\rangle+\frac{N(N-1)}{N^2}\langle f\rangle^2-\langle f\rangle^2 \\
&= \frac{1}{N}\langle(f-\langle f\rangle)^2\rangle.
\end{aligned}
$$

The quantity $\langle(f-\langle f\rangle)^2\rangle$ is completely determined since it is given by the integral:

$$\langle(f-\langle f\rangle)^2\rangle=\int_R (f(\mathbf{x})-\langle f\rangle)^2\phi(\mathbf{x})\mathrm{d}^n x.$$

In order to get a good numerical algorithm for the integral we need a calculable unbiased estimator whose variance can be made as small as we like. The estimator $\bar{f}$ is perfect for this purpose. Its variance is $1/N$ times a fixed quantity, and we can therefore make this variance as small as we wish by making N sufficiently large.

The integration algorithm therefore proceeds by generating a large number of independent points x_i, $i=1\ldots N$, evaluating $f_i=f(\mathbf{x}_i)$ on these points, and calculating $\bar{f}=\sum_i f_i/N$. Then $\bar{f}$ is an estimator for the integral

$$Z^{-1}\int_R \mathrm{d}^n x\, w(\mathbf{x})f(\mathbf{x}).$$

This algorithm is called *Monte Carlo integration*. We note that the Monte Carlo integration technique is a significant improvement on the Riemann sum technique we first described. It can produce unbiased estimators of integrals normalized by an appropriate factor Z with just a single function evaluation, and the unbiased estimators improve as the number of function evaluations increases. The technique has two basic problems however. First, it does not allow a direct evaluation of the normalization factor Z. For statistical mechanics purposes this is unfortunate, since it means that a direct evaluation of the partition function or its logarithm, and therefore the free energy, is not possible with Monte Carlo integration. Second, while it is possible, in principle, to find a direct numerical algorithm to generate points in a large-dimensional phase space distributed according to a particular probability density, in practice this is usually impossible to achieve with a realistic computer and consequently Monte Carlo integration of the particular kind so far described is only used in very special cases.

5.2 Markov processes

The simple Monte Carlo integration we have described above depends on being able to generate points within the multidimensional region we are integrating over and distributed according to a given probability distribution. In a statistical mechanics system, different points represent different possible states of the system. For a system which has a phase space parameterized by $3N$ momentum variables, $\mathbf{p}_i$, $i = 1 \dots N$, and $3N$ coordinate variables $\mathbf{x}_i$, $i = 1 \dots N$, a state is specified by giving values to all these $6N$ variables. The integral to be performed for such a system is then

$$\langle A \rangle = Z_N^{-1} \int \mathrm{d}^{3N} p \, \mathrm{d}^{3N} x \, \mathrm{e}^{-\beta H} A$$

$$Z_N = \int \mathrm{d}^{3N} p \, \mathrm{d}^{3N} x \, \mathrm{e}^{-\beta H}$$

where $H \equiv H(p, x)$ is the Hamiltonian. Our numerical problem is to generate states distributed with probability density

$$\phi(p, x) \mathrm{d}^{3N} p \, \mathrm{d}^{3N} x = Z_N^{-1} \mathrm{e}^{-\beta H(p,x)} \mathrm{d}^{3N} p \, \mathrm{d}^{3N} x.$$

In the last section we have described the difficulties to produce workable algorithms to generate independent states with this distribution. What is possible, however, is to generate a sequence of states each one of which is generated in a probabilistic way from the immediately preceding state in the sequence. By carefully adjusting the probabilistic transitions between neighbors in the sequence it is possible to adjust the weighting of states within the sequence so that they are distributed according to any desired probability density. The process which controls this probabilistic transition is called a *Markov process*, and it is the basis of all numerical integration techniques in statistical mechanics.

To simplify the discussion of Markov processes, we will assume for the moment that our statistical mechanics system has only a finite number of distinct states which we label by an index $i = 1 \dots N_s$ where N_s is the total number of states in the system. For the system with continuous coordinates $\mathbf{p}_i$, $\mathbf{x}_i$, we can arrange this by dividing the total phase space into hypercubes of volume $\Delta p^{3N} \Delta x^{3N}$ and we can specify states by specifying the identity of the particular hypercube into which the momenta and coordinates of that state fall. So this assumption does not limit the applicability of the techniques we are about to describe in any way.

The problem we face is to find an algorithm which generates states from the set of all states with a given probability distribution. Let ϕ_i be the required probability for state i. Since the set of states $i = 1 \dots N_s$ includes all possible states in the system we have

$$\sum_{i=1}^{N_s} \phi_i = 1.$$

The evaluation of the expectation $\langle A \rangle$ reduces to the evaluation of the sum

$$\langle A \rangle = \sum_{i=1}^{N_s} A(i)\phi_i$$

where $A(i)$ is the value of A on state i.

For the hypercube discrete system introduced above, for example, ϕ_i is given by

$$\phi_i = \frac{1}{Z_N} \exp(-\beta H) \Delta p^{3N} \Delta x^{3N} \Big|_i$$

where the notation $|_i$ denotes that the function on the right-hand side is to be evaluated on the ith hypercube. The relation $\sum \phi_i = 1$ then becomes

$$\sum_i \frac{1}{Z_N} \exp(-\beta H) \Delta p^{3N} \Delta x^{3N} \Big|_i = 1$$

and the definition for the expectation value of A becomes

$$\langle A \rangle = \sum_i A|_i \left(\frac{1}{Z_N} \exp(-\beta H) \Delta p^{3N} \Delta x^{3N} \Big|_i \right)$$

These latter two equations are just discretized versions of the continuum definitions of Z_N and $\langle A \rangle$ given at the beginning of this section and can reasonably be expected to approach their continuum values when Δp and Δx become small.

The basic idea underlying our integration technique to evaluate $\langle A \rangle$ is to generate a sequence of states each one of which is produced in a probabilistic way from the preceding state in the sequence. Start with any convenient initial state which we label i_0. Execute a transition to produce state i_1. Execute a second transition to produce i_2. This can be repeated over and over to produce a sequence $i_0, i_1, \ldots, i_\tau, i_{\tau+1}, \ldots$ Since states in the sequence are time-ordered during the computation, the state i_0 is produced first, then the state i_1, etc. It is convenient to think of the label τ specifying position in the sequence as a time index.

Suppose the sequence is in state i_τ at time τ. A completely general transition rule to the next state in the sequence can be defined by specifying a matrix M whose elements $M_{ji}(\tau)$ specify the transition probabilities that state i at time τ jumps to state j at time $\tau + 1$. The indices and argument of M allows for the possibility that the transition probabilities can differ from step to step, that they can differ depending on the state i at time τ, and that more than one possible state j is allowed for the result of the transition. A moment's thought shows, however, that M must satisfy two fundamental constraints. First, if the values of M are to represent transition probabilities, they must all be real and positive.

$$M_{ji}(\tau) \geq 0 \quad \text{for all } i, j, \tau.$$

Second, we require that all transitions must produce an allowed state of the system, so that the probability that state i at time τ jumps to j when summed over all possible states j must be exactly one.

$$\sum_{j=1}^{N_s} M_{ji}(\tau) = 1.$$

These two properties will play an important role in the following and we therefore introduce the following definitions:

Definition 5.1 A *Markov matrix* M is a matrix which satisfies the two conditions: *(i)* $M_{ji} \geq 0$ for all i, j, and *(ii)* $\sum_j M_{ji} = 1$ for all i.

Definition 5.2

(1) A process which uses $M_{ji}(\tau)$ to generate a sequence of states is called a *stochastic process*.
(2) A *Markov process* is a stochastic process which uses the same transition matrix at all times, i.e. M_{ji} is independent of τ.

Markov processes are the stochastic processes normally used to integrate statistical mechanics averages. The restriction that M be τ independent allows us to develop a connection between the process defined by M_{ji} and the underlying state probabilities ϕ_i which we wish to model. For the more general case this connection is much more difficult to make, and so much harder to actually implement in a computer program. In the following we therefore concentrate on Markov processes defined by a Markov matrix M.

The problem we are now faced with is to establish the link between the probability density of a given statistical system and the corresponding Markov matrix M. We begin by describing the conditions on M for the existence of a corresponding probability density. This is contained in the following theorem.

Theorem 5.1 All eigenvalues λ of the transition probability matrix M have modulus $|\lambda| \leq 1$. Furthermore, if there exists a finite integer $k \geq 1$ such that $0 < (\mathsf{M}^k)_{ji} < 1$ for all i, j then there is a unique eigenvector of M with eigenvalue $\lambda = 1$. This eigenvector can be normalized so that all its components are non-negative, and sum to unity.

Proof. If $v_i, i = 1 \dots N_s$ is an eigenvector of M with eigenvalue λ, then

$$\sum_i M_{ji} v_i = \lambda v_j.$$

Summing over the index j, we find

$$\sum_j \lambda v_j = \sum_j \sum_i M_{ji} v_i = \sum_i v_i$$

where we have used property *(ii)* in Definition 5.1 to produce the rightmost term. This equation implies either $\lambda = 1$ or $\sum_j v_j = 0$. If we take the modulus of both sides above we find

$$\sum_i |M_{ji}||v_i| = \sum_i M_{ji}|v_i| \geq |\lambda||v_j|.$$

Again summing over j, and using property *(ii)*, this becomes

$$\sum_j \sum_i M_{ji}|v_i| = \sum_i |v_i| \geq |\lambda| \sum_j |v_j|.$$

This equation implies $|\lambda| \leq 1$ and proves the first part of our theorem.

We now proceed to prove the existence of an eigenvector with eigenvalue 1 and to show that this eigenvalue is non-degenerate. Existence of $\lambda = 1$ is equivalent to the vanishing of the determinant of the matrix

$$\mathsf{Q} = \begin{pmatrix} M_{11} - 1 & M_{12} & \cdots M_{1N_s} \\ M_{21} & M_{22} - 1 & \cdots M_{2N_s} \\ \cdots & & \\ M_{N_s 1} & M_{N_s 2} & \cdots M_{N_s N_s} - 1 \end{pmatrix}.$$

This however follows from the observation that the elements of each row of Q sum up to zero by condition *(ii)*, which implies that not all N_s rows of Q are linearly independent.

On the other hand if ϕ is an eigenvector of M with eigenvalue 1, then clearly $\mathsf{M}^k \phi = \phi$. Now taking again the modulus of both sides we have

$$\sum_j (M^k)_{ij} |\phi_j| \geq |\phi_i|.$$

If we furthermore sum over the index i and use that $\sum_i (M^k)_{ij} = 1$ we find

$$\sum_{ij} (M^k)_{ij} |\phi_j| = \sum_i |\phi_i|,$$

which together with the previous equation implies that $\sum_j (M^k)_{ij} |\phi_j| = |\phi_i|$. Thus there exists an eigenvector, ϕ, with eigenvalue 1 and such that $\phi_i \geq 0$ for all i. In fact since $(M^k)_{ij} > 0$, for all i and j, the equation $\mathsf{M}^k \phi = \phi$ implies that $\phi_i > 0$

for all i. If we now assume that there exists a second eigenvector ψ with eigenvalue 1 for which we can assume without restricting the generality that $\psi_i > 0$ for all i, then, by linearity, $\phi + \alpha\psi$ with α a real number, is also an eigenvector with eigenvalue 1. We can thus choose the value α such that one component of $\phi + \alpha\psi$ vanishes with the non-zero components being positive. This however contradicts our previous conclusion that all components of this eigenvector have to be strictly positive. Thus the eigenvector with eigenvalue 1 is unique. □

The Markov matrix M is said to be *connected* if the condition $0 < (\mathsf{M}^k)_{ji} < 1$ is satisfied for some finite k and for all i, j. In concrete terms this condition means that there is always a finite but not unit probability to jump from any state i to any other state j providing a sufficient number of transitions are allowed. Requiring connectivity allows us to exclude non-generic Markov matrices M which do not allow access to all possible states of the system from a given starting state. Examples of such non-generic matrices include the identity matrix $M_{ji} = \delta_{ji}$, and permutation matrices which just cycle between different states in a deterministic manner.

If M is a connected transition matrix, we have shown there exists a unique vector ϕ with components $\phi_i \geq 0$ such that

$$\sum_i \phi_i = 1$$

$$\sum_i M_{ij}\phi_j = \phi_i .$$

Since the ϕ_i's are all positive, and since they sum to unity, these components can be interpreted as probabilities. Furthermore, since all other eigenvalues of M have magnitude less than unity, repeated multiplication of an arbitrary vector **v** by M will project out the component of **v** which is parallel to ϕ and suppress all other eigenvectors in **v**. Under multiplication by M the component of **v** parallel to ϕ will replicate itself. All other components will be suppressed by at least a factor equal in magnitude to that largest eigenvalue of M which is strictly less than unity. The vector ϕ represents therefore the equilibrium which is reached after repeated multiplications by M.

Definition 5.3 The unique eigenvector ϕ of a connected Markov matrix M which has eigenvalue $\lambda = 1$ and which is normalized so that its components are positive and sum to unity is called the *equilibrium probability distribution* of M.

Our original problem was to find a technique to generate states distributed according to some specified probability distribution. What we have just shown is that any given connected Markov matrix has an equilibrium probability distribution associated to it, so we now need to consider how to reverse this association and find a Markov matrix which has our specified probability distribution as its equilibrium distribution. For this it is convenient to replace the requirement of an equilibrium distribution by the following stronger condition:

Definition 5.4 A Markov matrix M_{ji} is said to satisfy *detailed balance* for a probability density ϕ with components ϕ_i, $i = 1 \ldots N_s$ if

$$M_{ji}\phi_i = M_{ij}\phi_j \quad \text{for all } i, j \text{ (no index sum implied).}$$

The relation between this condition and the existence of an equilibrium probability distribution is contained in the following theorem:

Theorem 5.2 If M satisfies detailed balance for the probability density ϕ and if M is connected, then ϕ is the equilibrium probability distribution for M.

Proof. Summing the detailed balance defining relation over index i, we find

$$\sum_i M_{ji}\phi_i = \sum_i M_{ij}\phi_j = \phi_j$$

where we have used property *(ii)* of Markov matrices. Thus ϕ is an eigenvector of M with eigenvalue 1. Also by definition ϕ is a probability distribution, and so must be the unique probability distribution which is M's equilibrium distribution. □

Detailed balance is a sufficient but not necessary condition that a transition matrix M has a given probability distribution ϕ as its equilibrium probability distribution (see Problem 5.5). Other less constraining conditions are possible but, as we shall see in the next section, detailed balance has the very nice feature that it can be imposed "on the fly" on an element by element basis when implemented in a computer algorithm. Less constraining conditions on M require information about many more of the elements of M for their implementation, and since M is a very large matrix even for the simplest of systems, actual computation with these conditions is impossible practically.

Transition matrices M satisfying detailed balance have considerable structure which is useful for us to develop. Define the matrix Φ with components Φ_{ji} as

$$\Phi = \text{diag}(\phi_1, \cdots, \phi_i, \cdots)$$

that is, Φ is a diagonal matrix with components along the diagonal given by the components of the probability density ϕ. The matrix Φ can be used to rewrite the detailed balance condition as:

$$\begin{aligned}
\left(\Phi^{-\frac{1}{2}}\mathsf{M}\Phi^{\frac{1}{2}}\right)_{ij} &= \phi_i^{-\frac{1}{2}} M_{ij}\phi_j\phi_j^{-\frac{1}{2}} \\
&= \phi_i^{-\frac{1}{2}}\phi_i M_{ji}\phi_j^{-\frac{1}{2}} \quad \text{(using detailed balance)} \\
&= \phi_i^{\frac{1}{2}} M_{ji}\phi_j^{-\frac{1}{2}} \\
&= \left[(\Phi^{-\frac{1}{2}}\mathsf{M}\Phi^{\frac{1}{2}})^T\right]_{ij}.
\end{aligned}$$

Therefore if M satisfies detailed balance, then $\mathsf{S} \equiv \Phi^{-\frac{1}{2}}\mathsf{M}\Phi^{\frac{1}{2}}$ is real and symmetric. On the other hand real symmetric matrices have the property that all eigenvalues of a symmetric matrix are real and furthermore that all eigenvectors can be organized into a complete orthonormal basis for the vector space on which the matrix acts.

The eigenvector of S with eigenvalue 1 has the components $(\phi_1^{\frac{1}{2}}, \ldots, \phi_i^{\frac{1}{2}}, \ldots)$. More generally for a symmetric matrix S with eigenvalues λ_i, $i = 1 \ldots N_s$, and if Y_i are orthogonal projection operators onto the one-dimensional spaces spanned by the eigenvectors of S, then these two properties imply that S can be expressed as a direct weighted sum

$$\mathsf{S} = \sum_i \lambda_i \mathsf{Y}_i,$$

and as a result

$$\mathsf{M} = \sum_i \lambda_i \mathsf{Z}_i\,, \qquad \mathsf{Z}_i = \Phi^{\frac{1}{2}}\mathsf{Y}_i\Phi^{-\frac{1}{2}}.$$

Note also that the matrices Z_i are orthogonal projection operators, since $\mathsf{Y}_i\mathsf{Y}_j = \delta_{ij}\mathsf{Y}_j$ immediately implies $\mathsf{Z}_i\mathsf{Z}_j = \delta_{ij}\mathsf{Z}_i$.

We have therefore arrived at the result that a connected transition matrix M which satisfies detailed balance for a probability distribution ϕ can be expanded as a weighted sum of orthogonal projection operators. The weights in this sum are the eigenvalues of M, and the projection operators project onto the orthogonal eigenvectors of M. If we label the eigenvalues of M so that $\lambda_1 = 1$, then we can write

$$\mathsf{M} = \mathsf{Z}_1 + \sum_{i \neq 1} \lambda_i \mathsf{Z}_i,$$

where the λ_i, $i \neq 1$ are real and have modulus $|\lambda_i| < 1$, and Z_1 has the components $(Z_1)_{ij} = \phi_i$, for all j.

Consider now a Markov process generated by a connected transition matrix M which satisfies detailed balance. This process generates a sequence of states by repeatedly applying M to an initial state. Suppose the initial state is i_0. Applying

M once produces a new state i_1. The probability that i_1 is a particular state j is the probability that M generated a transition from i_0 to j. This probability is M_{ji_0} since this is just what the components of M are. Applying M a second time produces a state i_2. The probability that i_2 is a particular state j is also calculable, but it is a little more complicated than for i_1. To determine it we must sum all the possible ways that two successive transitions can arrive in the state j. The possibilities are that i_0 could go to k for any k in the first transition, then that k could go to j in the second transition. The probability of this occurring is therefore given by $\sum_k M_{jk} M_{ki_0} = (\mathsf{M}^2)_{ji_0}$. The pattern produced by repeated application of M to starting state i_0 is now clear. If i_a is the state which results after a steps, then the probability that i_a is a particular state j is $(\mathsf{M}^a)_{ji_0}$.

If we exclude the initial state i_0, then τ applications of M will generate the sequence $\{i_1, i_2, \ldots, i_\tau\}$. The preceding analysis allows us to calculate the expected number of times, $N(j, \tau)$, that state j will appear in this sequence of τ states. This expected number is just the sum of the probabilities that each different state in the sequence is equal to j and is given by

$$N(j, \tau) = (\mathsf{M})_{ji_0} + (\mathsf{M}^2)_{ji_0} + \cdots + (\mathsf{M}^\tau)_{ji_0}.$$

The representation of M in terms of projection operators allows us to evaluate this expression. We have, for example,

$$\begin{aligned}
\mathsf{M}^2 &= \left(\mathsf{Z}_1 + \sum_{i \neq 1} \lambda_i \mathsf{Z}_i\right)\left(\mathsf{Z}_1 + \sum_{j \neq 1} \lambda_j \mathsf{Z}_j\right) \\
&= \mathsf{Z}_1 + \sum_{i \neq 1} \lambda_i^2 \mathsf{Z}_i,
\end{aligned}$$

since $\mathsf{Z}_i^2 = \mathsf{Z}_i$, and $\mathsf{Z}_i \mathsf{Z}_j = 0$ if $i \neq j$. Similar expressions hold for other powers of M, so we can express $N(j, \tau)$ as

$$\begin{aligned}
N(j, \tau) &= \sum_{a=1}^{\tau} (\mathsf{M}^a)_{ji_0} \\
&= \sum_{a=1}^{\tau} \left(\mathsf{Z}_1 + \sum_{i \neq 1} \lambda_i^a \mathsf{Z}_i\right)_{ji_0} \\
&= \tau (\mathsf{Z}_1)_{ji_0} + \sum_{i \neq 1} \frac{\lambda_i (1 - \lambda_i^\tau)}{1 - \lambda_i} (\mathsf{Z}_i)_{ji_0}.
\end{aligned}$$

Suppose now that $A(i)$ is some function defined on the states i of our statistical mechanics system. Our overall aim is to evaluate expectation values of A on a given probability distribution Φ. These expectation values are defined to be

$$\langle A \rangle = \sum A(i) \phi_i.$$

We now want to compare this with a numerical estimate of the expectation value of A in terms of the Markov matrix. For this we define

Definition 5.5 The Markov integration estimate for $\langle A \rangle$ is the value obtained when A is averaged over the states in the Markov sequence $\{i_1, \ldots, i_\tau\}$. If we denote this estimate by $\bar{A}_\tau$ we have

$$\bar{A}_\tau = \frac{1}{\tau} \sum_{a=1}^{\tau} A(i_a) .$$

In any given computer calculation the value of $\bar{A}_\tau$ will depend, of course, on the state i_0 used to start the Markov process, on the particular transition matrix M used for the process, and on the number of steps τ in the process. It will also depend on the particular transitions which are generated as the process is executed. Remember that the matrix M specifies transition probabilities only. When M is actually applied at any given step, some particular transition must be chosen from all those allowed. The particular transition chosen will depend on random numbers generated within the computer during the course of the calculation. As a result the actual value of $\bar{A}_\tau$ generated by any given computer calculation is not a predictable quantity. This is acceptable however since the whole calculation process is probabilistic, and we can predict the expected value of $\bar{A}_\tau$ if the results of a large number of independent calculations are averaged. If we denote the process of averaging over calculations by $\langle \cdots \rangle_{\text{Calc}}$, then we have

$$\langle \bar{A}_\tau \rangle_{\text{Calc}} = \frac{1}{\tau} \sum_j A(j) N(j, \tau),$$

since $N(j, \tau)$ counts the average number of times that state j will appear in a sequence of length τ. Using our result for $N(j, \tau)$ above we find

$$\begin{aligned} \langle \bar{A}_\tau \rangle_{\text{Calc}} &= \sum_j A(j) (\mathsf{Z}_1)_{j i_0} + \frac{1}{\tau} \sum_j A(j) \left(\sum_{i \neq 1} \frac{\lambda_i (1 - \lambda_i^\tau)}{1 - \lambda_i} (\mathsf{Z}_i)_{j i_0} \right) \\ &= \sum_j A(j) \phi_j + \frac{1}{\tau} \sum_j A(j) \left(\sum_{i \neq 1} \frac{\lambda_i (1 - \lambda_i^\tau)}{1 - \lambda_i} (\mathsf{Z}_i)_{j i_0} \right) . \end{aligned}$$

To arrive at this last expression we have used the fact that Z_1 has components $(Z_1)_{ji} = \phi_j$ for all i.

In the limit $\tau \to \infty$ of a system with finite dimensional Markov matrix M, the only term in this expression which survives is the term $\sum_j A(j)\phi_j$. This term is exactly the expectation value we set out to calculate. The remaining terms are all annihilated by the factor $1/\tau \to 0$ as $\tau \to \infty$. Although we have used detailed balance in the derivation of this result, it is not hard to show that it holds for any

Markov matrix. Thus our Markov integral estimate, averaged over independent calculations, $\langle \bar{A}_\tau \rangle_{\text{Calc}}$, produces the statistical mechanics expectation we need in the limit $\tau \to \infty$. By a calculation similar to the one we have just completed it is also possible to show that the fluctuation in $\bar{A}_\tau$ between independent calculations goes to zero as $\tau \to \infty$. This is of course to be expected since, as we have seen in Chapter 3, fluctuations in the average of τ different measurements of a given variable normally vanish like $1/\sqrt{\tau}$ for τ large. It has the nice consequence, however, that

$$\bar{A}_\tau \to \langle A \rangle \quad \tau \to \infty$$

even when not averaged over independent calculations. This relation is our final defining relation for Monte Carlo integration with Markov processes. It states that the average of A evaluated on a sequence of states generated by a Markov process will approach the statistical mechanics expectation of A as the number of states in the Markov sequence goes to infinity.

To complete our task we still need to give a construction of a Markov matrix for a given equilibrium distribution. In the next two sections we will see that such matrices can be quite easily constructed for concrete systems making use of the detailed balance condition.

5.3 The Ising model

As a simple example of the application of Markov integration techniques we will consider the two-dimensional Ising model. This model describes a system of spins which are arranged on the vertices of a square grid. Spins are identified by their position on the square grid, and can take values ± 1 only. We label the spin which is located at the intersection of the ith vertical and jth horizontal grid line by $s_{i,j}$, where i and j take integer values in the range $1 \leq i, j \leq L$, and where L is the length of the grid. Spins can interact with their nearest neighbors and with an externally applied magnetic field B which is normally assumed to be constant over the whole grid.

Any given spin will have four nearest neighbors. For $s_{i,j}$ these nearest neighbors are $s_{i-1,j}$, $s_{i+1,j}$, $s_{i,j-1}$, and $s_{i,j+1}$. We impose periodic boundary conditions on the system when evaluating nearest neighbors. This is easily implemented by requiring that the arithmetic which evaluates neighbors be executed modulo the length L of the grid. Thus the nearest neighbors to spin $s_{L,L}$ are $s_{L-1,L}$, $s_{1,L}$, $s_{L,L-1}$, and $s_{L,1}$ for example.

The Hamiltonian for the Ising model is

$$H = -g \sum_{i,j} (s_{i,j} s_{i+1,j} + s_{i,j} s_{i,j+1}) - B \sum_{i,j} s_{i,j} .$$

States in the system are specified by specifying values for all L^2 spins. Since each spin has exactly two possible values, there are exactly 2^{L^2} states possible, and the Ising model therefore is an example of a statistical mechanics system which has a finite discrete state space.

Function expectations in the canonical ensemble are given by

$$\langle A \rangle = Z_N^{-1} \sum_{\text{states}} \exp(-\beta H) A$$

$$Z_N = \sum_{\text{states}} \exp(-\beta H)$$

and typical functions A include the Hamiltonian H of the system, and the magnetization $S = \sum_{i,j} s_{i,j}$. Our goal in this section is to find a numerical algorithm to evaluate these expectation values. To do this we must find a Markov matrix M for this system. This transition matrix must be connected so that we can access all states of the system from a given starting state. Following our discussion in the last section we will furthermore impose detailed balance for this transition matrix.

Consider first the possible states of the Ising model. These are specified by giving values for all the possible spins in the system. Transitions between states can be made by flipping (changing sign) of one or more of the spins in a given state. The simplest possible transition is to flip a single spin, so we will consider now how to build a Markov matrix using detailed balance which attempts to execute this simple transition.

Below we will explain two approaches to construct a Markov matrix for this system. First we introduce a simple procedure which leads to the so-called *Metropolis Markov matrix*. A more sophisticated ansatz introduced in the second part will lead to a *heat bath Markov matrix*.

In the present case, we have chosen a particular spin, spin $s_{i,j}$ for example, and wish to find a transition matrix $\mathsf{M}(i, j)$ which is to decide whether to flip that spin. Let q denote the state of our system before the transition occurs, and let r denote the state which results if spin $s_{i,j}$ is flipped. Detailed balance requires that

$$M(i, j)_{rq}\phi_q = M(i, j)_{qr}\phi_r$$

where $\phi_q = \exp(-\beta E_q)/Z_N, \phi_r = \exp(-\beta E_r)/Z_N$ and E_q and E_r are the energies of states q and r respectively. Note first that the factors Z_N appearing in ϕ_q and ϕ_r play no role in this equation. They are common to both sides and therefore immediately cancel. Our detailed balance equation therefore becomes

$$M(i, j)_{rq} \exp(-\beta E_q) = M(i, j)_{qr} \exp(-\beta E_r) .$$

A naive transition rule which we could now adopt is simply to always flip the spin $s_{i,j}$. This rule corresponds to the choice $M(i, j)_{rq} = 1 = M(i, j)_{qr}$. Note that this choice is symmetric in the states q and r. This rule is certainly easy to implement but it satisfies detailed balance only if the states q and r have the same energy. If the energies of the two states are different, then detailed balance is not satisfied. However, if the energies are different, then it is possible to distinguish q and r since one of these states will have an energy greater than the other.

The Metropolis accept/reject algorithm uses this distinguishability to correct detailed balance. The algorithm rule is to always accept a transition if the new state has lower energy than the old, and to accept a transition with probability $\exp(-\beta(E_{\text{new}} - E_{\text{old}}))$ if the new state has greater energy than the old. This rule adds an extra probability component to the Markov matrix defining the spin flip transition. The probability of generating a transition now becomes the probability of attempting a transition times the probability of accepting that attempt. Suppose, for example, that we have $E_r < E_q$. An attempted transition from q to r will always succeed, while an attempted transition from r to q will succeed only some of the time. The actual probabilities for both of these transitions are then

$$\begin{aligned} M_{rq}(i,j) &= 1 \qquad E_r < E_q \\ M_{qr}(i,j) &= \exp(-\beta(E_q - E_r)) \qquad E_r < E_q \end{aligned}$$

Condition (ii) for a Markov matrix then requires that $M(i,j)_{qq} = 0$ and $M(i,j)_{rr} = 1 - M_{qr}(i,j)$ respectively. Detailed balance is trivially satisfied since

$$\begin{aligned} M_{rq}(i,j)\phi_q &= \exp(-\beta E_q) \\ &= \exp(-\beta(E_q - E_r))\exp(-\beta E_r) \\ &= M_{qr}(i,j)\phi_r. \end{aligned}$$

If the opposite case holds, that is $E_r > E_q$, then the transition probabilities are different

$$\begin{aligned} M_{rq}(i,j) &= \exp(-\beta(E_r - E_q)) \qquad E_r > E_q \\ M_{qr}(i,j) &= 1 \qquad E_r > E_q \end{aligned}$$

where the diagonal elements are again determined by condition (ii). However, detailed balance again holds.

A Metropolis algorithm attempts to execute a transition $q \to r$ which may or may not succeed. The possible results of the transition therefore are that $q \to r$ or $q \to q$. If the system under consideration is such that there are more than two possible final states after the transition (for instance if every individual spin can take more than two values) then the Metropolis algorithm is usually defined by the Markov matrix

$$M_{rq}(i,j) = W_{rq} \times \begin{cases} 1 & E_r < E_q \\ \exp(-\beta(E_r - E_q)) & E_r > E_q \end{cases}$$

where $W_{rq} = W_{qr}$ is has to be adjusted such that the total probability equals one (condition (ii)).

An alternative algorithm to employ is the heat bath algorithm. This algorithm allows multiple possible final states but tries to adjust the transition probabilities to these final state so that an accept/reject step is not required. For a spin-flipping transition the two possible final states are q and r. The heat bath algorithm chooses the transition probability to state r to be proportional to $\exp(-\beta E_r)$ and that to q to be proportional to $\exp(-\beta E_q)$. Since only two final states are possible we have therefore:

$$\begin{aligned} M_{qq} &= \exp(-\beta E_q)/(\exp(-\beta E_q) + \exp(-\beta E_r)) \\ M_{rq} &= \exp(-\beta E_r)/(\exp(-\beta E_q) + \exp(-\beta E_r)). \end{aligned}$$

To check detailed balance we also need the probabilities for transitions which start in state r. The possibilities are either that $r \to r$ or $r \to q$. The corresponding probabilities are given by

$$\begin{aligned} M_{rr} &= \exp(-\beta E_r)/(\exp(-\beta E_q) + \exp(-\beta E_r)) \\ M_{qr} &= \exp(-\beta E_q)/(\exp(-\beta E_q) + \exp(-\beta E_r)). \end{aligned}$$

Comparing probabilities, it is simple to see that detailed balance is trivially satisfied in this case.

Both the Metropolis and heat bath algorithms provide implementations of a Markov matrix $\mathsf{M}(i, j)$ which attempts transitions involving the flip of a single spin $s_{i,j}$. By construction these transition matrices will satisfy detailed balance. However, since $\mathsf{M}(i, j)$ allows only a single spin to flip, it is not a connected Markov matrix. Starting in any given state a single $\mathsf{M}(i, j)$ will only allow transitions between this state and one other state in the system. All remaining states of the system will remain inaccessible.

There are many possible solutions to this final problem. The simplest is to use a compound transition matrix which is the tensor product of the single-spin Markov matrices we have already generated. For example we can define the compound matrix to be

$$\mathsf{M} = \mathsf{M}(1, 1) \otimes \cdots \otimes \mathsf{M}(L, L).$$

This matrix is Markov, and has the same equilibrium probability distribution as the individual spin-flip matrices $\mathsf{M}(i, j)$ (see Problem 5.6).

The compound matrix M is easily seen to be connected. It allows all possible spin flips, so all states of the system are accessible from any given starting state. Since M is a connected Markov matrix with the correct equilibrium probability distribution, the sequence of states it generates can be used to evaluate the expectation averages for the Ising model in the canonical ensemble.

5.4 Implementation of the algorithm

Let us now describe how to implement the Metropolis algorithm for the two-dimensional Ising model numerically on a computer. The first task is to set up an array of L^2 variables to parametrize the possible configurations. To continue one programs an algorithm performing the following steps:

(1) Set up a counter counting from $n = 1, \cdots, L^2$ with each step representing a position in the array.
(2) Choose an initial configuration, that is, initial values (± 1) for the L^2 variables using a random number generator.
(3) Evaluate the energy of the initial configuration using the discrete Hamiltonian given at the beginning of Section 5.3.
(4) Set the initial value of the counter to $n = 1$.
(5) Evaluate the energy of the configuration obtained after changing the sign of the nth variable in the array.
(6) If the energy difference between the new configuration and the previous configuration is less than or equal to zero, then the program should update the array by changing the sign of the variable in question, increment the counter, $n \to n + 1$ and then return to step (5). If, however, the energy difference is bigger than zero one chooses a random number, x, between 0 and 1, using the random number generator. Then, if $\exp(-\beta(E_{\text{new}} - E_{\text{old}})) \geq x$ the array will be updated before incrementing the counter and the program returns to step (5). If $\exp(-\beta(E_{\text{new}} - E_{\text{old}})) < x$ then the array is not updated and the program returns to step (5) after incrementing the counter.
(7) Once the counter has reached the last variable in the array ($n = L^2$) the program evaluates the observable of interest (e.g. the energy) and returns to step (4).

In fact, the precise initial configuration is not particularly relevant since the Markov process loses its memory after a few steps. However, ideally the initial configuration should not be chosen to be totally atypical in order to save computer time. An alternative to updating the spins in the array in order of increasing n, is to choose a site randomly and repeat this process L^2 times to generate a new configuration. It would appear that another way to speed up the process is to update all spins before doing the Metropolis test. However, this drastically reduces the acceptance rate and should therefore be avoided.

The program set up in this way can now be used to evaluate the expectation values of the various observables by evaluating the value of the observable in each update of the complete array. In the case of the Ising model these expectation values will depend on the temperature β and the external magnetic field B. Care should be taken to take into account that when these external parameters are modified the program will take a certain number of steps before settling near the new equilibrium configuration. In the case at hand, the external parameters are $t \equiv 1/\beta g$ and $b \equiv$

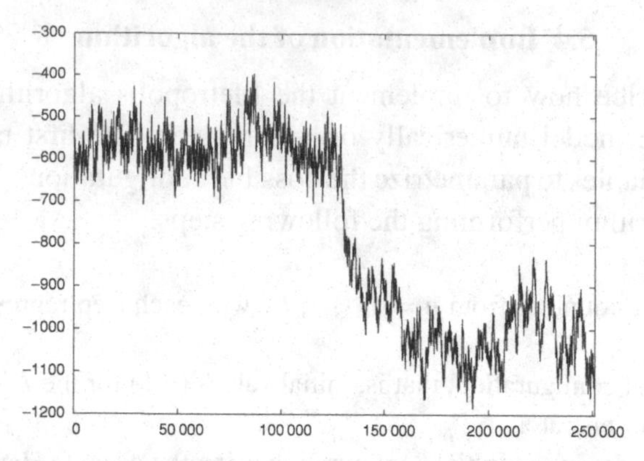

Figure 5.1 Fluctuations of the internal energy as a function of the time after increasing the inverse temperature from 1 to 1.5 in numerical units.

βB. As an example we plot in Figure 5.1 the "measured" energy as a function of the time after lowering the temperature. Taking this into account, the sampling period necessary for measuring the various quantities can be shortened significantly if the sampling is started after the system has settled into the new equilibrium state.

After the program has been "tuned", that is the parameters have been set such that fluctuations between different "runs", i.e. different measurements of energy, are reasonably small we can start with the quantitative analysis of the system. As an example let us measure the specific heat, c_V, as a function of temperature. This quantity is interesting since it can be used to detect signatures of a phase transition. We recall that c_V is given in terms of the internal energy U by $c_V = \partial\langle U\rangle/\partial T$, where

$$\begin{aligned} \langle U\rangle &= \frac{1}{Z_N}\sum_{\text{states}} H\,\exp(-\beta H) \\ &= -\frac{1}{Z_N}\frac{\partial Z_N}{\partial\beta}\,. \end{aligned}$$

Thus,

$$c_V = k\beta^2\left(\sum_{\text{states}}\frac{H^2\exp(-\beta H)}{Z_N} - \langle U\rangle^2\right).$$

But this last expression is just $k\beta^2$ times the variance of U introduced in Section 5.1,

$$\mathrm{Var}(U) = \langle (U - \langle U\rangle)^2\rangle\,.$$

Thus we have the simple result, $c_V = k\beta^2\mathrm{Var}(U)$. In order to compute c_V in the Metropolis algorithm we then evaluate U and U^2 by sampling the Hamiltonian $H(i_a)$ over the sequence of states i_a, $a = 1, \cdots \tau$ with the program set up above. We plot the results as function of t in Figure 5.2.

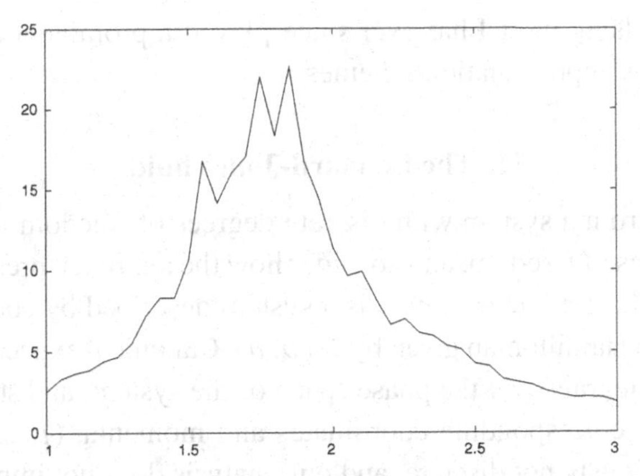

Figure 5.2 The specific heat, c_V, as a function of temperature on 32×32 lattice and a sampling period of 1 million steps at each temperature.

The interpretation of this result is interesting. We see that c_V is peaked near the temperature t_c. On the other hand we have seen above that c_V is a measure of the variance of U. Thus we see that the fluctuations in the energy grow near t_c. This is the typical behaviour of a second-order phase transition (see Chapter 13). Thus Figure 5.2 suggests that the two-dimensional Ising model undergoes a second-order phase transition at the critical temperature $t_c \simeq 1.7$. If we take the values for the spins to be $\pm 1/2$ instead of ± 1, then t_c is rescaled by a factor $(1/2)^2$ so that $t_c \simeq 0.43$.

Note that the plot of c_V in Figure 5.2 is not at all smooth, in spite of having averaged over a substantial number of updates. This is for two reasons: first since c_V itself is a variance, the fluctuations in c_V are a measure of the "variance of a variance". This is one source of the problem, since variances of variances are generally big. Second the algorithm we have used is not ideal for this purpose. The result can be improved considerably by using a so-called "cluster" algorithm, where a large block of spins is grown and then flipped in unison. Another possible modification is to replace the canonical sampling by a micro canonical sampling where a set of "demons" circulates around the lattice trying to flip spins. Each carries a sack of energy ranging. Any energy change associated with a spin flip is compensated by a change in this sack. If the demon's sack cannot accommodate the change, the flip is rejected. In this mode the temperature is not fixed, but calculated from the average demon energy. We refer the reader to the literature at the end of this chapter for further details about these algorithms. We finish this section by noting that the two-dimensional Ising model has been solved exactly in a veritable "tour de force" by L. Onsager in 1944. He did not care to publish the derivation and so it was not until 1952 that C. N. Yang gave the first published derivation. The two-dimensional Ising model has ever since played a prominent role as a testing ground for various approximation schemes.

5.5 The Lennard-Jones fluid

Let us turn now from a system with discrete degrees of freedom to a system with continuous degrees of freedom, and consider how the approach presented so far has to be modified. The generic case now is a system described by coordinates q_i and momenta p_i, with Hamiltonian given by $H(p, q)$. Canonical expectation values are now defined as integrals over the phase space of the system, and states are defined by specifying the corresponding coordinates and momenta, (p, q). The space of states is now obviously not discrete, and our analysis does not immediately apply. However, we have already described a simple approximation which determines how to proceed. This approximation is to divide the continuous phase space of the system into a sequence of small discrete elements. For example we could divide the phase space into a large number of hypercubes which have principal axes lying along the momentum and coordinate axes of phase space. If we define the length of a hypercube to be Δp_i on the edge parallel to the p_i momentum axis, and Δq_i on the edge parallel to the q_i axis, then each hypercube will have volume $\prod_i \Delta p_i \Delta q_i$. Each hypercube will contain very many different states of our system but, if the hypercube is sufficiently small, we can expect that all these states are approximately equivalent. As a result, we can adopt the view that each hypercube defines a single distinct state with coordinates and momenta given, for example, by the center point of the hypercube.

Mathematically, this approximation is nothing more than the statement that we can approximate an integral over some volume by a sum of contributions from sub-volumes. Thus

$$\begin{aligned}\langle A \rangle &= Z_N^{-1} \int \prod_i \frac{\mathrm{d}p_i \mathrm{d}q_i}{2\pi\hbar} \mathrm{e}^{-\beta H} A \\ &\simeq Z_N^{-1} \sum_s \prod_i \frac{\Delta p_i \Delta q_i}{2\pi\hbar} \left(\mathrm{e}^{-\beta H} A\right)\Big|_s ,\end{aligned}$$

where

$$\begin{aligned}Z_N &= \int \prod_i \frac{\mathrm{d}p_i \mathrm{d}q_i}{2\pi\hbar} \mathrm{e}^{-\beta H} \\ &\simeq \sum_s \prod_i \frac{\Delta p_i \Delta q_i}{2\pi\hbar} \left(\mathrm{e}^{-\beta H}\right)\Big|_s .\end{aligned}$$

The index s here runs over the individual hypercubes which make up the states of the system in this approximation, and the notation $|_s$ denotes that the corresponding factors are evaluated on the state s.

In this discrete formulation, we can now apply all our results about Markov processes. States in our system are labeled by an index s. The equilibrium probability for a state in a hypercube centered at s is given by

$$\phi_s = Z_N^{-1}\, \mathrm{e}^{-\beta H}\Big|_s \prod_i \frac{\Delta p_i \Delta q_i}{2\pi\hbar}$$

and the detailed balance condition we need to impose on a Markov matrix to generate this equilibrium probability is the standard one

$$M(t \leftarrow s)\phi_s = M(s \leftarrow t)\phi_t.$$

Here $M(t \leftarrow s)$ is the probability that state s jumps to state t. In our discrete approximation, state t is a hypercube, and we expect therefore that the probability that s jumps to t should depend on the volume of the hypercube t. Thus we find that $M(t \leftarrow s)$ should have the form

$$M(t \leftarrow s) = \left(\prod_i \frac{\Delta p_i \Delta q_i}{2\pi\hbar}\right)\Bigg|_t \mathcal{M}(t \leftarrow s)$$

where $\mathcal{M}$ is now a function. In deriving this formula, we have made explicit the dependence of M on the hypercube volume, by factorizing it out. The new function $\mathcal{M}$ so introduced is therefore a probability density function.

The correct generalization of the Markov method to systems with continuous degrees of freedom is now clear. The equilibrium probability vector ϕ_i becomes a probability density function $\phi(s)$. The Markov transition matrix M also becomes a density function, $\mathcal{M}(t \leftarrow s) \equiv \mathcal{M}(t, s)$. The variables s and t here are continuous, and take values in the phase space of the system. If $\mathrm{d}\Gamma$ is an infinitesimal volume element in this phase space, then $\phi(s)\ \mathrm{d}\Gamma_s$ is the equilibrium probability density that the system occupies an infinitesimal volume $\mathrm{d}\Gamma_s$ about the point s in phase space, and $\mathcal{M}(t, s)\mathrm{d}\Gamma_t$ represents the probability that the Markov transition matrix generates a jump from a point s in phase space to an infinitesimal volume element $\mathrm{d}\Gamma_t$ about the point t.

The basic properties of the equilibrium probability vector ϕ_i and of a Markov transition matrix M for the Lennard-Jones fluid are then obtained from those of a discrete system in the following way.

$$\phi_i \geq 0 \Longrightarrow \phi(s) \geq 0$$
$$\sum_i \phi_i = 1 \Longrightarrow \int \mathrm{d}\Gamma\phi = 1$$
$$\mathcal{M}_{ji} \geq 0 \Longrightarrow \mathcal{M}(t, s) \geq 0$$
$$\sum_j M_{ji} = 1 \Longrightarrow \int \mathrm{d}\Gamma_t \mathcal{M}(t, s) = 1$$
$$M_{ji}\phi_i = M_{i,j}\phi_j \Longrightarrow \mathcal{M}(t, s)\phi(s) = \mathcal{M}(s, t)\phi(t).$$

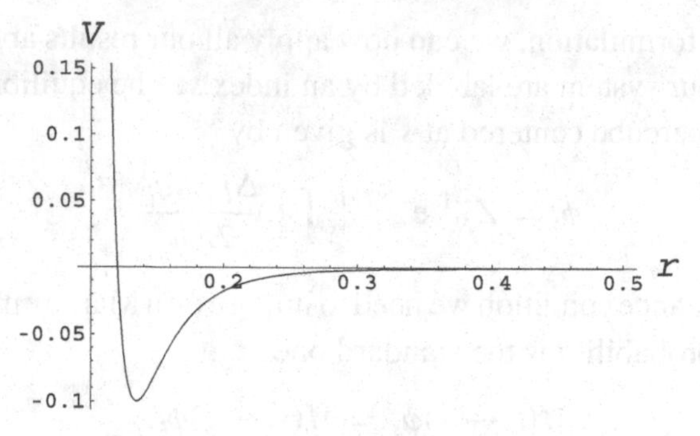

Figure 5.3 Lennard-Jones potential

We will now consider how to apply our understanding of how Markov processes are generated for systems with discrete state spaces to generate a Markov integration process for a system with continuous degrees of freedom. The system which we will analyze is that of a *Lennard-Jones fluid.* This is a system of N atoms or molecules which interact pairwise. The momenta and positions of the atoms are $\mathbf{p}_i$, and $\mathbf{x}_i$ with $i = 1, \ldots, N$. The Hamiltonian for the system is

$$H = \sum_i \frac{1}{2m} |\mathbf{p}|^2 + \sum_{i<j} V_{ij}$$

with an interaction potential between the ith and jth atom

$$V_{ij} \equiv V(r_{ij}), \quad r_{ij} = \left|\mathbf{x}_i - \mathbf{x}_j\right|.$$

A Lennard-Jones fluid is a fluid with potential

$$V(r) = 4\epsilon\left(\left(\frac{\sigma}{r}\right)^{12} - \left(\frac{\sigma}{r}\right)^{6}\right).$$

The shape of this potential is of the form shown in Figure 5.3. It is a heuristic potential to approximate the interatomic potential with repulsion between two atoms at short distance and a weak attractive force at intermediate distances representing the chemical binding energy. As mentioned in Chapter 4 this potential has the added feature that the first virial coefficient can be computed exactly.

For the calculation of the partition function, the quantity βH will be relevant. In all numerical calculation, it is useful to transform from dimensional quantities like β, H, $\mathbf{p}_i$ and $\mathbf{x}_i$ to dimensionless quantities before generating any computer programs. By so transforming we can arrange that the typical numbers which we work with are quantities of $O(1)$ rather than the very large or very small numbers with which we had to deal if working in m-kg-s units. In the Lennard-Jones system, we have four dimensional parameters, ϵ which has units of energy, σ which has units of length, m which has units of mass, and β which also has units of energy. The standard approach to rescaling physical quantities is to define a numerical length scale, S_x, a numerical time scale, S_t, and a numerical mass scale S_m, and re-express all variables in these new scales, that is

$$\mathbf{x}_i^{(n)} = S_x^{-1}\mathbf{x}_i, \quad t^{(n)} = S_t^{-1}t, \quad m^{(n)} = S_m^{-1}m$$

where $\mathbf{x}$, t and m are expressed in m-kg-s units. Since momenta is defined in terms of mass times velocity, it also scales and we find that momentum in time units, $\mathbf{p}_i^{(n)}$, is given by

$$\mathbf{p}_i^{(n)} = m^{(n)}\frac{\mathrm{d}\mathbf{x}_i^{(n)}}{\mathrm{d}t^{(n)}} = \frac{S_t}{S_m S_x}\mathbf{p}_i.$$

We can now apply these changes of units in βH and consider how we can best choose the scale factors S_x, S_t, and S_m. A convenient choice is

$$S_x = \sigma, \quad S_m = m, \quad \text{and} \quad \frac{S_m S_x^2}{\epsilon S_t^2} = 1.$$

With this choice, $m^{(n)} = 1$, and the Lennard-Jones Hamiltonian takes the form,

$$\beta E_{LJ} = \frac{1}{T}\left(\sum_i \frac{1}{2}|\mathbf{p}_i|^2 + \sum_{i<j} V_{ij}\right)$$

where

$$V_{ij} = V(r_{ij}) = 4\left(r_{ij}^{-12} - r_{ij}^{-6}\right)$$

and we have defined $T \equiv 1/\beta\epsilon$. This form of the Lennard-Jones system is the standard form used in all numerical calculations, and we have dropped the suffixes (n) on the coordinates and momenta here since we will presume that we are working in numerical units from now on. Note the very important result, however, that the Lennard-Jones system is defined by a single parameter T, which is known as the reduced temperature. Further note that this parameter is dimensionless. The apparently independent variables β, ϵ, m, and σ with which we began have been reduced to just a single dimensionless variable T.

One further point about this Hamiltonian needs to be made. The canonical ensemble integrations which we need to perform have the momenta degrees of freedom completely unconstrained. The position degrees of freedom, however, are constrained to lie within the volume V in which the system is confined. In principle, it is quite simple to implement such a constraint in any numerical calculation. In practice, numerical calculations are very computer intensive, and we tend to work with only very small systems as a result. In a very small system, the edge effects resulting from the very small volumes in which the particles are confined can be very large indeed. The normal approach is therefore to impose periodic boundary conditions on the system rather than confining it directly in a volume. Thus if we have N particles in a volume V, we view the system as if there are infinitely many copies of this volume replicated over and over in all three dimensions. An atom close to the edge of our original volume can then interact with atoms within the replicated volumes surrounding the original system.

Let us turn now to the problem of generating a Markov process which will allow the calculation of expectation values of operators in the canonical ensemble of the Lennard-Jones system. The basic problem is to find a Markov transition matrix for this system. The phase space for the Lennard-Jones system is $6N$-dimensional, with N momenta, $\mathbf{p}_i$ and N coordinates, $\mathbf{x}_i$. A state for the system is described by giving values for all $6N$ momenta and coordinates, and a Markov transition for the system must begin in some state and transit to another state with probability which satisfies the detailed balance condition. Clearly, attempting a direct transition in which all $6N$ coordinates and momenta change together is impossible. Our experience with the Ising model however provides the solution. We shall build a general Markov transition matrix as a product of simpler Markov matrices each of which modify just a few of the degrees of freedom. In the Ising model, these simpler transition matrices flipped only a single spin at a single site. This suggests that we look for two basic kinds of transition matrix. The first kind should change a single momentum degree of freedom, the second should change a single position degree of freedom.

We consider first finding a transition matrix which changes a momentum degree of freedom, $\mathbf{p}_i$, to a new value $\mathbf{p}'_i$. We denote this transition matrix, $\mathcal{M}(\mathbf{p}'_i \leftarrow \mathbf{p}_i)$. In designing this matrix, we need to make sure to satisfy detailed balance. This takes the form

$$\mathcal{M}(\mathbf{p}'_i \leftarrow \mathbf{p}_i)\mathrm{e}^{-\beta H} = \mathcal{M}(\mathbf{p}_i \leftarrow \mathbf{p}'_i)\mathrm{e}^{-\beta H'}$$

where H' denotes the value taken by the Hamiltonian after $\mathcal{M}$ has acted, and H denotes the value taken by the Hamiltonian before $\mathcal{M}$ has acted. The action of $\mathcal{M}$ here is simply to change the value of one momentum degree of freedom, $\mathbf{p}'_i \leftarrow \mathbf{p}_i$. This momentum degree of freedom contributes to only one term in the Hamiltonian. All other terms are left unchanged. Thus the detailed balance equation can be reduced to a very simple form,

$$\mathcal{M}(\mathbf{p}'_i \leftarrow \mathbf{p}_i)\mathrm{e}^{-\frac{1}{2T}|\mathbf{p}_i|^2} = \mathcal{M}(\mathbf{p}_i \leftarrow \mathbf{p}'_i)\mathrm{e}^{-\frac{1}{2T}|\mathbf{p}'_i|^2}.$$

The simplest implementation of a Markov which satisfies this detailed balance condition is to choose

$$\mathcal{M}(\mathbf{p}'_i \leftarrow \mathbf{p}_i) = \frac{1}{z} e^{-\frac{1}{2T}|\mathbf{p}'_i|^2}$$

where

$$z = \int d^3 p'_i \, e^{-\frac{1}{2T}|\mathbf{p}'_i|^2} .$$

Remember that $\mathcal{M}$ is a probability density function for systems with continuous degrees of freedom. What this equation tells us to do is to generate $\mathbf{p}'_i$ with the Gaussian density function given by this last equation. There exist many ways to generate Gaussian random numbers. Note that in generating $\mathbf{p}'_i$ this way we do not need to know the value of $\mathbf{p}_i$. It drops everywhere from the equations.

Let us turn now to the problem of finding a transition matrix which changes a position degree of freedom, $\mathbf{x}_i$, to a new value, $\mathbf{x}'_i$. We denote this transition matrix by $\mathcal{M}(\mathbf{x}'_i \leftarrow \mathbf{x}_i)$. Again, this matrix is required to satisfy detailed balance,

$$\mathcal{M}(\mathbf{x}'_i \leftarrow \mathbf{x}_i) e^{-\beta H} = \mathcal{M}(\mathbf{x}_i \leftarrow \mathbf{x}'_i) e^{-\beta H'}$$

where, as before, H' denotes the value taken by the Hamiltonian after $\mathcal{M}$ has acted, and H denotes the value taken by the Hamiltonian before $\mathcal{M}$ has acted. The action of $\mathcal{M}$, now, is simply to change the value of one position degree of freedom, $\mathbf{x}'_i \leftarrow \mathbf{x}_i$. This position degree of freedom contributes to only the potential term in the Hamiltonian. The kinetic energy is left unchanged. Thus the detailed balance equation can again be reduced to

$$\mathcal{M}(\mathbf{x}'_i \leftarrow \mathbf{x}_i) e^{-\frac{1}{2T}\sum_{j\neq i} V(|\mathbf{x}_i - \mathbf{x}_j|)} = \mathcal{M}(\mathbf{x}_i \leftarrow \mathbf{x}'_i) e^{-\frac{1}{2T}\sum_{j\neq i} V(|\mathbf{x}'_i - \mathbf{x}_j|)} ,$$

but now the form is not so simple, since the potential energy terms are more complicated than the kinetic energy terms. The ansatz we made for the momentum Markov transition matrix is thus now not appropriate. In particular, although the choice

$$\mathcal{M}(\mathbf{x}'_i \leftarrow \mathbf{x}_i) = \frac{1}{z} e^{-\frac{1}{2T}\sum_{j\neq i} V(|\mathbf{x}'_i - \mathbf{x}_j|)} , \quad z = \int d^3 x'_i \, e^{-\frac{1}{2T}\sum_{j\neq i} V(|\mathbf{x}'_i - \mathbf{x}_j|)}$$

is a valid Markov matrix, it is not the best choice since there does not exist any direct way to generate new positions $\mathbf{x}'_i$ according to this density function. In the momentum case, the density function involved was a Gaussian. There exist many ways to generate Gaussian random numbers, and we did not have this problem. However, a simple form for $\mathcal{M}$ for the position update is found using the Metropolis idea. This involves two steps. First we generate $\mathbf{x}'_i$ by adding a random vector $\mathbf{a}$ to

$\mathbf{x}_i$. Then we accept the new vector $\mathbf{x}'_i$ with the probability $\mathcal{A}$ where

$$\mathcal{A} = \begin{cases} 1 & e^{-\Delta H} > 1 \\ e^{-\Delta H} & e^{-\Delta H} < 1 , \end{cases}$$

with

$$\Delta H \equiv H' - H = \frac{1}{2T} \sum_{j \neq i} (V(|\mathbf{x}'_i - \mathbf{x}_j|) - V(|\mathbf{x}_i - \mathbf{x}_j|)).$$

In order that this transition rule satisfies detailed balance, the random vector **a** must be symmetrically distributed about the origin.

We have now defined two basic sets of transition matrices which implement transitions for momenta, and for coordinates. The generation of a sequence of states via these Markov matrices is now straightforward. We begin by specifying some initial state i_0 in which we arbitrarily assign coordinates and momenta. Then we generate a state i_1 by applying in turn the transitions $\mathcal{M}(\mathbf{p}'_i \leftarrow \mathbf{p}_i)$ for $i = 1, \ldots, N$, and $\mathcal{M}(\mathbf{x}'_i \leftarrow \mathbf{x}_i)$ for $i = 1, \ldots, N$. There are in all $2N$ applications of these transitions required to change all $2N$ vector momenta $\mathbf{p}_i$ and vector coordinates $\mathbf{x}_i$ in our starting configuration. After these $2N$ applications are made, a new state i_1 is generated.

This process is then repeated to generate states i_2, i_3, etc. Once we have a sufficiently large set of different states we can begin to calculate expectations of operators. These expectations are given by averaging the operators over the states generated by the Markov transition we have defined.

Algorithm: Let us now describe how to implement the Metropolis algorithm for the Lennard-Jones fluid numerically on a computer. First we set up an array of $2N$ vectors to parametrize the possible configurations in phase space. The program then has to perform the following steps:

(1) Choose an initial configuration, that is, initial values for the N position vectors using a random number generator and for the N momentum vectors using a Gaussian normal distribution, $f(\mathbf{p})$ with variance $\sigma = T$, and centered at the origin, that is

$$f(\mathbf{p}) = \frac{1}{\sqrt{2\pi T}} e^{-\frac{\mathbf{p}^2}{2T}} .$$

(2) Evaluate the kinetic and potential energy separately for this configuration.
(3) Update the first position vector by a random vector symmetrically distributed about the origin.
(4) Evaluate the difference in potential energy before and after the update. If the energy difference is less than or equal to zero, then the update for this position vector is accepted. If not, choose a random number, x, between 0 and 1, using the random number

generator and update the position if $\exp(-\beta(V_{\text{new}} - V_{\text{old}})) \geq x$. Otherwise leave this position vector unchanged.
(5) Repeat the previous step for all position vectors.
(6) Measure the quantity of interest for this configuration.
(7) Update the N momentum vectors using a Gaussian normal distribution, $f(\mathbf{p})$.
(8) Return to step (3).

The comments made after the suggested algorithm in Section 5.4 could be repeated here. In particular the concrete initial configuration is not important as long as it is not in some completely irrelevant part of the phase space. Furthermore updating all coordinates at once should be avoided since the resulting configuration is unlikely to satisfy the Metropolis criterion.

Note that for many measurements, such as the total energy or the pair distribution function, which is an order parameter for the melting transition (see Section 6.4), step (7) can be omitted since the average of the kinetic energy is determined by the temperature alone. Furthermore the momenta and the coordinates are updated independently so that there is no correlation between the two updates.

Problems

Problem 5.1 If a computer can execute one function evaluation in 10^{-9} seconds, how long would it take to evaluate a 600-dimensional integral when $n = 10$.

Problem 5.2 Calculate the probability that

$$(\bar{f} - \langle f \rangle)^2 < \epsilon(\langle (f - \langle f \rangle)^2 \rangle)$$

for ϵ fixed.

Problem 5.3 Consider the Markov matrix

$$\mathsf{M} = \begin{pmatrix} a & 0 & 0 \\ \frac{1}{2}(1-a) & 0 & 1 \\ \frac{1}{2}(1-a) & 1 & 0 \end{pmatrix}.$$

Show that $\mathsf{M}^n \mathbf{v}$ approaches a periodic orbit exponentially fast and thus the Markov process does not converge.

Problem 5.4 Prove that any transition matrix for a two-state system must satisfy detailed balance.

Problem 5.5 Prove that for a system with three or more states, detailed balance is a sufficient but not necessary condition for M to have ϕ as its equilibrium probability distribution. (Hint: count degrees of freedom.)

Problem 5.6 Prove that the tensor product of the single-spin Markov matrices defined in Section 5.3 has the same equilibrium probability distribution as the individual spin-flip matrices $\mathsf{M}(i, j)$.

Problem 5.7 Show that the compound Markov matrix M of Problem 5.6 does not satisfy detailed balance. Suggest a modified Markov matrix which does satisfy detailed balance.

Problem 5.8 Show that in the heat bath algorithm M satisfies the Markov condition $\sum_j M_{ji} = 1$.

Problem 5.9 Consider a circle of radius $r = 1$ centered on the origin which is inscribed in a square with side length $l = 2$ also centered on the origin. The area of the circle is π. The area of the square is 4. The ratio, $\pi/4$, of the two areas can be expressed as an integral

$$\frac{\pi}{4} = \frac{\int_{|x|<1,|y|<1} \mathrm{d}x\mathrm{d}y f(x, y)}{\int_{|x|<1,|y|<1} \mathrm{d}x\mathrm{d}y}$$

where $f(x, y)$ takes the value 1 if (x, y) is within the circle, and the value 0 otherwise. Define a Monte Carlo integration algorithm to evaluate this ratio. How many different pairs of points will you have to generate in order to have a 90% probability of calculating π correctly up to 10 decimal places with your algorithm.

Problem 5.10 Following the rules for implementing the Metropolis algorithm given in Section 5.4 write a program in your preferred language for a two-dimensional Ising model on a lattice with 20×20 sites. The program should have the following features

(1) It should allow for a variety of initial conditions.
(2) The temperature should be a controllable parameter in the program.
(3) It should be able to "measure" the spin-spin correlation function $< s_i s_j >$.

Use this program to estimate the internal energy as well as the expectation value of the magnetization as a function of the temperature for vanishing external field. Plot these quantities as a function of temperature. Using the spin-spin correlator estimate the magnetic susceptibility as a function of temperature and discuss the existence and order of the phase transition. Estimate the critical temperature, T_c.

Further reading

The Metropolis algorithm was first published in 1953 by N. Metropolis, A. Rosenbluth, M. Rosenbluth, A. Teller and E. Teller. Since then a number of variations of this approach have been suggested. A collection of articles on Monte Carlo methods is contained in K. Binder, *Monte Carlo Methods in Statistical Physics*, Springer (1979). In particular, the first article by K. Binder presents a nice summary of Monte Carlo methods. A good text on Monte Carlo methods together with sample codes can be found in S. E. Koonin, *Computational Physics*, Benjamin-Cummings (1986). For a recent graduate textbook see D. Landau and K. Binder, *A Guide to Monte Carlo Simulations in Statistical Physics*, Cambridge University Press (2000). An essential reference for numerical simulation of classical systems is M. P. Allen and D. J. Tildesley, *Computer Simulation of Liquids*, Clarendon Press (1986). A good and concise introduction to Monte Carlo methods in statistical physics can be found in D. Chandler, *Introduction to Modern Statistical Mechanics*, Oxford University Press (1987) and also in J. Zinn-Justin, *Quantum Field Theory and Critical Phenomena*, Clarendon Press (1993). The "cluster" algorithm is described in R. H. Swendsen and J.-S. Wang, *Phys. Rev. Letters* **58**, 86 (1987). The canonical sampling using "demons" can be found in M. Creutz, *Phys. Rev. Letters* **50**, 1411 (1983). Finally, the exact solution of the two-dimensional Ising model, due to Onsager, can be found, for instance, in C. J. Thompson, *Mathematical Statistical Mechanics*, Princeton University Press (1972).

6
Numerical molecular dynamics

In the last chapter, we described a numerical technique which allows us to calculate canonical ensemble partition functions. We saw there that the canonical ensemble required us to evaluate a large multidimensional integral, and our numerical approach was to attempt to construct representative numerical sampling methods. In this chapter we address the issue of how to evaluate expectation values in the micro canonical ensemble. This often turns out to be a much more economical way in actual numerical computations.

6.1 Equations of motion and the micro canonical ensemble

In the canonical ensemble we consider all possible states of the system when generating expectation averages. Different states are, however, weighted with different exponential factors, $\exp(-\beta H)$, which depend on their energy H. In the micro-canonical ensemble, we limit our attention only to states which have fixed energy (Figure 6.1). We saw in Chapter 3 that these two different ensembles produce the same results when applied to macroscopic systems where the number of degrees of freedom of the system is very large relative to the typical microscopic length scales. The defining formula for expectations of operators $A(p, q)$ in the micro canonical ensemble is

$$\langle A \rangle = \frac{\int_{E<H<E+\Delta} \mathrm{d}\Gamma\, A}{\int_{E<H<E+\Delta} \mathrm{d}\Gamma}.$$

As before, the calculation of $\langle A \rangle$ requires that we calculate a ratio of integrals. We face however a fundamental problem in defining these integrals because we must confine the integral to a shell of energy E and width Δ in phase space. For almost all systems, defining this shell in a useful way is not possible. The perfect gas is one of the few exceptions, and we have already discussed this exceptional case in Chapter 5.

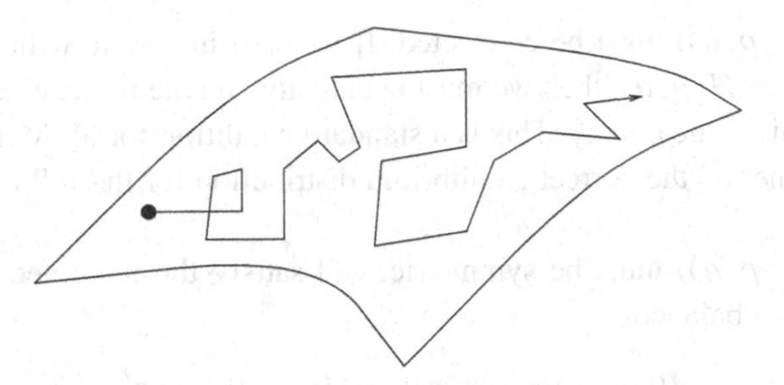

Figure 6.1 Sketch of a micro canonical sampling.

Let us first consider applying the numerical techniques we developed in the last chapter to the problem at hand. Monte Carlo methods require that we choose some random starting state for our system, then continually make Markov transitions from that starting state to new states. In order that we generate a correctly weighted sequence of states we need to arrange that the Markov transition probability $M(i \leftarrow j)$ from one state j to the next state i satisfies detailed balance:

$$M(i \leftarrow j)\mathrm{e}^{-\beta H}|_j = M(j \leftarrow i)\mathrm{e}^{-\beta H}|_i.$$

Markov methods are as applicable to the calculation of the micro canonical expectations as to canonical expectations, since a ratio of integrals is required in both cases. However, because we are in the micro canonical ensemble, the states involved, and the detailed balance condition are changed somewhat. In the micro canonical ensemble, we include only states with a fixed energy E. This means that the exponential factors in the detailed balance equation are always given by $\exp(-\beta E)$ independently of the state, and automatically cancel. For a system with a finite number of discrete states of fixed energy E, therefore, a satisfactory Markov transition matrix is one which is connected, and which is symmetric ($M_{i \leftarrow j} = M_{j \leftarrow i}$).

We have also applied Markov techniques to systems described by continuous variables in the last chapter. Here states represent points in phase space, and we define a state by specifying the coordinates of the corresponding phase space points, $(p, q) = (p_1, \ldots, p_n, q_1, \ldots, q_n)$. Markov transitions, which take states to new states or equivalently take points to new points in phase space, are mappings on phase space. A Markov transition matrix, $M((p', q') \leftarrow (p, q))$ which generates a micro canonical ensemble of states for a system with Hamilton $H(p, q)$ must be a mapping on phase space, which satisfies three conditions:

(1) $M((p', q') \leftarrow (p, q))$ must preserve energy so that $H(p', q') = H(p, q)$. This condition is necessary so that our transition probability only produces states with the correct energy required in the integrals we have to perform.

(2) $M((p', q') \leftarrow (p, q))$ must be connected. If we start in a state with coordinates p, q and energy $E = H(p, q)$, then we must eventually be able to generate all other states which have this same energy. This is a standard condition for all Markov processes if they are to generate the correct equilibrium distribution for the full phase space being studied.

(3) $M((p', q') \leftarrow (p, q))$ must be symmetric, and satisfy the restricted micro canonical form of detailed balance,

$$M((p', q') \leftarrow (p, q)) = M((p, q) \leftarrow (p', q')).$$

The first and the third condition suggest that we consider the natural energy-conserving mapping on phase space generated by the *Hamiltonian flow*, that is, by evolving any point for some time interval t under the equations of motion. Thus we define $(p', q') = (P(t, p, q), Q(t, p, q))$ where $P_i(t)$ and $Q_i(t)$ satisfy Hamilton's equations of motion with initial conditions $(P(0), Q(0)) = (p, q)$. We clearly get a different mapping on phase space for each different choice of t.

We first recall some important properties of Hamiltonian flows. If $A(p, q)$ is some function of the coordinates and momenta, and if H is the Hamiltonian for the system, then we have

$$\begin{aligned}
\frac{\mathrm{d}}{\mathrm{d}t} A &= \sum_i \frac{\partial A}{\partial q_i} \frac{\mathrm{d}q_i}{\mathrm{d}t} + \sum_i \frac{\partial A}{\partial p_i} \frac{\mathrm{d}p_i}{\mathrm{d}t} \\
&= -\left(\sum_i \frac{\partial H}{\partial q_i} \frac{\partial A}{\partial p_i} - \sum_i \frac{\partial H}{\partial p_i} \frac{\partial A}{\partial q_i} \right) \\
&= -\left(\sum_i \frac{\partial H}{\partial p_i} \frac{\partial}{\partial q_i} - \sum_i \frac{\partial H}{\partial q_i} \frac{\partial}{\partial p_i} \right) A.
\end{aligned}$$

This equation involves two fundamental objects which play important roles in classical mechanics. The first of these is the Poisson Bracket. The second is the Lie derivative.

Definition 6.1 If $A(p, q)$ and $B(p, q)$ are functions of the coordinates and momenta, then the *Poisson bracket* of A and B is denoted $\{A, B\}$, and is defined to be

$$\{A, B\} = \sum_i \left(\frac{\partial A}{\partial q_i} \frac{\partial B}{\partial p_i} - \frac{\partial A}{\partial p_i} \frac{\partial B}{\partial q_i} \right).$$

Definition 6.2 If $A(p, q)$ and $B(p, q)$ are functions of the coordinates and momenta, then the *Lie derivative* of B with respect to A is denoted $\mathcal{L}(A)B$, and is defined by the relation,

$$\mathcal{L}(A)B = \{A, B\}$$

which immediately implies

$$\mathcal{L}(A) = \left(\sum_i \frac{\partial A}{\partial p_i} \frac{\partial}{\partial q_i} - \sum_i \frac{\partial A}{\partial q_i} \frac{\partial}{\partial p_i} \right).$$

Inspection of these definitions immediately leads to the following properties

$$\{A, B\} = -\{B, A\}$$
$$\mathcal{L}(\lambda A + \mu B) = \lambda \mathcal{L}(A) + \mu \mathcal{L}(B)$$

where A and B are functions of momenta and coordinates, and λ and μ are constants. We will need one further important result about Poisson brackets and Lie derivatives. We state it as a theorem.

Theorem 6.1 If $A(p, q)$, $B(p, q)$, and $C(p, q)$ are functions of momenta and coordinates, then

$$\{A, \{B, C\}\} + \{B, \{C, A\}\} + \{C, \{A, B\}\} = 0.$$

This result is known as the Jacobi identity.

Proof. The proof of this theorem is completely mechanical. We simply replace the Poisson brackets in the statement of the theorem with their formula in terms of derivatives as given in the definition of the Poisson bracket. Rearrangement of the resulting expressions then gives the result. □

Corollary 6.2

$$[\mathcal{L}(A), \mathcal{L}(B)] = \mathcal{L}(\{A, B\})$$

In words, the commutator of the Lie derivatives with respect to two functions A and B is equal to the Lie derivative with respect to the Poisson bracket of the two functions A and B.

Proof. To prove this result we apply $[\mathcal{L}(A), \mathcal{L}(B)]$ to a third function C.

$$\begin{aligned} [\mathcal{L}(A), \mathcal{L}(B)]\, C &= (\mathcal{L}(A)\mathcal{L}(B) - \mathcal{L}(B)\mathcal{L}(A))\, C \\ &= \mathcal{L}(A)\, (\mathcal{L}(B) C) - \mathcal{L}(B)\, (\mathcal{L}(A) C) \\ &= \mathcal{L}(A)\, (\{B, C\}) - \mathcal{L}(B)\, (\{A, C\}) \\ &= \{A, \{B, C\}\} - \{B, \{A, C\}\} \\ &= \{\{A, B\}, C\} \quad \text{using Jacobi identity} \\ &= \mathcal{L}(\{A, B\}) C \end{aligned}$$

Since C is arbitrary, this establishes the corollary. □

Poisson brackets and Lie derivatives allow us to write the equations defining the time derivative of a function A in a compact form

$$\frac{\mathrm{d}}{\mathrm{d}t}A = -\{H, A\} = -\mathcal{L}(H)A.$$

This in turn leads us to define the exact *time-stepping operator* as

$$A|_{t+h} = \mathrm{e}^{-h\mathcal{L}(H)}A|_t.$$

This time-stepping operator represents exact evolution according to the Hamiltonian H. Thus we expect that it should keep the Hamiltonian H constant. Indeed, we have

$$\begin{aligned} H|_{t+h} &= \mathrm{e}^{-h\mathcal{L}(H)}\, H|_t \\ &= \left(1 - h\mathcal{L}(H) + \frac{1}{2}h^2\mathcal{L}(H)\mathcal{L}(H) + \ldots\right)H\bigg|_t . \end{aligned}$$

All terms except the first in the expansion of the exponential involve one or more applications of $\mathcal{L}(H)$ to H. But

$$\mathcal{L}(H)H = \{H, H\} = 0$$

so we see immediately that, when acting on H, the time-stepping operator gives

$$H|_{t+h} = H|_t$$

proving that H remains constant.

Equations of motion therefore generate a mapping on phase space which is energy conserving and satisfies the first condition needed for a mapping to be a Markov mapping generating a micro canonical ensemble. The second condition required is that the mapping be connected in the Markov sense. What we require is that the trajectory generated by the equations of motion visit all points in phase space which have energy E, before it returns to the point at which it starts. Systems governed by equations of motion which have this property are said to be *ergodic*. In Markov theory, connectedness and ergodicity are often used interchangeably as the name describing this particular property. Proving connectedness or ergodicity for classical mechanics systems is extremely difficult, and is the subject of much research. Some classical systems are known definitely to be non-ergodic or non-connected, and some are known to be ergodic or connected. In Figure 6.2 we give a schematic picture of a phase space with non-ergodic time evolution. We will consider this issue in a little more detail in Chapter 12. For the moment, we leave the issue hanging, and accept that the ergodicity is a requirement which we cannot determine simply, and which may or may not hold for any given system.

The final condition for a mapping to be a Markov transition which generates a micro canonical ensemble of configurations is that of symmetry. In the current

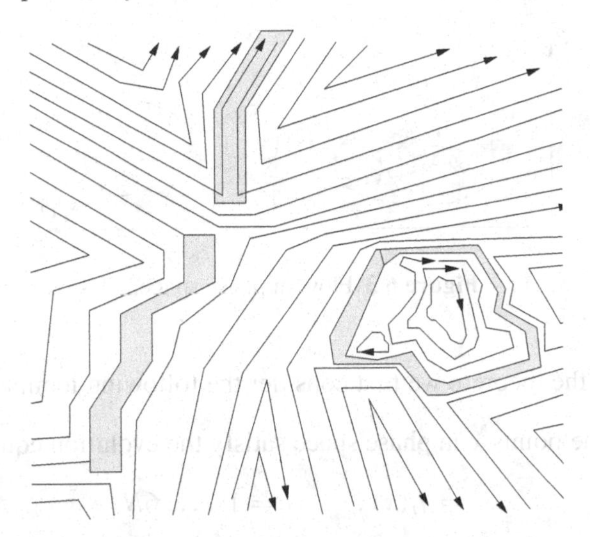

Figure 6.2 Sketch of a non-ergodic system. In an ergodic sytem all trajectories would be different sections of a single, long trajectory.

context, this condition needs some care to implement. In normal Markov processes, the detailed balance condition imposes a condition on the probability density that a state i is mapped to some infinitesimal volume element about a state j. For mappings generated by equations of motion, there is no probability at all involved. A single initial state goes to a single final state, so the size of an infinitesimal volume element about the state j seems to play no role. However, we can address the question in a slightly different way. Rather than mapping states to states, we can consider mapping volumes to volumes. A volume represents a collection of states. Now, since the Hamiltonian flow is reversible the final volume will be mapped again into the initial volume upon time reversion. If we furthermore accept that the size of the volume being mapped is a count of the number of independent states contained in that volume, then symmetry requires that the number of states remain unchanged under the mapping (see Figure 6.3). Imposing symmetry of the Markov transition matrix then implies that we must impose a condition of volume preservation on the mapping on the phase space which is implementing the transition. This, however is guaranteed by Liouville's theorem.

Theorem 6.3 The mapping on phase space generated by time evolution under Hamilton's equations of motion preserves phase space volume. If $D(0)$ is some region in phase space which is mapped to a region $D(t)$, then the volume in phase space occupied by $D(t)$ is equal to the volume occupied by $D(0)$. Explicitly,

$$\int_{D(t)} \prod_i \mathrm{d}P_i(t)\mathrm{d}Q_i(t) = \int_{D(0)} \prod_i \mathrm{d}P_i(0)\mathrm{d}Q_i(0).$$

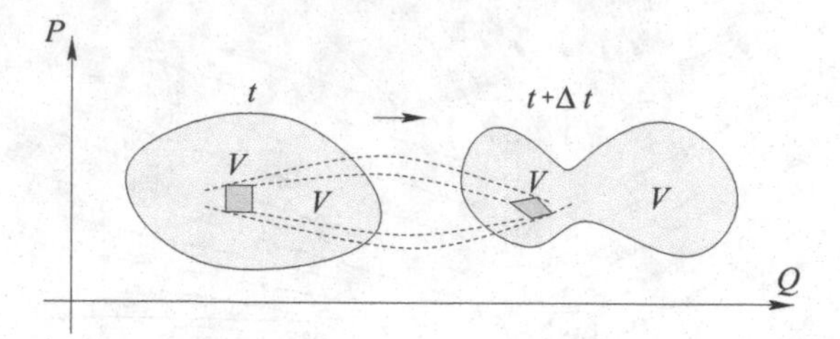

Figure 6.3 Flow in phase space.

Proof. To prove the theorem we first consider the following lemma:

Lemma 6.4 If the points x in phase space satisfy the evolution equation

$$\dot{x}_i = f_i(x)\,, \qquad i = 1, \ldots, 6N$$

then we have

$$\frac{\mathrm{d}}{\mathrm{d}t} D|_{t=0} = \int_D \left(\sum_{i=1}^{6N} \frac{\partial}{\partial x_i} f_i(x) \right) \mathrm{d}\omega$$

where $\mathrm{d}\omega = \mathrm{d}x_1 \ldots \mathrm{d}x_{6N}$ is the infinitesimal volume element in phase space.

Proof. Let $x(t) = g^t(x)$ be the time evolution of x. Then

$$D(t) = \int_{D(t)} \mathrm{d}\omega_t$$

$$= \int_{D(0)} \det \left(\frac{\partial g^t(x)_i}{\partial x_j} \right) \mathrm{d}\omega$$

where $\det(\cdots)$ is the Jacobian of the change of variable $g^t(x) \to x$. Now, for small t we can expand

$$(g^t(x))_i = x_i + f_i(x)t + O(t^2).$$

Thus

$$\frac{\partial g^t(x)_i}{\partial x_j} = \delta_{ij} + \frac{\partial f_i}{\partial x_j} t + O(t^2)$$

$$= \delta_{ij} + A_{ij}t + O(t^2).$$

On the other hand expanding the determinant as

$$\det(I + At) = 1 + t\,\mathrm{Tr}(A) + O(t^2)$$

we end up with

$$\frac{\mathrm{d}}{\mathrm{d}t}D|_{t=0} = \int_D \mathrm{Tr}(A)\,\mathrm{d}\omega$$

$$= \int_D \left(\sum_{i=1}^{6N} \frac{\partial}{\partial x_i} f_i(x)\right) \mathrm{d}\omega.$$

□

To prove the theorem we then notice that for a Hamiltonian evolution

$$\dot{x}_i = \begin{pmatrix} \dot{q}_\alpha \\ \dot{p}_\beta \end{pmatrix} = \begin{pmatrix} \frac{\partial H}{\partial p_\alpha} \\ \frac{-\partial H}{\partial q_\alpha} \end{pmatrix}, \quad \alpha, \beta = 1, \ldots, 3N.$$

Thus

$$\sum_{i=1}^{6N} \frac{\partial}{\partial x_i} f_i(x) = \sum_{\alpha=1}^{3N} \left(\frac{\partial^2 H}{\partial q_\alpha \partial p_\alpha} - \frac{\partial^2 H}{\partial p_\alpha \partial q_\alpha} \right) = 0.$$

It then follows from the lemma that

$$\frac{\mathrm{d}}{\mathrm{d}t}D|_{t=0} = 0.$$

We can repeat this proof without modification for any $t = t_0 \neq 0$. This, however, implies that $\mathrm{d}/\mathrm{d}t\, D = 0$ for all t so that

$$D(t) = D(0) \quad \text{for all } t.$$

□

Mappings on phase space which preserve phase space volume are called *symplectic mappings*. Mappings generated by evolution for any Hamiltonian function $H(p, q)$ are symplectic. But the reverse is not true. Not all symplectic mappings can be generated by evolution according to the equations of motion for some Hamiltonian.

What we have finally shown in the preceding analysis is that, so long as we are dealing with an ergodic system, the mapping on phase space generated by equations of motion satisfies all the conditions needed for it to be a Markov process which generates a micro canonical ensemble. If we initiate the Markov process with some state with coordinates $(p(0), q(0))$, and define the transition between states as being evolution for a time interval h, then the sequence of states generated by this Markov system is the set of states which occur along the evolution trajectory which begins with $(p(0), q(0))$, and which are separated by time intervals h. The states so included are therefore $(p(t_i), q(t_i))$ where $t_i = ih$ and i is a zero or positive integer.

The Monte Carlo formula for the expectation value of an operator A is then given in the usual way,

$$\lim_{N\to\infty} \frac{1}{N} \sum_{i=1}^{N} A(p(t_i), q(t_i)).$$

This formula does not place any constraints on h except that the resulting transitions between states should be connected in the Markov sense. If the system is ergodic, then the only condition we need to impose is that h is not a rational fraction of the time taken for the trajectory in phase space which begins at state $(p(0), q(0))$ to return to $(p(0), q(0))$. Since h is (almost) arbitrary, we can take the limit as $h \to 0$. To take this limit, we first rewrite our formula for the expectation value as a ratio of two sums

$$\langle A \rangle = \lim_{Nh\to\infty} \frac{h \sum_{i=1}^{N} A(p(t_i), q(t_i))}{hN}.$$

Taking the limit $h \to 0$ is now straightforward,

$$\langle A \rangle \to \lim_{T\to\infty} \frac{1}{T} \int \mathrm{d}t\, A(p(t), q(t)).$$

We find therefore that expectations in the micro canonical ensemble can be evaluated by averaging for a long time over a classical trajectory. Physically this is not a surprising result. Our physical view of expectations is that they represent values of macroscopic properties of statistical mechanics systems. These values are measured in macroscopic time. The underlying dynamics of the system, however, evolves in microscopic time, so any operator measured in macroscopic time must represent the average of that operator over a very long timescale relative to the typical interaction times of the atomic constituents of the system. Qualitatively therefore the expectation value of a macroscopic variable is given by the time average of the instantaneous values that that variable takes due to the microscopic motion of the system. If we assume that that microscopic motion is governed by classical equations of motion, then we arrive at the time-averaging formula we have just developed. Micro canonical ensemble averages become time averages.

6.2 Numerical integration

Let us turn now to the numerical problem of estimating time averages of operators over trajectories. We want to evaluate

$$\langle A \rangle = \lim_{T\to\infty} \frac{1}{T} \int \mathrm{d}t\ A(p(t), q(t)).$$

Clearly the work involved here is to actually generate the trajectories themselves. We need an algorithm which will allow us to find the momenta $p_i(t)$ and coordinates $q_i(t)$ as a function of t over some time interval T which, in principle, needs to be made infinitely large. Once we have these momenta and coordinate values, the calculation of the average of A is quite straightforward.

The equations of motion which govern a trajectory are Hamilton's equations. The simplest algorithm to solve ordinary differential equations is *Euler's algorithm*. This is generated by replacing the time derivatives on the left-hand side of Hamilton's equations with finite differences. Introducing a small time interval h, we have

$$\frac{\mathrm{d}p_i(t)}{\mathrm{d}t} = \frac{p_i(t+h) - p_i(t)}{h} + O(h)$$
$$\frac{\mathrm{d}q_i(t)}{\mathrm{d}t} = \frac{q_i(t+h) - q_i(t)}{h} + O(h).$$

If we substitute these finite difference approximations into Hamilton's equations and rearrange slightly, we find

$$p_i(t+h) = p_i(t) - h\frac{\partial H}{\partial q_i} + O(h^2)$$
$$q_i(t+h) = q_i(t) + h\frac{\partial H}{\partial p_i} + O(h^2).$$

These approximate equations represent the Euler-discretized version of our equations of motion. The Euler numerical integration method now is to choose a value of h which is sufficiently small so that the $O(h^2)$ contributions to these equations are negligible. The time interval h is called the *step size* for the numerical integration algorithm. We then have a system of equations which is simple to implement in a computer program. We initialize this system by choosing some random starting configuration of the system. This involves fixing $(p(0), q(0))$ and corresponds directly to choosing an initial configuration for a Markov integration in the canonical ensemble. The initial configuration and step size represent all the information we need to calculate the right-hand sides of the Euler-discretized equations for $t = 0$, and therefore allows us to calculate the left-hand side also. We generate a new configuration $(p(h), q(h))$. This process is obviously repeatable, and we generate a whole sequence of configurations at times $t = ih$ for $i = 0, 1, \ldots$

The Euler algorithm is only one of a large class of algorithms which can be used to solve ordinary differential equations. Standard text books on numerical analysis develop many other possibilities. However, all the various algorithms possible follow basically the pattern of the Euler algorithm. We begin by choosing some starting configuration. Then we evolve this configuration in some manner which approximates the evolution generated by the exact equations of motion. This generates a second configuration, and we can repeat this process over and over to generate a large sequence of configurations which should approximate the configurations we would have generated if we had integrated the equations of motion exactly. When attempting to generate a computer program to simulate the micro canonical ensemble, the major problem we face therefore is to decide which of the different algorithms we should adopt to implement the integration of the equations of motion. Once we have decided on an algorithm, the mechanics of its implementation follows almost exactly that pattern for the Euler algorithm.

6.3 Choices of algorithm

In order to choose between the various possible algorithms, we must develop some understanding of what possibilities exist, and of the advantages and disadvantages of each possibility. Let us consider therefore a generic set of coupled ordinary differential equations for functions $x_i(t)$ where $i = 1, \ldots, n$

$$\frac{\mathrm{d}}{\mathrm{d}t} x_i(t) = f_i(x).$$

The usual applications in molecular dynamics involve Hamiltonians which have no explicit time dependence, so the functions f_i depend explicitly only on $(x_1(t), \ldots, x_n(t))$. The solution of these differential equations gives us the functions $x_i(t)$.

Consider some function $A(x)$ which depends explicitly on these solutions $x_i(t)$, and therefore implicitly depends on t. At some time t the function A will have the value

$$A(x)|_t \equiv A(x(t))$$

and we can relate the values taken by A at different times by generating Taylor expansions,

$$\begin{aligned} A(x)|_{t+h} &= A(x)|_t + h \left.\frac{\mathrm{d}}{\mathrm{d}t} A(x)\right|_t + \left.\frac{1}{2!} h^2 \frac{\mathrm{d}^2}{\mathrm{d}t^2} A(x)\right|_t + \cdots \\ &= \left.\left(1 + h\frac{\mathrm{d}}{\mathrm{d}t} + \frac{1}{2} h^2 \frac{\mathrm{d}^2}{\mathrm{d}t^2} + \cdots\right) A(x)\right|_t \\ &= \left. \mathrm{e}^{h\, \mathrm{d}/\mathrm{d}t} A(x)\right|_t . \end{aligned}$$

The last expression here is a formal expression only. The operator $h\mathrm{d}/\mathrm{d}t$ acts on functions of t. We define the exponential of this operator by specifying how it acts on functions of t. This action is given by the power series expansion of the exponential function. So long as we act on functions which are infinitely differentiable and for which the series resulting from applying the exponential converges, then this formal definition is well defined. If, however, we act on non-differentiable functions, or if the resulting series does not converge, then problems will arise, and our formula for $A(x)|_{t+h}$ in terms of an exponential operator will not apply. For dynamical systems satisfying Hamilton's equations of motion, we will normally not have to worry, since the coordinates and momenta defining a trajectory are almost always infinitely differentiable, and the corresponding Taylor series expansions almost always converge.

The formal Taylor series expansion we have just derived can be written in a slightly different way by noticing that the differential equations defining the functions $x_i(t)$ allow us to partially evaluate the action of $\mathrm{d}/\mathrm{d}t$ on functions A. We have

$$\begin{aligned}\frac{\mathrm{d}}{\mathrm{d}t}A(x) &= \sum_i \left(\frac{\partial}{\partial x_i}A(x)\right)\frac{\mathrm{d}}{\mathrm{d}t}x_i\\ &= \sum_i f_i(x)\frac{\partial}{\partial x_i}A(x)\\ &= \left(f(x)\cdot\frac{\partial}{\partial x}\right)A(x).\end{aligned}$$

Thus the action operator $\mathrm{d}/\mathrm{d}t$ is seen to be equivalent to the action of the operator $f\cdot\partial/\partial x$. If we now replace $\mathrm{d}/\mathrm{d}t$ in our Taylor expansion formula, we find

$$A(x)|_{t+h} = \mathrm{e}^{hf\cdot\partial/\partial x}A(x)\Big|_t.$$

Functions x_i on the left-hand side of this equation are evaluated at $t+h$. Functions on the right-hand side are evaluated at t.

The operator $\exp(hf\cdot\partial/\partial x)$ is seen from these equations to be the time evolution operator for our system. Acting with this operator on an arbitrary function $A(x)$ at time t gives us the same function at the later time $t+h$. We see therefore that this operator is actually a time-stepping operator. It steps functions evaluated at time t to the same functions evaluated at time $t+h$. Further, to actually execute this step, we see that we need only the values of the functions $x_i(t)$ at t, the originating time. The effects of multiple steps with the stepping operator is also calculable. We have for example,

$$A(x)|_{t+h_1+h_2} = \mathrm{e}^{h_2 f\cdot\frac{\partial}{\partial x}}A(x)\Big|_{t+h_1}.$$

Note, however, that the right-hand side here is a function of $x_i(t+h_1)$ evaluated at time $t+h_1$, so we can step this again to find,

$$A(x)|_{t+h_1+h_2} = \mathrm{e}^{h_2 f\cdot\frac{\partial}{\partial x}}\mathrm{e}^{h_1 f\cdot\frac{\partial}{\partial x}}A(x)\Big|_t$$

All functions x_i on the right-hand side are evaluated now at time t, and we have therefore generated an expression which gives us $A(x)|_{t+h_1+h_2}$ as the action of two stepping operators on the function $A(x)|_t$.

Another important property of our stepping operator arises from the following observation

$$A(x)B(x)|_{t+h} = A(x)|_{t+h}\ B(x)|_{t+h}.$$

This immediately implies

$$e^{hf\cdot\frac{\partial}{\partial x}}\left(A(x)B(x)\right) = \left(e^{hf\cdot\frac{\partial}{\partial x}}A(x)\right)\left(e^{hf\cdot\frac{\partial}{\partial x}}B(x)\right).$$

This result can also be proved by expanding the exponential operator on the left in a power series, and noting that

$$f\cdot\frac{\partial}{\partial x}\left(A(x)B(x)\right) = \left(f\cdot\frac{\partial}{\partial x}A(x)\right)B(x) + A(x)\left(f\cdot\frac{\partial}{\partial x}B(x)\right).$$

The result has an even more general form,

$$A(x)|_{t+h} = A(x|_{t+h})$$

which implies

$$e^{hf\cdot\frac{\partial}{\partial x}}A(x) = A\left(e^{hf\cdot\frac{\partial}{\partial x}}x\right).$$

It is important to note however that both of these results require that we include all the terms in the infinite series expansion of the exponential stepping operator.

Let us turn now to the Euler numerical algorithm to integrate the coupled differential equations for x_i. This algorithm is defined by replacing the first-order derivatives in the differential equations with approximate differences. We have therefore that

$$\frac{\mathrm{d}x_i(t)}{\mathrm{d}t} \to \frac{1}{h}\left(x_i(t+h) - x_i(t)\right) + O(h).$$

This substitution followed by some rearrangement gives the Euler form of the differential equations,

$$\begin{aligned} x_i(t+h) &= x_i(t) + hf_i(x(t)) + O(h^2) \\ &= \left(1 + hf(x(t))\cdot\frac{\partial}{\partial x}\right)x_i(t) + O(h^2) \end{aligned}$$

or rewriting in the notation we have developed,

$$x_i|_{t+h} = \left(1 + hf\cdot\frac{\partial}{\partial x}\right)x_i\Big|_t + O(h^2).$$

Comparison of the exact stepping formula and the Euler approximate stepping formula now explains what is involved in the numerical integration of our system of coupled first order equations. The exact stepping operator is

$$e^{hf\cdot\frac{\partial}{\partial x}} = 1 + hf \cdot \frac{\partial}{\partial x} + \frac{1}{2!}\left(hf \cdot \frac{\partial}{\partial x}\right)^2 + \cdots$$

Note that there are an infinite number of terms in this series. The Euler stepping operator is given by

$$1 + hf \cdot \frac{\partial}{\partial x}.$$

Clearly, the Euler stepping formula is an approximation to the exact stepping operator. The steps in the Euler algorithm involve acting with this approximate stepping operator on the fundamental functions x_i,

$$x_i|_{t+h,\text{Euler}} = \left(1 + hf \cdot \frac{\partial}{\partial x}\right) x_i \bigg|_t .$$

If $A(x)$ is some function of these fundamental functions x_i, then we have

$$A(x)|_{t+h,\text{Euler}} \neq \left(1 + hf \cdot \frac{\partial}{\partial x}\right) A(x) \bigg|_t .$$

For example, consider $A(x) = \sum_i x_i x_i$. If we execute an Euler step, then we find

$$\begin{aligned}\sum_i x_i x_i &\to \sum_i \left(\left(1 + hf \cdot \frac{\partial}{\partial x}\right) x_i\right)\left(\left(1 + hf \cdot \frac{\partial}{\partial x}\right) x_i\right)\\ &\neq \left(1 + hf \cdot \frac{\partial}{\partial x}\right) \sum_i (x_i x_i).\end{aligned}$$

We thus see that the truncations involved in generating the Euler algorithm do very fundamental damage to the properties we would like a time-stepping operator to have.

The problems we posed at the beginning of this section were to describe the possible algorithms and to find a way to choose among those possibilities. The first of these problems is now seen to break down into describing the various ways in which we might approximate the exact stepping operator. There are two basic possibilities. We can either approximate the exact stepping operator by truncating the power series at some point or we can approximate by writing the exact operator as a product of simpler operators. Of course, we can combine the two possibilities. The Euler algorithm is typical of the truncation possibility. The other well-known differential equation algorithm, the *Runge–Kutta* method, is also of this type.

The splitting possibility we suggest here is to approximate the exponential of an operator as a product of other operators. If a and b are commuting numbers then we can split the exponential of their sum into a product of two exponentials,

$$\mathrm{e}^{a+b} = \mathrm{e}^{a}\mathrm{e}^{b} \quad a, b \text{ commuting numbers.}$$

If a and b are operators, then this result does not hold. The exponential of a sum of operators is not, in general, equal to the products of the exponentials of the individual operators.

$$\mathrm{e}^{a+b} \neq \mathrm{e}^{a}\mathrm{e}^{b} \quad a, b \text{ operators.}$$

This is disappointing, since it precludes us making the obvious splitting of our step operator,

$$\mathrm{e}^{hf\cdot\partial/\partial x} = \mathrm{e}^{h\sum_i f_i \partial/\partial x_i} \neq \mathrm{e}^{hf_1\partial/\partial x_1} \ldots \mathrm{e}^{hf_n\partial/\partial x_n}.$$

To illustrate the problems which arise when we replace commuting numbers with operators, let us consider in detail the product of exponentials of two operators a and b. In our current application these operators take the form:

$$a = \sum_i f_i(x)\frac{\partial}{\partial x_i} \quad b = \sum_j g_h(x)\frac{\partial}{\partial x_j}$$

where $f_i(x)$ and $g_i(x)$ are functions of x_i. In evaluating the exponentials we will need to take products of a and b. For example, we have

$$\begin{aligned} ab &= \left(\sum_i f_i(x)\frac{\partial}{\partial x_i}\right)\left(\sum_j g_j(x)\frac{\partial}{\partial x_j}\right) \\ &= \sum_{ij}\left(f_i g_j \frac{\partial^2}{\partial x_i \partial x_j} + f_i \frac{\partial g_j}{\partial x_i}\frac{\partial}{\partial x_j}\right) \end{aligned}$$

while

$$\begin{aligned} ba &= \left(\sum_j g_j(x)\frac{\partial}{\partial x_j}\right)\left(\sum_i f_i(x)\frac{\partial}{\partial x_i}\right) \\ &= \sum_{ij}\left(f_i g_j \frac{\partial^2}{\partial x_i \partial x_j} + g_j \frac{\partial f_i}{\partial x_j}\frac{\partial}{\partial x_i}\right). \end{aligned}$$

Comparison shows that $ab \neq ba$. This is the defining difference between operators and commuting numbers. Operators in general do not commute. Their noncommutativity is encoded in commutation relations,

$$[a, b] = ab - ba$$

which gives the differences which result when the operators act in different orders.

The non-commutativity of the operators a and b is responsible for the fact that the exponential of a sum is not equal to the product of exponentials for operators. For operators a more complicated relationship results. The relationship is encoded in the Baker–Campbell–Hausdorff relation which we state in the form of a theorem.

Theorem 6.5 If a and b are operators, then

$$e^a e^b = e^{K_2(a,b)}$$

$$K_2(a,b) = a + b + \frac{1}{2}[a,b] + \frac{1}{12}[a,[a,b]] + \frac{1}{12}[b,[b,a]] + \cdots$$

The exact formula for K_2 is an infinite series, but the terms listed here represent all terms which have three or less powers of the operators a or b.

Proof. The proof of this relationship involves expanding the exponentials on the left-hand side, and rearranging the resulting terms. To keep track of powers of operators, we insert a commuting number factor h multiplying a and b on the left-hand side. Thus,

$$\begin{aligned} e^{ha}e^{hb} &= \left(1 + ha + \frac{1}{2}h^2a^2 + O(h^3)\right)\left(1 + hb + \frac{1}{2}h^2b^2 + O(h^3)\right) \\ &= 1 + h(a+b) + h^2(a^2 + 2ab + b^2) + O(h^3) \\ &= 1 + h(a+b) + \frac{1}{2}h^2(a^2 + 2ab + b^2) + O(h^3) \\ &= 1 + h(a+b) + \frac{1}{2}h^2(a+b)^2 + \frac{1}{2}h^2(ab - ba) + O(h^3) \\ &= 1 + h(a+b) + \frac{1}{2}h^2(a+b)^2 + \frac{1}{2}h^2[a,b] + O(h^3). \end{aligned}$$

This formula includes all terms up to $O(h^2)$. We can perform a similar expansion for the right-hand side.

$$\begin{aligned} e^{K_2(ha,hb)} &= 1 + K_2(ha,hb) + \frac{1}{2}K_2(ha,hb)^2 + O(K_2)^3 \\ &= 1 + h(a+b) + \frac{1}{2}h^2(a+b)^2 + \frac{1}{2}h^2[a,b] + O(h^3). \end{aligned}$$

Comparing these two expansions we see that the identity is exact to $O(h^2)$. Since we have inserted h to keep track of powers of the operators a and b, we see that the formula is exact up to all terms which contain two or fewer powers of these operators. To show the exactness up to higher order, we need to keep track of more powers of h in both equations. This is straightforward but tedious and thus we do not repeat it here. □

Corollary 6.6 If a and b are operators, then

$$e^{\frac{1}{2}a}e^{b}e^{\frac{1}{2}a} = e^{K_3(a,b)}$$

$$K_3(a,b) = a + b - \frac{1}{24}[a,[a,b]] + \frac{1}{12}[b,[b,a]] + \cdots$$

This formula is similar to that for K_2 in that it is an infinite series, but again the terms listed here represent all terms which have three or less powers of the operators a or b. An important point to note about this particular formula is that on the right-hand side no quadratic term, $[a, b]$, appears.

Proof. We have, according to our theorem,

$$e^{\frac{1}{2}a}e^{b}e^{\frac{1}{2}a} = e^{\frac{1}{2}a}e^{K_2(b,\frac{1}{2}a)}$$
$$= e^{K_2(\frac{1}{2}a,K_2(b,\frac{1}{2}a))}.$$

Evaluating $K_2(\frac{1}{2}a, K_2(b, \frac{1}{2}a))$ gives the formula listed here for $K_3(a, b)$. □

Of course, the usual commuting number formula for exponential products will hold if the operators involved commute with each other. In particular if $[a, b] = 0$, then $K_2(a, b) = a + b$. All other terms in K_2 involve commutators of a and b, and so are zero. Since an operator always commutes with itself, we find that our time evolution operator, $\exp(hf \cdot \partial/\partial x)$, can be arbitrarily split into multiple products in at least one obvious way:

$$e^{hf\cdot\frac{\partial}{\partial x}} = \left(e^{\frac{h}{n}f\cdot\frac{\partial}{\partial x}}\right)^n.$$

Even though an operator $f \cdot \partial/\partial x$ is involved here, we do not have to worry about the ordering of the products on the right-hand side of this equation, since the result is order independent because the operator commutes with itself.

Before leaving this generic formulation of the numerical solution of coupled first-order differential equations, let us consider Liouville's theorem once again. In Hamiltonian dynamics, this theorem states that a volume in phase space is conserved if points in phase space are evolved according to Hamilton's equations of motion. In our generic formulation, phase space coordinates are the functions $x_1(t), \ldots, x_n(t)$. A region in this phase space is denoted by $D(x(t))$, and the volume occupied by this region is $\int_{D(x)} dx_1 \ldots dx_n$. We now wish to ask what conditions are required on these mappings so that volume in phase space is preserved. For this we consider the mapping,

$$x'_i = (1 + hf \cdot \partial_x)\, x_i.$$

This mapping will take the region $D(x)$ to $D(x')$, and the volume of this new region in phase space is $\int_{D(x')} dx'_1 \ldots dx'_n$. Reviewing our previous proof of Liouville's theorem, we see that this mapping will preserve volume in phase space if the Jacobian, J, of the mapping has value one. We have seen in the proof of Liouville's theorem that

$$J = \mathrm{Det}\left(\frac{\partial x'}{\partial x}\right)$$

with

$$\frac{\partial x'_i}{\partial x_j} = \delta_{ij} + h\frac{\partial f_i}{\partial x_j}.$$

The matrix $\partial x'/\partial x$ is therefore of the form $1 + hA$, where 1 is the identity matrix, and A has elements $A_{ij} = \partial f_i/\partial x_j$. Expanding the determinant in terms of h

$$\mathrm{Det}(1 + hA) = 1 + h\,\mathrm{Tr}(A) + \frac{1}{2}h^2\left(\mathrm{Tr}(A)^2 - \mathrm{Tr}(A^2)\right) + O(h^3)$$

we find

$$J = 1 + h\left(\sum_i \frac{\partial f_i}{\partial x_i}\right) + \frac{1}{2}h^2\left(\left(\sum_i \frac{\partial f_i}{\partial x_i}\right)^2 - \left(\sum_{ij} \frac{\partial f_i}{\partial x_j}\frac{\partial f_j}{\partial x_i}\right)\right) + O(h^3).$$

If now we have

$$\sum \frac{\partial f_i}{\partial x_i} = 0$$

then the term of order h here is zero, $J = 1 + O(h^2)$, and

$$\int_{D(x')} \mathrm{d}x'_1 \ldots \mathrm{d}x'_n = \int_{D(x)} J\mathrm{d}x_1 \ldots \mathrm{d}x_n$$
$$= \int_{D(x)} \mathrm{d}x_1 \ldots \mathrm{d}x_n + O(h^2).$$

The $O(h^2)$ term in this formula shows immediately that the mapping generated by the operator $1 + hf \cdot \partial/\partial x$ does not conserve phase space volume so long as h is finite. Thus the Euler algorithm, which involves steps with finite h using this operator, is not phase space volume conserving. However, the exact exponential mapping $\exp(hf \cdot \partial/\partial x)$ is phase space volume conserving. This is so because we can write this exponential mapping as a limit,

$$\mathrm{e}^{hf\cdot\frac{\partial}{\partial x}} = \lim_{n\to\infty}\left(\mathrm{e}^{\frac{h}{n}f\cdot\frac{\partial}{\partial x}}\right)^n = \lim_{n\to\infty}\left(1 + \frac{h}{n}f\cdot\frac{\partial}{\partial x}\right)^n.$$

Each of the operators $1 + (h/n)f \cdot \partial/\partial x$ here conserves phase space to $O(h^2/n^2)$ according to our result above. In the limit $n \to \infty$, the total error we make in ignoring terms of $O(h^2/n^2)$ is $nO(h^2/n^2)$ which goes to zero, and phase space volume is exactly preserved, provided of course that $\sum_i \partial f_i/\partial x_i = 0$. Note that this last condition is an essential requirement. If it did not hold, then the individual terms would preserve phase space only to $O(h/n)$ rather than $O(h^2/n^2)$, which does not go to zero fast enough to guarantee phase space conservation in the limit $n \to \infty$.

We therefore have arrived at an important result about the varieties of numerical integration algorithm possible. Algorithms involving stepping operators which are exact exponential operators will preserve phase space volume if $\sum_i \partial f_i/\partial x_i = 0$. The Euler stepping operator will not conserve phase space volume for any finite h. More generally, any algorithm based on truncation of the infinite power series of an exponential stepping operator will not preserve phase space volume. This is because the Jacobian of such a mapping will have many terms in it involving different orders of h, and it is usually not possible to arrange that all these terms cancel. Higher-order Euler algorithms and Runge–Kutta algorithms fall into this latter class, and generally do not preserve phase space volume as a result. Truncation methods also fail to preserve energy. Under an Euler step for example, we find

$$\begin{aligned} H(p|_{t+h,\text{Euler}}, q|_{t+h,\text{Euler}}) &= H((1 - h\mathcal{L}(H))p, (1 - \mathcal{L}(H))q) \\ &\neq (1 - h\mathcal{L}(H))H(p, q) = H(p, q). \end{aligned}$$

Thus the Euler step algorithm changes the value of H from step to step. More generally, we expect that all approximations based on truncation will also fail to preserve H.

The basic alternative we have suggested to truncation methods is to split the exponential stepping operator into a product of terms which are also exponential. Exponential stepping operators generate exact time transformations of our system and generally evaluating this transformations exactly is impossible. In fact, this is the very reason we are led to consider numerical integration techniques. However, there are special cases when exact time transformations are simple to evaluate. Two important cases occur. If H is a function of the momenta only, $H(p, q) = T(p)$, then Hamilton's equations reduce to

$$\begin{aligned} \frac{\mathrm{d}q_i}{\mathrm{d}t} &= \frac{\mathrm{d}}{\mathrm{d}p_i} T(p) \implies q_i(t + h) = q_i(t) + h \frac{\mathrm{d}}{\mathrm{d}p_i} T(p(t)) \\ \frac{\mathrm{d}p_i}{\mathrm{d}t} &= 0 \implies p_i(t + h) = p_i(t). \end{aligned}$$

On the other hand, if H is a function of coordinates only, $H(p,q) = V(q)$, then Hamilton's equations reduce to

$$\frac{\mathrm{d}q_i}{\mathrm{d}t} = 0 \implies q_i(t+h) = q_i(t)$$

$$\frac{\mathrm{d}p_i}{dt} = -\frac{\mathrm{d}}{\mathrm{d}q_i}V(q) \implies p_i(t+h) = p_i(t) - h\frac{\mathrm{d}}{\mathrm{d}q_i}V(q(t)).$$

Both of these sets of equations can be simply implemented in computer programs. The first involves shifting the coordinates by a term proportional to the derivative of $T(p)$, while leaving the momenta fixed. The second involves shifting the momenta by a term proportional to the force, while leaving the coordinates fixed. In terms of stepping operators, we see therefore that exact computer implementations exist for the two stepping operators

$$\mathrm{e}^{-h\mathcal{L}(T(p))} \qquad \mathrm{e}^{-h\mathcal{L}(V(q))}.$$

Since these are exact rather than truncated approximations we immediately find that these individual stepping operators preserve phase space volume, and also preserve energy. This is in marked contrast to the situation for truncated step operators as used in Euler's algorithm.

Hamiltonians of the form $H = T(p)$ or $H = V(q)$ are of course quite atypical of real physical systems. The more usual situation is that $H(p,q) = T(p) + V(q)$, and our problem is to find some numerical method to implement the stepping operator

$$\mathrm{e}^{-h\mathcal{L}(T+V)} = \mathrm{e}^{-h\mathcal{L}(T)-h\mathcal{L}(V)}.$$

Now, $\mathcal{L}(T)$ and $\mathcal{L}(V)$ are operators. If they commuted, we could write the exponential of their sum as a product of exponentials, and we would be done. However, they do not commute, but splitting the exponential into products should be a good approximation if h is not too large. We therefore consider a numerical integration scheme which uses exact operators to approximate the evolution according to the true Hamiltonian $H = T + V$. Many possibilities occur. The simplest include

$$\mathrm{e}^{-h\mathcal{L}(T)}\mathrm{e}^{-h\mathcal{L}(V)}$$
$$\mathrm{e}^{-h\mathcal{L}(V)}\mathrm{e}^{-h\mathcal{L}(T)}$$
$$\mathrm{e}^{-\frac{h}{2}\mathcal{L}(V)}\mathrm{e}^{-h\mathcal{L}(T)}\mathrm{e}^{-\frac{h}{2}\mathcal{L}(V)}$$
$$\mathrm{e}^{-\frac{h}{2}\mathcal{L}(T)}\mathrm{e}^{-h\mathcal{L}(V)}\mathrm{e}^{-\frac{h}{2}\mathcal{L}(T)}$$

but clearly there are many other splittings possible also. Such operators can be considered compound stepping operators, and are classified according to the number of stepping operators combined to make the compound stepping operator. The first two examples here are two-step operators, the second two are three-step operators.

Let us analyze the first two-step operator, $\exp(-h\mathcal{L}(T))\exp(-h\mathcal{L}(V))$. This operator is implemented on the computer by first applying the operator $\exp(-h\mathcal{L}(V))$ to evolve the coordinates and then by applying the operator $\exp(-h\mathcal{L}(T))$ to evolve the momenta. Note the actual order in which the operations occur. We can analyze the effect of the combined operator by using the Baker–Campbell–Hausdorff formula,

$$\mathrm{e}^{-h\mathcal{L}(T)}\mathrm{e}^{-h\mathcal{L}(V)} = \mathrm{e}^{K_2(-h\mathcal{L}(T),-h\mathcal{L}(V))}$$

where

$$\begin{aligned}K_2(-h\mathcal{L}(T), -h\mathcal{L}(V)) &= -h\mathcal{L}(T) - h\mathcal{L}(V) + \frac{1}{2}[-h\mathcal{L}(T), -h\mathcal{L}(V)] + O(h^3)\\ &= -h\left(\mathcal{L}(T) + \mathcal{L}(V) + \frac{1}{2}[\mathcal{L}(T), \mathcal{L}(V)] + O(h^2)\right).\end{aligned}$$

We can now recast this expression as

$$K_2(-h\mathcal{L}(T), -h\mathcal{L}(V)) = -h\mathcal{L}(T + V + h\{T, V\} + O(h^2)),$$

and we therefore find that

$$\mathrm{e}^{-h\mathcal{L}(T)}\mathrm{e}^{-h\mathcal{L}(V)} = \mathrm{e}^{-h\mathcal{L}(H_e)},$$

where

$$H_e(p, q, h) = T(p) + V(q) + \frac{1}{2}h\{T(p), V(q)\} + O(h^2).$$

Consequently, the product of our two stepping operators is actually equivalent to a single stepping operator with a different Hamiltonian. This Hamiltonian is H_e, and can be considered as an effective Hamiltonian. This effective Hamiltonian depends on the coordinates and momenta as usual, but it also depends on h. In fact, because the Baker–Campbell–Hausdorff theorem generates an infinite series for the function K_2, the effective Hamiltonian involves an infinite series of terms. We have only calculated the first few terms in the formula above. Other terms arise from commutators involving more factors of $\mathcal{L}(T)$ and $\mathcal{L}(V)$. However, we can then appeal to the result that the commutator of two Lie derivatives is equal to the Lie derivative of the corresponding Poisson bracket to show that all these higher-order commutators simply add more terms to our effective Hamiltonian.

If we consider our effective Hamiltonian more closely, we find that its leading h independent contribution is $T(p) + V(q)$. This particular term is the exact Hamiltonian for which we wanted an integration algorithm. The next term in the effective

Hamiltonian is given by

$$\frac{1}{2}h\{T, V\} = -\frac{1}{2}h\sum_i \frac{\partial T}{\partial p_i}\frac{\partial V}{\partial q_i}.$$

Since this term has a factor h multiplying it, we expect that we can make its effects small by making h small.

We can repeat the analysis which produces the effective Hamiltonian for the other suggested splittings in our list to see how changing the splitting effects the resulting effective Hamiltonian. In all cases we find that the effective Hamiltonian is given by the exact Hamiltonian, $T + V$, plus corrections of $O(h)$ or $O(h^2)$. The results are as follows,

$$\begin{aligned}
\mathrm{e}^{-h\mathcal{L}(T)}\mathrm{e}^{-h\mathcal{L}(V)} &= \mathrm{e}^{-h\mathcal{L}(H_e^{(1)})} \\
\mathrm{e}^{-h\mathcal{L}(V)}\mathrm{e}^{-h\mathcal{L}(T)} &= \mathrm{e}^{-h\mathcal{L}(H_e^{(2)})} \\
\mathrm{e}^{-\frac{h}{2}\mathcal{L}(V)}\mathrm{e}^{-h\mathcal{L}(T)}\mathrm{e}^{-\frac{h}{2}\mathcal{L}(V)} &= \mathrm{e}^{-h\mathcal{L}(H_e^{(3)})} \\
\mathrm{e}^{-\frac{h}{2}\mathcal{L}(T)}\mathrm{e}^{-h\mathcal{L}(V)}\mathrm{e}^{-\frac{h}{2}\mathcal{L}(T)} &= \mathrm{e}^{-h\mathcal{L}(H_e^{(4)})}
\end{aligned}$$

where

$$\begin{aligned}
H_e^{(1)}(p, q, h) &= T(p) + V(q) + \frac{1}{2}h\{T, V\} + O(h^2) \\
H_e^{(2)}(p, q, h) &= T(p) + V(q) + \frac{1}{2}h\{T, V\} + O(h^2) \\
H_e^{(3)}(p, q, h) &= T(p) + V(q) + \frac{1}{2}h^2\{T, V\} + O(h^3) \\
H_e^{(4)}(p, q, h) &= T(p) + V(q) + \frac{1}{2}h^2\{T, V\} + O(h^3).
\end{aligned}$$

Note that the last two of these stepping operators involve effective Hamiltonians which have leading $O(h^2)$ corrections to the exact Hamiltonian rather than $O(h)$ corrections of the first two stepping operators. These are therefore better approximate stepping operators. In fact, it is possible to force the corrections to any given power in h by further splitting the evolution operator, but we shall not follow this route here. Effective Hamiltonians with leading $O(h^2)$ corrections have another nice feature in that they are reversible. If we act with these operators to step backward in time, we can undo an equivalent step forward. For example a forward step of size h with

$$\mathrm{e}^{-\frac{h}{2}\mathcal{L}(T)}\mathrm{e}^{-h\mathcal{L}(V)}\mathrm{e}^{-\frac{h}{2}\mathcal{L}(T)}$$

is canceled by a backward step of size $-h$ with

$$\mathrm{e}^{-\frac{(-h)}{2}\mathcal{L}(T)}\mathrm{e}^{-(-h)\mathcal{L}(V)}\mathrm{e}^{-\frac{(-h)}{2}\mathcal{L}(T)}.$$

The backward step operator is identical to the forward step operator except that h is replaced by $-h$. This result does not hold for

$$e^{-h\mathcal{L}(T)}e^{-h\mathcal{L}(V)}.$$

The reverse of this operator is given by

$$e^{-(-h)\mathcal{L}(V)}e^{-(-h)\mathcal{L}(T)}$$

which is not obtained by replacing h with $-h$.

Compound stepping operators of the form given in our list satisfy two important properties not satisfied by truncated stepping operators such as the Euler stepping operator. They preserve volume in phase space, and they act consistently on the fundamental coordinates and momenta and on functions of those coordinates and momenta. Furthermore, we have found that they generate evolution according to an effective Hamiltonian which can be made as close to the true Hamiltonian as liked by making h sufficiently small. In the general case, they represent the best algorithm to generate classical trajectories.

6.4 The Lennard-Jones fluid

Let us consider now an application of the integration techniques which we have just developed to the Lennard-Jones fluid defined in the last chapter. This is a system on N atoms with momenta $\mathbf{p}_i$ and coordinates $\mathbf{x}_i$ for $i = 1, \ldots, N$. The Hamiltonian in numerical units is given by,

$$\begin{aligned} H_{LJ} &= T + V \\ T &= \sum_i \frac{1}{2} |\mathbf{p}_i|^2 \\ V &= \sum_{i<j} V_{ij} \end{aligned}$$

where

$$V_{ij} = V(r_{ij}) = 4\left(r_{ij}^{-12} - r_{ij}^{-6}\right)$$

and the temperature in numerical units is given by T. Momenta are considered unconstrained, while coordinates satisfy periodic boundary conditions as we had for the canonical system.

Our basic problem now is to generate classical trajectories for this Hamiltonian. We begin by arbitrarily assigning initial values to coordinates and momenta. To calculate classical trajectories we will adopt the integration method which uses

$\exp(-h\mathcal{L}(T))$ and $\exp(-h\mathcal{L}(V))$ as time stepping operators. We therefore need to calculate how these operators act on the coordinates and momenta. We define p_{ia} to the ath component of the vector $\mathbf{p}_i$, and similarly we define x_{ia} to be the ath component of the vector $\mathbf{x}_i$. We have

$$\begin{aligned} \mathrm{e}^{-h\mathcal{L}(T)} p_{ia} &= p_{ia} \\ \mathrm{e}^{-h\mathcal{L}(T)} x_{ia} &= x_{ia} + h p_{ia} \end{aligned}$$

while

$$\begin{aligned} \mathrm{e}^{-h\mathcal{L}(V)} p_{ia} &= p_{ia} - h \sum_i \frac{\partial V}{\partial x_{ia}} \\ &= h F_{ia} \\ \mathrm{e}^{-h\mathcal{L}(V)} x_{ia} &= x_{ia} \end{aligned}$$

where $F_{ia} = (\mathbf{F}_i)_a$ is the ath component of the force on atom i. These stepping equations are seen to be quite simple. The operator $\exp(h\mathcal{L}(T))$ shifts $\mathbf{x}_i$ by $h\mathbf{p}_i$ while leaving $\mathbf{p}_i$ unchanged. The operator $\exp(h\mathcal{L}(T))$ shifts $\mathbf{p}_i$ by $h\mathbf{F}_i$ while leaving $\mathbf{p}_i$ unchanged. These two shifts can be implemented in separate computer routines. The shift of the positions $\mathbf{x}_i$ is completely trivial and involves only a few lines of code. The shift of the momenta $\mathbf{p}_i$ involves calculating the net force which acts on atom i. This in turn involves a sum over all other atoms in the fluid, since each atom exerts a force on each other atom. The code to calculate this force term is also quite simple in principle. In practice, however, it is very expensive computationally. Thus almost all the computer time in integrating equations of motion is spent in calculating force terms.

Once the fundamental stepping operators have been designed, we need to choose a combination of these operators to implement a compound step which is to approximate evolution according to the true Hamiltonian for the system. The normal choice now is to select one of the three-step compound operators we have defined above. These are

$$\mathrm{e}^{-\frac{h}{2}\mathcal{L}(V)}\mathrm{e}^{-h\mathcal{L}(T)}\mathrm{e}^{-\frac{h}{2}\mathcal{L}(V)} \quad \text{or} \quad \mathrm{e}^{-\frac{h}{2}\mathcal{L}(T)}\mathrm{e}^{-h\mathcal{L}(V)}\mathrm{e}^{-\frac{h}{2}\mathcal{L}(T)}.$$

Either of these step operators are acceptable integrators. The algorithm corresponding to these two choices is often called the *Leapfrog algorithm*. The name leapfrog comes from one of the ways to write this algorithm, where positions and velocities "leap over" each other. Positions are defined at times $t = ih$, spaced at constant intervals, while the velocities are defined at times halfway in-between, $t = (i + 1/2)h$. Indeed consider for instance the repeated application of the operator $\mathrm{e}^{-\frac{h}{2}\mathcal{L}(V)}\mathrm{e}^{-h\mathcal{L}(T)}\mathrm{e}^{-\frac{h}{2}\mathcal{L}(V)}$ to an initial configuration $(\mathbf{x}_i(0), \mathbf{p}_i(0))$. This then generates

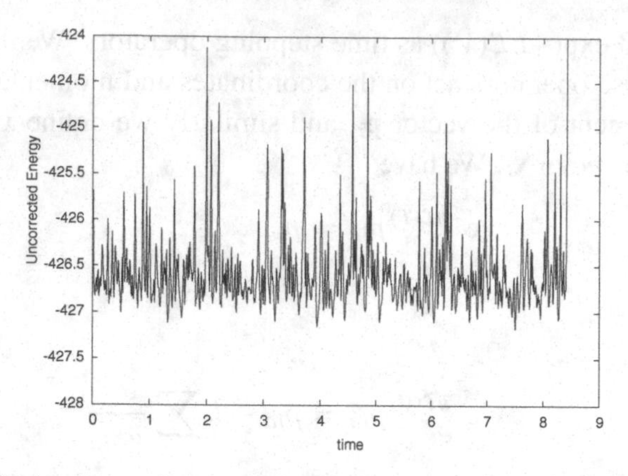

Figure 6.4 Fluctuations of the uncorrected energy H in numerical units as a function of time for a step size $\tau = 0.011$. In this simulation the fluctuations diverge eventually.

the sequence

$$\begin{aligned}
\mathbf{p}_i(0) &\rightarrow \mathbf{p}_i(0) + \tfrac{h}{2}\mathbf{F}_i(\mathbf{q}_i(0)) \equiv \mathbf{p}_i\left(\frac{1}{2}\right) \\
\mathbf{q}_i(0) &\rightarrow \quad \mathbf{q}_i(0) + h\mathbf{p}_i\left(\tfrac{1}{2}\right) \quad \equiv \mathbf{q}_i(1) \\
\mathbf{p}_i\left(\frac{1}{2}\right) &\rightarrow \mathbf{p}_i\left(\tfrac{1}{2}\right) + h\mathbf{F}_i(\mathbf{q}_i(1)) \equiv \mathbf{p}_i\left(\frac{3}{2}\right) \\
\mathbf{q}_i(1) &\rightarrow \quad \mathbf{q}_i(1) + h\mathbf{p}_i\left(\tfrac{3}{2}\right) \quad \equiv \mathbf{q}_i(2) \\
&\text{etc.}
\end{aligned}$$

We have shown that this algorithm is a *symplectic integrator*, that is, it preserves the volume in phase space.

There now remain two quantities which need to be determined before we can proceed with useful calculation. First, we need to choose a step size h to use when applying the compound stepping operator. Different choices of h lead to different effective Hamiltonians, and the larger h is, the larger will be the difference between the effective Hamiltonians and the true Hamiltonian which we want to approximate. This difference between effective and true Hamiltonians leads to fluctuations in the value of $H = T + V$ when this quantity is calculated on the states generated by our numerical integration algorithm. As h gets bigger, these fluctuations will also increase. Thus we must choose h so that these fluctuations remain at a reasonable level. Acceptable values for fluctuations are in the range $\leq 1\%$. In Figure 6.4 we plotted the fluctuations of the energy as a function of time using a symplectic integrator.

The other parameter which needs to be fixed is the temperature T. Temperature does not enter the equations of motion at any point. However, the trajectory of states generated when we integrate these equations of motion represents a micro canonical ensemble of states. We can therefore use our understanding of statistical mechanics to help in finding T. In particular, we note that equipartition of energy tells us that, in physical units,

$$\left\langle \frac{1}{2m} |\mathbf{p}_i|^2 \right\rangle = \frac{3}{2\beta}.$$

In the numerical units which we are using for coordinates and momenta in the Lennard-Jones system, this equation becomes

$$\left\langle \frac{1}{2} |\mathbf{p}_i|^2 \right\rangle = \frac{3T}{2}.$$

To determine the temperature T of our system therefore we measure $\langle 1/2\, |\mathbf{p}_i|^2 \rangle$ over a number of trajectories, and use the equipartition formula.

An important aspect of numerical molecular dynamics is the *equilibration*. The algorithm outlined above does not allow the energy to adjust to a prescribed value. In addition it will not immediately be in thermodynamic equilibrium. We can add or subtract energy by scaling the momenta appropriately. After a few such adjustments, we will arrive at a system which has the correct internal energy for the temperature we wish to study and, thereafter, the states we generate by evolving with our compound stepping operators will be states in the micro canonical ensemble for this temperature. To bring the system to equilibrium we need the equilibration phase. The equilibration phase is completed when the system has settled into definite mean values of the kinetic and potential energy and no systematic drift in these quantities can be observed anymore. In the case where the initial condition is a lattice, this configuration corresponds to minimal potential energy. If furthermore the kinetic energy is tuned to a temperature above the melting point, the potential energy will rise from a large negative value to the value typical for a liquid. In Figures 6.5 and 6.6 we show the behavior of the kinetic energy during equilibration. The initial temperature was chosen to be 2.7 in the numerical units introduced in Section 5.5. This is well above the melting temperature of the lattice. At the beginning of the simulation we observe a dramatic drop in the kinetic energy. This is due to the melting of the initial lattice. The simulation must at this stage transfer kinetic energy into potential energy. This can be seen in Figure 6.5. After the kinetic energy has reached a stable average, the momenta are scaled until the target temperature is reached in Figure 6.6.

To ensure that the system has lost all memory of the initial condition certain parameters can be followed. For instance, the degree of translational order can be

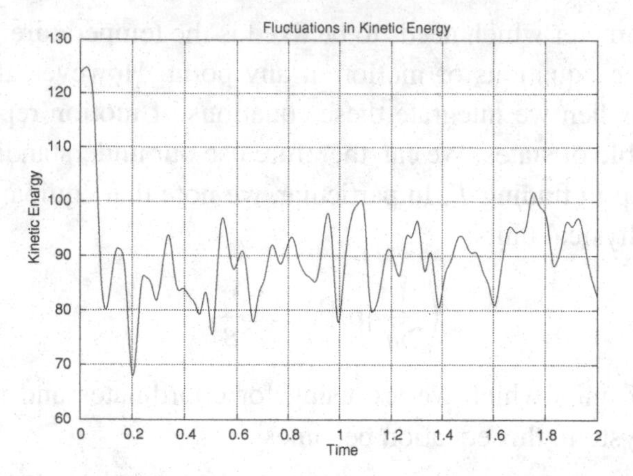

Figure 6.5 The kinetic energy drops as a result of melting the initial lattice. A sample simulation plotted in numerical units.

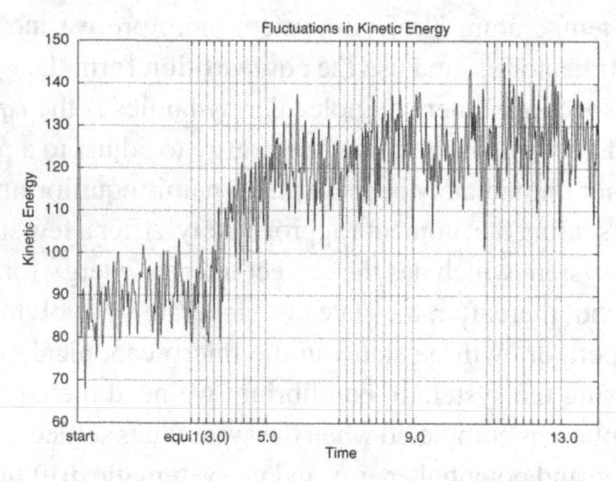

Figure 6.6 After the lattice has melted the kinetic energy is increased to match the target temperature. A sample simulation plotted in numerical units.

measured by the *translational order parameter*

$$\rho(\mathbf{k}) = \frac{1}{N} \sum_{i=1}^{N} \cos(\mathbf{k} \cdot \mathbf{q}_i) ,$$

where $\mathbf{q}_i$ is the position of the ith particle and $\mathbf{k}$ is a vector of the reciprocal lattice. For a solid $\rho(k)$ is of order unity while in the liquid phase $\rho(k)$ will fluctuate around zero with amplitude $O(N^{-1/2})$.

Problems

Problem 6.1 Show that the set of mappings generated by evolution for all possible times t, (both positive and negative) forms a group.

Problem 6.2 Prove that reversibility of a compound stepping operator implies that the corresponding effective Hamiltonian will contain terms with only even powers of h.

Problem 6.3 Prove the equations

$$e^{-h\mathcal{L}(T)} p_{ia} = p_{ia}$$

$$e^{-h\mathcal{L}(V)} p_{ia} = p_{ia} - h \sum_i \frac{\partial V}{\partial x_{ia}} = hF_{ia}.$$

by expanding the exponential step operators in infinite power series, then evaluating the action of each term in these series on p_{ia} and x_{ia}.

Problem 6.4 Show that the compound stepping operator $e^{-\frac{h}{2}\mathcal{L}(V)} e^{-h\mathcal{L}(T)} e^{-\frac{h}{2}\mathcal{L}(V)}$ leads to a leapfrog algorithm which is equivalent to the one described in the text.

Problem 6.5 Show that the leapfrog algorithm described in Section 6.4 is mathematically equivalent to the Verlet algorithm which is defined by the sequence

$$\mathbf{q}_i(t + \Delta t) = 2\mathbf{q}_i(t) - \mathbf{q}_i(t - \Delta t) + \tfrac{1}{m}\mathbf{F}_i(\mathbf{q}_i(t)).$$

Note that the Verlet algorithm does not involve velocities!

Problem 6.6 Consider the following combination between leapfrog and Metropolis algorithm which is used to sample the phase space of Lennard-Jones fluids in the canonical ensemble. In this algorithm one chooses a random initial configuration for $\mathbf{p}_i$ and $\mathbf{x}_i$. In step two the positions and momenta are updated using the leapfrog algorithm described in the text. This updating is then repeated n times, where n is a number between 1 and 1000, say, determined randomly. After that we do a Metropolis test, that is we evaluate $\Delta E = H(\mathbf{p}_i', \mathbf{x}_i') - H(\mathbf{p}_i, \mathbf{x}_i)$. If $\Delta E \leq 0$ the update is accepted, if $\Delta E > 0$ it is accepted with probability $\exp(-\Delta E/T)$. In case of rejection we return to step two. If the update is successful the momenta $\mathbf{p}_i'$ (but not the positions) are reconfigured using a Gaussian normal distribution centered at zero with variance T. This update is again subjected to a metropolis test. After that the program returns to step two.

As we have seen in the text the update using the leapfrog algorithm is volume preserving and reversible. Show that the Markov matrix constructed in this way ensures that detailed balance is satisfied. Note that detailed balance is satisfied for any step size in the leapfrog process. However, the acceptance rate will be decreased with increasing step size. In practical applications an acceptance rate of approximately 50% should be aimed for. The algorithm described here presents an improvement of the naive Metropolis algorithm described in Chapter 5 in that it has better decorrelation properties and converges faster to a new thermal equilibrium. This is due to the fact that it explores more "distant" configurations in phase space while keeping a high acceptance rate.

Program this algorithm on a computer for a Lennard-Jones fluid and plot the internal energy as a function of temperature. Compare the result with that of the Metropolis algorithm for the Lennard-Jones fluid in Chapter 5. Give a qualitative discussion of the acceptance rate as a function of the step size.

Further Reading

For a good text on numerical simulations see M. P. Allen and D. J. Tildesley, *Computer Simulation of Liquids*, Clarendon Press (1996). For a blend of tutorial and recipe collection see D. C. Rapaport, *The Art of Molecular Dynamics Simulation*, Cambridge University Press (2004). Various aspects of symplectic integrators are reviewed by H. Yoshida, "Recent Progress in the Theory and Application of Symplectic Integrators". *Celestial Mechanics and Dynamical Astronomy*, **56** (1993) 27. A discussion of alternative symplectic integrators can be found in M. P. Calvo and J. M. Sanz-Serna, *Numerical Hamiltonian Problems*, Chapman & Hall (1994).

7
Quantum statistical mechanics

So far we have considered systems of molecules which are governed by the laws of classical mechanics. On the other hand, we know that when the separation distance between molecules becomes small, quantum effects become important. Thus for molecules which are in a condensed state such as, for instance, electrons in a metal or electrons in a white dwarf star, it is expected that quantum effects cannot be ignored. We have thus to learn how to set up a statistical mechanics for such systems. In this chapter we will develop statistical mechanics for a system of N non-interacting identical quantum molecules of mass m. In Chapter 9 we will see how the approach can be generalized to take interactions into account.

7.1 Quantum mechanics

In quantum mechanics, the properties of a system are encoded in the state vector $|\psi\rangle$ of the system. This state vector satisfies the Schrödinger equation which in vector form is given by

$$\mathsf{H} \mid \psi(t)\rangle = i\hbar\frac{\partial}{\partial t} \mid \psi(t)\rangle$$

where the state vector is a vector in a complex Hilbert space $\mathcal{H}$ and the Hamiltonian H is a linear self-adjoint operator in $\mathcal{H}$. Let us first consider a single particle of mass m and momentum $\mathbf{p}$ moving in a potential $V(\mathbf{x})$, for instance. The quantum mechanical Hamilton operator is given by

$$\mathsf{H} = \frac{\mathsf{p}\cdot\mathsf{p}}{2m} + V(\mathsf{x})$$

where $[\mathsf{p}_a, \mathsf{x}_b] = \hbar/i\ \delta_{ab}$. The vector form of the Schrödinger equation has an equivalent representation as a partial differential equation for the wave functions $\psi(\mathbf{x}, t) \equiv \langle \mathbf{x} \mid \psi(t)\rangle$. Here the state vectors or kets $\{|x\rangle\}$ form a complete set of eigenvectors of the position operator x, normalized by the Dirac delta function. In

Dirac's Bra and Ket notation

$$\langle \mathbf{x} \mid \mathbf{y} \rangle = \delta^{(3)}(\mathbf{x} - \mathbf{y})$$
$$\int \mathrm{d}^3 x \mid \mathbf{x}\rangle\langle \mathbf{x} \mid = 1 .$$

The set $\{|x\rangle\}$ thus forms a generalized basis of the Hilbert space $\mathcal{H}$. The wave functions $\psi(\mathbf{x}, t)$ are just the "components" of $\langle \mathbf{x} \mid \psi(t)\rangle$ in this basis. The momentum and position operators are represented on wave functions as

$$\mathsf{p} \to -\mathrm{i}\hbar\nabla$$
$$\mathsf{X} \to \mathbf{x} .$$

The Schrödinger equation in our example is then the differential equation

$$\left[-\frac{\hbar^2}{2m}\nabla^2 + V(\mathbf{x})\right]\psi(\mathbf{x}, t) = \mathrm{i}\hbar\frac{\partial}{\partial t}\psi(\mathbf{x}, t).$$

Solving it for $\psi(\mathbf{x}, t)$ with appropriate boundary conditions determines all observable properties of the system. In particular, in the stationary regime $\mathsf{H} \neq \mathsf{H}(t)$ considered here we can separate the time dependence leading to the equation

$$\left[-\frac{\hbar^2}{2m}\nabla^2 + V(\mathbf{x})\right]\psi(\mathbf{x}, t) = E\psi(\mathbf{x}, t)$$

which determines the possible energies of the quantum mechanical system.

The generalization of the Schrödinger equation to a system of N identical *non-interacting* molecules of mass m then is simply

$$\mathrm{i}\hbar\frac{\partial}{\partial t}\psi(\mathbf{x}_1, \mathbf{x}_2, \ldots, \mathbf{x}_N, t) = -\left[\sum_{j=1}^{N}\frac{\hbar^2}{2m}\nabla^2_{\mathbf{x}_j}\right]\psi(\mathbf{x}_1, \mathbf{x}_2, \ldots, \mathbf{x}_N, t).$$

After separation of the time dependence we then obtain the time-independent many-particle Schrödinger equation

$$-\left(\sum_{j=1}^{N}\frac{\hbar^2}{2m}\nabla^2_{\mathbf{x}_j}\right)\psi_E\,(\mathbf{x}_1, \ldots, \mathbf{x}_N) = E\,\psi_E\,(\mathbf{x}_1, \ldots, \mathbf{x}_N)$$

where the energy is the sum of the energies of the individual particles, $E = \sum_{i=1}^{N}\epsilon_i$, where ϵ_i are the energy eigenvalues of the time independent, single particle Schrödinger equation

$$-\frac{\hbar^2}{2m}\,\nabla_i^2\,\psi_{\epsilon_i}\,(\mathbf{x}_i) = \epsilon_i\,\psi_{\epsilon_i}\,(\mathbf{x}_i)\,.$$

There is, however, a subtlety due to the fact that the N molecules cannot be distinguished in quantum mechanics: the statement that one molecule is at $\mathbf{x}_1$ and has energy ϵ_1 while a second molecule at $\mathbf{x}_2$ has energy ϵ_2 is meaningless. If, for instance, the ith and jth molecules are interchanged, the Hamiltonian $\mathsf{H}(\mathsf{p}_1, \ldots, \mathsf{p}_N)$ given by

$$\mathsf{H} = \sum_{j=1}^{N} \frac{\mathsf{p}_j{}^2}{2m}$$

remains unchanged. Let us assume that the energy eigenfunctions of the system are non-degenerate. This means that the wave function corresponding to any energy eigenvalue E is unique (up to a scale factor). Then the symmetry property under exchange of any pair of identical molecules of the system can be summarized in terms of the following operational statements.

(1) Introduce P_{ij}, the operator that interchanges the locations of the ith and jth molecules. Then
(2) $[\mathsf{H}, \mathsf{P}_{ij}] = 0$
(3) $\mathsf{P}_{ij}\, \psi_E(\mathbf{x}_1, \ldots, \mathbf{x}_N) = \lambda\, \psi_E(\mathbf{x}_1, \ldots, \mathbf{x}_N)$.

Since $\mathsf{P}_{ij}^2 = 1$, it follows that $\lambda^2 = 1$, therefore $\lambda = \pm 1$. Thus the wave function $\psi_E(\mathbf{x}_1, \ldots, \mathbf{x}_N)$ is either a totally symmetric function of $\mathbf{x}_1, \ldots, \mathbf{x}_N$, or it is a totally antisymmetric function of $\mathbf{x}_1, \ldots, \mathbf{x}_N$. Observe, further, that if two molecules are placed in the same quantum state, i.e. $\psi_{\epsilon_i}(\mathbf{x}_i) = \psi_{\epsilon_j}(\mathbf{x}_j)$ with $\epsilon_i = \epsilon_j$, $\mathbf{x}_i = \mathbf{x}_j$ then the corresponding wave function must vanish in the case where the wave function is totally antisymmetric. The N-particle wave function with these properties is given by the product of the 1-particle wave functions $\psi_{\epsilon_i}(\mathbf{x}_i)$ summed over all permutations σ of the positions $\mathbf{x}_i$ and multiplied by the signature of σ, that is

$$\psi_E(\mathbf{x}_1, \ldots, \mathbf{x}_N) = \frac{1}{\sqrt{N!}} \sum_{\sigma} \mathrm{sign}(\sigma) \psi_{\epsilon_1}\left(\mathbf{x}_{\sigma(1)}\right) \cdots \psi_{\epsilon_N}\left(\mathbf{x}_{\sigma(N)}\right).$$

An immediate consequence is that a given, totally antisymmetric quantum state can contain at most one molecule; a remarkable result. This result is known as the *Pauli exclusion principle* and was originally discovered to be a property of electrons by analyzing spectroscopic data. Systems for which this is true are known as *Fermi–Dirac* systems. It has been established that the Fermi–Dirac system describes the behavior of identical quantum molecules with half integer spin, in units of $\hbar$.

The situation where the wave function is totally symmetric places no restrictions on the occupation number of any given quantum state. Systems for which this is true are known as *Bose–Einstein* systems. Again it has been established that Bose–Einstein systems describe the behavior of identical quantum molecules having integer spin, in units of $\hbar$. The N-particle wave function is again given by the product of the 1-particle wave functions $\psi_{\epsilon_i}(\mathbf{x}_i)$ summed over all permutations but without multiplying by the signature

$$\psi_E(\mathbf{x}_1, \ldots, \mathbf{x}_N) = \frac{1}{\sqrt{N!}} \sum_\sigma \psi_{\epsilon_1}\left(\mathbf{x}_{\sigma(1)}\right) \cdots \psi_{\epsilon_N}\left(\mathbf{x}_{\sigma(N)}\right).$$

This elegant characterization of the two theoretical possibilities is one of the deep results of quantum mechanics. It is known as the *spin statistics theorem* and was first established by Pauli using general ideas of relativity and causality in the framework of quantum field theory. One intriguing aspect of Pauli's result is that it holds in three and higher space dimensions and is not true in two spatial dimensions!

Finally note that from $[\mathsf{H}, \mathsf{P}_{ij}] = 0$ and the observation that the Hamilton operator H is responsible for the time evolution of a wave function, it follows that a wave function which is initially in a symmetric (antisymmetric) state will continue to maintain its symmetry property for all time.

To complete our task we need to determine the possible energy eigenvalues ϵ_i. Since we wish to study a statistical mechanics problem, we confine the N identical molecules to a cubic box of side L and volume $V = L^3$, and for simplicity we choose periodic boundary conditions on ψ. Periodicity means that a shift of any molecule coordinate by an integer multiple of L leaves ψ unchanged. Explicitly, if $\mathbf{r}_i \equiv (r_{i,x}, r_{i,y}, r_{i,z})$, $i = 1, \ldots, N$, are vectors with integer components, then periodicity requires

$$\psi(\mathbf{x}_1 + L\mathbf{r}_1, \mathbf{x}_2 + L\mathbf{r}_2, \ldots, \mathbf{x}_N + L\mathbf{r}_N, t) = \psi(\mathbf{x}_1, \mathbf{x}_2, \ldots, \mathbf{x}_N, t).$$

Other boundary conditions are also possible, such as for example requiring $\psi = 0$ at the boundary. But so long as the volume and number of particles is large, the particular boundary conditions on the wave function are not important. We will come back to this point in Chapter 12. The allowed energy levels ϵ_i of a single molecule are obtained by solving the time independent equation

$$-\frac{\hbar^2}{2m} \nabla^2 \psi(\mathbf{x}) = \epsilon\, \psi(\mathbf{x}),$$

subject to periodic boundary conditions for ψ. This eigenvalue problem is then easily solved with

$$\epsilon = \left(\frac{p_1{}^2}{2m} + \frac{p_2{}^2}{2m} + \frac{p_3{}^2}{2m}\right)$$

where $(i = 1, 2, 3)$

$$p_i = \frac{2\pi\hbar}{L} m_i, \quad m_i = 0, \pm 1, \ldots$$

Thus the momenta and energies take discrete values parameterized by L and integers m_1, m_2, m_3. Hence each quantum state can be characterized by a set of three integers (m_1, m_2, m_3). Notice that the quantum states can also be characterized by the physical momentum variables p_1, p_2, p_3 and that the number of quantum states between $p_1 + \Delta p_1$, $p_2 + \Delta p_2$, $p_3 + \Delta p_3$, and p_1, p_2, p_3 can be counted from

$$\Delta p_i = \frac{2\pi \hbar}{L} \Delta m_i .$$

If the volume of the box containing the molecules is very big we can approximate the sum over all possible quantum states by an integral for momenta as follows:

$$\sum_{m_1,m_2,m_3} \rightarrow \frac{V}{(2\pi\hbar)^3} \int (\Delta p_1)\,(\Delta p_2)\,(\Delta p_3),$$

where $V = L^3$ is the volume of the cubic box containing the molecules.

Since, by assumption, the molecules do not interact, we can write the energy of our N molecule system as the sum over all possible energies of the individual molecules times the number of molecules having this energy, i.e.

$$E = \sum_i n_i \epsilon_i, \quad \text{with} \quad \sum n_i = N.$$

Hence the total energy of a collection of N non-interacting molecules is given by specifying how many molecules n_i have energy ϵ_i. From our discussions we know that $n_i = 0$ or 1 for a Fermi–Dirac system, while $n_i = 0, 1, \ldots, \infty$ for a Bose–Einstein system.

7.2 The quantum partition function

After this digression into quantum mechanics let us return to the way a statistical mechanics description of such a system is to be carried out. We recall that the grand canonical ensemble Z_Ω was defined as

$$\begin{aligned} Z_\Omega &= \mathrm{e}^{-\beta\Omega} \\ &= \sum_N \sum_{E_N} C(E_N)\mathrm{e}^{-\beta(E_n - \mu N)}, \qquad \beta = \frac{1}{kT} \end{aligned}$$

where N was the number of molecules, E_N the energy of the N molecules and $C(E_N)$ the phase space factor counting the number of configurations with the same energy E_N (see Fig. 3.2).

For a quantum system all of these physical quantities are well-defined. Indeed for the system of identical non-interacting molecules we can write, as we have just shown

$$E_N = \sum_i n_i \epsilon_i \, , \qquad N = \sum_i n_i$$

where the sum is over the possible energy eigenvalues ϵ_i that a single molecule can have. Here $C(E_N)$ counts the degeneracy of the energy eigenvalue E_N. We note that $\sum_N \sum_{E_N}$ can be replaced by an unrestricted sum over $\{n_i\}$. The grand canonical partition function is thus just the sum over the occupation number of each possible energy eigenstate

$$\begin{aligned} Z_\Omega &= \sum_{n_1, n_2, \ldots} e^{-\beta \sum_i (\epsilon_i - \mu) n_i} \\ &= \prod_i \left(\sum_{n_i} e^{-\beta(\epsilon_i - \mu) n_i} \right) . \end{aligned}$$

At this stage we have to distinguish the Fermi–Dirac statistics where $n_i = 0$ or 1 and the Bose–Einstein statistics where $n_i = 0, 1, 2, \ldots, \infty \; \forall i$. We begin with the Fermi–Dirac statistics in the next section. Note also that the label i goes over all possible quantum states, some of which may have the same energy ϵ_i but differ by some other quantum number such as the spin. This takes into account the degeneracy of the energy eigenvalues and therefore the phase space factor $C(E)$ is already included in this sum.

7.3 Fermi–Dirac system

As explained in the last section, in the case of Fermi–Dirac statistics each quantum state can have occupation number either zero or one. This simplifies the expression for the partition function since

$$\sum_{n_i} e^{-\beta(\epsilon_i - \mu) n_i} = 1 + e^{-\beta(\epsilon_i - \mu)} \, .$$

Substitution into our expression for the partition function at the end of the last section gives

$$Z_\Omega^{FD} = \prod_i \left(1 + e^{-\beta(\epsilon_i - \mu)} \right) .$$

We then take the logarithm to arrive at the expression for the grand canonical potential

$$\Omega^{FD} = -\frac{1}{\beta} \sum_i \ln \left(1 + e^{-\beta(\epsilon_i - \mu)} \right) .$$

Once this link with the thermodynamic function Ω has been made, it is possible to calculate all thermal properties of the quantum system. Some of these properties

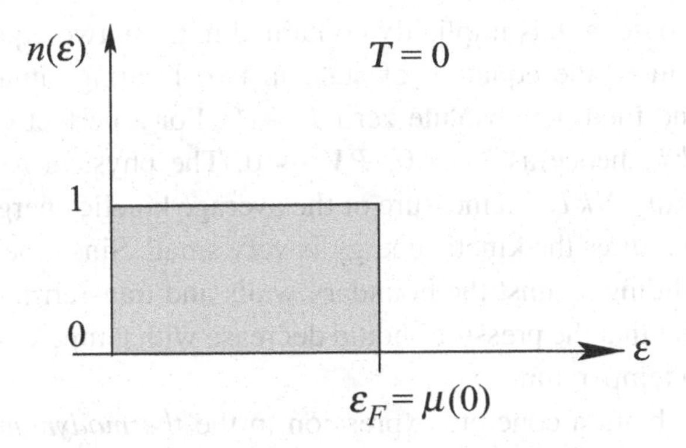

Figure 7.1 All energy levels up to the Fermi energy are filled at zero temperature.

are very different from the corresponding properties of the classical system. As an immediate application of the above formula we compute the expectation value for the occupation numbers $\langle n_i \rangle$ (see Fig. 7.1)

$$\begin{aligned}\langle n_i \rangle &= \frac{\partial}{\partial \epsilon_i}\left(\frac{-1}{\beta}\ln Z_\Omega^{FD}\right)\\ &= \frac{1}{1+\mathrm{e}^{\beta(\epsilon_i-\mu)}}\,.\end{aligned}$$

Let us now determine the equation of state for a quantum mechanical system of Fermi–Dirac particles. For this we first recall that for the present case

$$\sum_i \cdots = \sum_{\alpha=1}^{g}\sum_{\mathbf{p}} \cdots = g\sum_{\mathbf{p}},$$

where the sum over α counts the additional quantum numbers which do not change the energy and the second sum is over the momenta. We then have

$$\Omega^{FD}(T,V,\mu) = -\frac{g}{\beta}\sum_{\mathbf{p}}\ln\left(1+\mathrm{e}^{-\beta(\epsilon(\mathbf{p})-\mu)}\right).$$

To continue we use $\Omega^{FD} = -PV$ which follows from the definition of Ω and $U - TS - \mu N$. Thus

$$PV = \frac{g}{\beta}\sum_{\mathbf{p}}\ln\left(1+\mathrm{e}^{-\beta(\epsilon(\mathbf{p})-\mu)}\right)$$

In order to obtain the equation of state we need to express the chemical potential μ in terms of P, V, T, N. This can be done using

$$N = -\left(\frac{\partial \Omega^{FD}}{\partial \mu}\right)_{V,T}$$

The equation of state is thus implicitly contained in these two equations. We will calculate the form of the equation of state in two limiting situations. First the $T \to \infty$ limit and then the absolute zero $T \to 0$. For a perfect classical gas we know $PV = NkT$, hence as $T \to 0$, $PV \to 0$. The physical reason for this is simple. The quantity NkT is a measure of the average kinetic energy of the system and at low temperatures the kinetic energy is very small. Since the pressure is due to molecules colliding against the boundary walls and transferring kinetic energy to them, we expect that the pressure should decrease with temperature if the kinetic energy falls with temperature.

We will now obtain a concrete expression in the *thermodynamic limit*; that is we let the volume V and the number of particles tend to infinity in such a way that the particle density remains finite. Recalling the formula for the continuum approximation obtained at the end of Section 7.1

$$\sum_{\mathbf{p}} \to \frac{V}{(2\pi\hbar)^3} \int \mathrm{d}^3 p\,,$$

we then have

$$\Omega^{FD} = -\frac{g}{\beta} \frac{V}{(2\pi\hbar)^3} \int \mathrm{d}^3 p \ln\left(1 + \mathrm{e}^{-\beta\epsilon(p)} z\right), \quad z = \mathrm{e}^{\beta\mu}.$$

The quantity z introduced here is the *fugacity*. Similarly,

$$N = \frac{g}{\beta} \frac{V}{(2\pi\hbar)^3} \int \mathrm{d}^3 p \frac{\partial}{\partial\mu} \ln\left(1 + \mathrm{e}^{-\beta\epsilon(p)} z\right).$$

For a non-relativistic Fermi–Dirac gas the one-particle energy is

$$\epsilon(p) = \frac{|\mathbf{p}|^2}{2m}.$$

Changing the variables $|\mathbf{p}| \to \epsilon(\mathbf{p})$ we get

$$\Omega^{FD} = -g\sqrt{\beta} \frac{2V}{\sqrt{\pi}\lambda^3} \int \mathrm{d}\epsilon\, \epsilon^{1/2} \ln(1 + z\mathrm{e}^{-\beta\epsilon}),$$

where $\lambda = h/\sqrt{2\pi mkT}$ is the thermal wavelength introduced in Chapter 2. Upon integration by parts we then find

$$\Omega^{FD} = -\frac{2}{3} A \int_0^\infty \frac{\epsilon^{\frac{3}{2}}\,\mathrm{d}\epsilon}{(\mathrm{e}^{\beta\epsilon} z^{-1} + 1)},$$

$$N = A \int_0^\infty \frac{\epsilon^{\frac{1}{2}}\,\mathrm{d}\epsilon}{(\mathrm{e}^{\beta\epsilon} z^{-1} + 1)},$$

where

$$A = g\frac{2V\beta^{\frac{3}{2}}}{\sqrt{\pi}\lambda^3} = g\frac{V2^{\frac{5}{2}}m^{\frac{3}{2}}\pi}{h^3}.$$

We note that the factor $(e^{\beta\epsilon} z^{-1} + 1)^{-1}$ represents the average number of molecules of energy ϵ at temperature T. Furthermore $A\,\epsilon^{1/2}\,d\epsilon$ represents the density of states with energies between ϵ and $\epsilon + d\epsilon$. This provides a clear interpretation of the integral representing the average number of molecules N at temperature T. With this interpretation in mind we turn to the expression for Ω^{FD}. Rewriting Ω^{FD} as

$$\Omega^{FD} = -\frac{2}{3}\int_0^\infty A\,\epsilon^{\frac{1}{2}}\left(\frac{\epsilon}{e^{\beta\epsilon}\frac{1}{z}+1}\right)d\epsilon\,,$$

it becomes clear that $\Omega^{FD} = -\frac{2}{3}U$, where U is the average energy of the system. But $\Omega^{FD} = -PV$ so that we have the general result

$$PV = \frac{2}{3}U\,.$$

This result is valid for all values of the temperature. As we saw in Chapter 2 this result also holds for a gas of non-interacting non-relativistic classical molecules, so we have here an equation which is true for both a classical as well as a quantum system. It represents the essential features of a non-interacting system. It is not hard to see that the above identity also holds for a system with Bose–Einstein statistics. We leave the verification of this to the reader.

7.3.1 High temperature, classical limit

As explained in the introduction to this chapter we expect quantum effects to be small when the molecules are far from each other. For a quantum system this condition means that the thermal wavelength is much smaller than the distance between the molecules or $\lambda^3/(V/N) \ll 1$. We can achieve this either by increasing the temperature for a given density or by reducing the density for given temperature. In order to find the equation of state we once more change the integration variable to $x = \beta\epsilon$. Let us start with the expression for N

$$N = g\frac{2V}{\sqrt{\pi}\lambda^3}\int_0^\infty x^{\frac{1}{2}} z\,e^{-x}\left(\frac{1}{1+z\,e^{-x}}\right)dx\,.$$

This integral can be evaluated as a power series in z

$$N = g\frac{V}{\lambda^3}\frac{2}{\sqrt{\pi}}\int_0^\infty x^{\frac{1}{2}}\,z\,e^{-x}(1 - z\,e^{-x} + \cdots)\,dx.$$

The remaining x-integrals are then recognized as the integral representation of the Gamma function

$$\Gamma(n) = \int_0^\infty \mathrm{d}t\, t^{n-1}\,\mathrm{e}^{-t}\,,$$

with $\Gamma[\frac{3}{2}] = \sqrt{\pi}/2$. We thus end up with

$$\frac{N}{V} = \frac{g}{\lambda^3} f_{\frac{3}{2}}(z), \quad f_{\frac{3}{2}}(z) = -\sum_{\ell \geq 1} \frac{(-z)^\ell}{\ell^{\frac{3}{2}}}\,.$$

In particular, we see that $\lambda^3/(V/N) \ll 1$ implies $z \ll 1$. The expansion in z is thus meaningful in the classical limit. We can now repeat this procedure for Ω^{FD} in an analogous manner with the result

$$\Omega^{FD} = -g\frac{V}{\beta\lambda^3} f_{\frac{5}{2}}(z), \quad f_{\frac{5}{2}}(z) = -\sum_{\ell \geq 1} \frac{(-z)^\ell}{\ell^{\frac{5}{2}}}\,.$$

Finally we need to express z in terms of V, T, N. Starting with the expression for N we have, for small z

$$\left(\frac{N}{V}\lambda^3\right)\frac{1}{g} \simeq z - \frac{z^2}{2^{\frac{3}{2}}}\,.$$

For $\left(\frac{N}{V}\lambda^3\right) \ll 1$ we can then expand this equation perturbatively up to order z^2. This gives $z = z_0 + z_1$, where

$$z_0 = \left(\frac{N}{V}\lambda^3\right)\frac{1}{g}, \quad z_1 = \frac{1}{2^{\frac{3}{2}}}\, z_0^2\,.$$

Substituting this in Ω^{FD} we then find

$$\Omega^{FD} = -g\frac{V}{\beta\lambda^3}\left[z_0 + \frac{z_0^2}{2^{\frac{5}{2}}}\right].$$

Finally replacing z_0 by the above expression and recalling that $\Omega^{FD} = -PV$ we find

$$PV = NkT\left[1 + \frac{1}{4\sqrt{2}g}\left(\frac{N\lambda^3}{V}\right)\right].$$

The pressure approaches the perfect gas law $PV = NkT$ in the limit where $N\lambda^3/V \to 0$. This is the classical regime. Note that for a Fermi–Dirac system the first quantum correction to the classical equation of state increases the pressure compared to the perfect gas law.

7.3.2 Equation of state at $T \to 0$

Let us now consider the opposite limit when quantum effects are expected to become important. We consider the quantum gas in the region $T \to 0$. For this we note that the average number of Fermi–Dirac molecules $n(\epsilon)$ in the energy state $\epsilon(p)$ as $T \to 0, \beta \to \infty$ has a rather simple structure, namely

$$n(\epsilon) = \frac{1}{e^{\beta(\epsilon-\mu)}+1} \to \begin{cases} 1 & \text{if } \epsilon < \mu; \\ 0 & \text{if } \epsilon > \mu. \end{cases}$$

As a consequence of this, the expression for N becomes

$$\begin{aligned} N &= A \int_0^{\mu} \epsilon^{\frac{1}{2}} \, d\epsilon \\ &= \frac{2}{3} A \mu^{\frac{3}{2}} \end{aligned}$$

while Ω^{FD} takes the form

$$\begin{aligned} \Omega^{FD} &= -\frac{2}{3} A \int_0^{\mu} \epsilon^{\frac{3}{2}} d\epsilon \\ &= -\frac{4}{15} A \mu^{\frac{5}{2}}. \end{aligned}$$

Let us now return to equation of state, $PV = -\Omega^{FD}$. We see that since Ω^{FD} does not vanish at zero temperature, neither does the pressure. More precisely

$$PV = \frac{4}{15} A \left(\frac{3N}{2A}\right)^{\frac{5}{3}}.$$

The reason for this is simple. Unlike a Bose–Einstein system where an arbitrary number of molecules can settle down to the same lowest energy state, a Fermi–Dirac system has to move up the energy ladder as more and more molecules are introduced. As a consequence all the energy levels below μ are fully occupied as $T \to 0$ (Fig. 7.1). If the number of molecules in the system is large, μ can be large (recall $\mu \propto N^{\frac{2}{3}}$), so that the Fermi–Dirac system, even at $T \to 0$, may contain very energetic molecules. It is for this reason that a Fermi–Dirac system has pressure even in the limit $T \to 0$. This phenomenon, which is a direct consequence of the Pauli exclusion principle, has very important consequences for the stability of matter. Indeed the world as we see it could not exist if electrons and protons were not Fermi–Dirac particles satisfying the exclusion principle!

7.3.3 Thermal properties at low temperature

We now turn to the thermal properties of a Fermi–Dirac gas at low temperatures. In order to do this we need to determine the chemical potential μ and the thermodynamic potential Ω^{FD} for low (but non-zero) values of the absolute temperature. We recall that in the thermodynamic limit

$$\Omega^{FD} = -\frac{2}{3}A\int_0^\infty \frac{\epsilon^{\frac{3}{2}}\mathrm{d}\epsilon}{\mathrm{e}^{\beta(\epsilon-\mu)}+1}$$

while

$$N = A\int_0^\infty \frac{\epsilon^{\frac{1}{2}}\mathrm{d}\epsilon}{\mathrm{e}^{\beta(\epsilon-\mu)}+1}.$$

In order to deal with these expressions for small values of the temperature we will make use of the following mathematical result

Lemma 7.1 If $I(\mu) = \int_0^\infty \frac{f(\epsilon)\mathrm{d}\epsilon}{\mathrm{e}^{\beta(\epsilon-\mu)}+1}$, with $f(\epsilon)$ a non-singular function and $\mu > 0$, then

$$I(\mu) \cong \int_0^\mu \mathrm{d}\epsilon f(\epsilon) + 2(kT)^2 f'(\mu)\int_0^\infty \left(\frac{x}{\mathrm{e}^x+1}\right)\mathrm{d}x + O((kT)^4).$$

Proof. Let us write

$$I(\mu) = \int_0^\mu \frac{f(\epsilon)\mathrm{d}\epsilon}{\mathrm{e}^{\beta(\epsilon-\mu)}+1} + \int_\mu^\infty \frac{f(\epsilon)\mathrm{d}\epsilon}{\mathrm{e}^{\beta(\epsilon-\mu)}+1}.$$

In the first term we set $x = \beta(\mu - \epsilon)$, so that

$$\begin{aligned}\int_0^\mu \frac{f(\epsilon)}{\mathrm{e}^{\beta(\epsilon-\mu)}+1}\mathrm{d}\epsilon &= \int_0^{\mu\beta} \frac{\mathrm{d}x}{\beta}\frac{f(\mu-\frac{x}{\beta})}{\mathrm{e}^{-x}+1} \\ &= \int_0^{\mu\beta}\frac{\mathrm{d}x}{\beta}f\left(\mu-\frac{x}{\beta}\right)\frac{\mathrm{e}^x}{\mathrm{e}^x+1} \\ &= \int_0^{\mu\beta}\frac{\mathrm{d}x}{\beta}f\left(\mu-\frac{x}{\beta}\right)\left[1-\frac{1}{\mathrm{e}^x+1}\right],\end{aligned}$$

while in the second term we set $x = \beta(\epsilon - \mu)$ to get

$$\int_\mu^\infty \frac{f(\epsilon)}{\mathrm{e}^{\beta(\epsilon-\mu)}+1}\mathrm{d}\epsilon = \int_0^\infty \frac{f(\mu+\frac{x}{\beta})}{\mathrm{e}^x+1}\frac{\mathrm{d}x}{\beta}.$$

We now observe that

$$\int_0^{\mu\beta}\frac{\mathrm{d}x}{\beta}f\left(\mu-\frac{x}{\beta}\right) = \int_0^\mu f(\epsilon)\mathrm{d}\epsilon,$$

which is the first term on the right-hand side of $I(\mu)$. On the other hand, for large $\beta = 1/kT$ we have up to corrections of $O((kT)^4)$

$$\int_0^\infty \frac{f\left(\mu + \frac{x}{\beta}\right)}{e^x + 1}\frac{dx}{\beta} - \int_0^{\mu\beta} \frac{f\left(\mu - \frac{x}{\beta}\right)}{e^x + 1}\frac{dx}{\beta} = \int_0^\infty \frac{f\left(\mu + \frac{x}{\beta}\right) - f\left(\mu - \frac{x}{\beta}\right)}{e^x + 1}\frac{dx}{\beta}$$
$$= 2\,(kT)^2\, f'(\mu) \int_0^\infty dx \frac{x}{e^x + 1}\,.$$

This then establishes the lemma. □

Using this result we can write

$$N = A\Big[\int_0^\mu \epsilon^{\frac{1}{2}} d\epsilon + 2\,(kT)^2 \left(\frac{d}{d\mu}\mu^{\frac{1}{2}}\right)\frac{\pi^2}{12} + \cdots\Big]$$

where $\cdots$ denote higher orders in kT and we have used

$$\int_0^\infty dx \frac{x}{e^x + 1} = \frac{\pi^2}{12}.$$

Setting $\mu = \mu_0 + (kT)^2\,\mu_1 + O((kT)^4)$ we can solve for μ_0 and μ_1. We get

$$\mu_0 = \left(\frac{3N}{2A}\right)^{\frac{2}{3}}, \qquad \mu_1 = -\frac{\pi^2}{12}\frac{1}{\mu_0}.$$

Thus, the chemical potential decreases with increasing temperature (see Fig. 7.2). We can now use the lemma again to the effect

$$\Omega^{FD} = -\frac{2}{3}A\Big[\int_0^\mu \epsilon^{\frac{3}{2}} d\epsilon + 2\,(kT)^2 \left(\frac{d}{d\mu}\mu^{\frac{3}{2}}\right)\frac{\pi^2}{12} + \cdots\Big]$$
$$= -\frac{4}{15}A\mu^{\frac{5}{2}} + \frac{\pi^2}{4}\,(kT)^2\,\mu^{\frac{1}{2}}\left(-\frac{2}{3}A\right) + \cdots$$

We then substitute the expression for μ obtained above to get

$$\Omega^{FD} = -\frac{4}{15}A\mu_0^{\frac{5}{2}} - \frac{\pi^2}{9}(kT)^2 A\mu_0{}^{\frac{1}{2}} + \cdots$$

Having expressed Ω^{FD} in terms of V and T we are now in a position to determine the thermodynamic properties of the system. The equation of state of the gas is given by

$$PV = \frac{4}{15}A\left(\frac{3N}{2A}\right)^{\frac{5}{3}} + \frac{\pi^2}{9}\,(kT)^2\,A\left(\frac{3N}{2A}\right)^{\frac{1}{3}} + \cdots$$

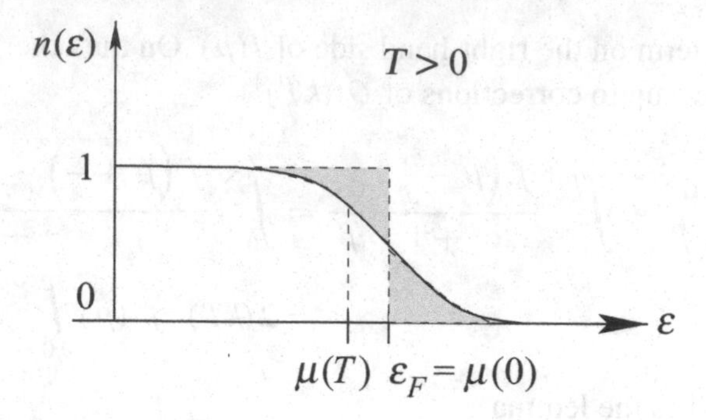

Figure 7.2 Occupation number and chemical potential as a function of temperature.

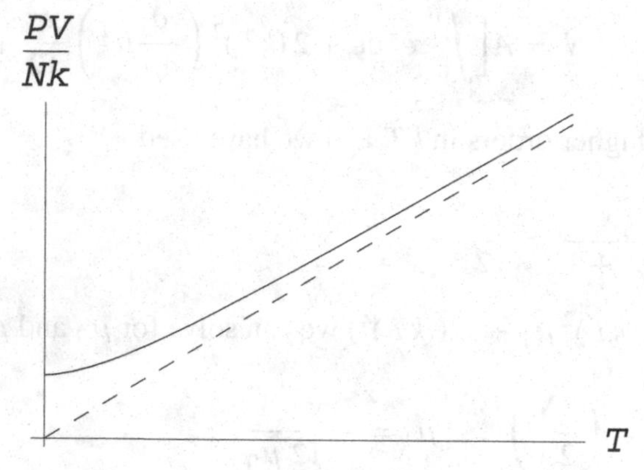

Figure 7.3 Sketch of the equation of state $PV(N, T)$ for the classical ideal gas (dashed) and the ideal Fermi gas (full line).

Combining this result with the first correction to the classical result obtained in Section 7.3.1 we get an equation of state as sketched in Figure 7.3. On the other hand, the entropy S given by

$$S = -\left(\frac{\partial \Omega^{FD}}{\partial T}\right)_{V,\mu} = \frac{k^2 \pi^2 A \mu_0^{\frac{1}{2}}}{3} T ,$$

is found to vanish linearly in T as $T \to 0$. Thus, by taking quantum effects into account, statistical mechanics leads to results that are consistent with the third law of thermodynamics. Finally the specific heat becomes

$$\begin{aligned} c_V &= T \left(\frac{\partial S}{\partial T}\right)_{V,N} \\ &\simeq \frac{N k^2 \pi^2}{2\mu_0} T \end{aligned}$$

as $T \to 0$, which is qualitatively different from the classical result $c_V = 3/2\, Nk$.

7.3.4 Ultra-relativistic gas

A curious consequence of our result for the chemical potential μ is that it is possible to have μ so that $\epsilon \leq \mu$ represents an extremely relativistic fermion, even though the system has temperature close to the absolute zero. We are thus led to consider an extremely relativistic Fermi–Dirac gas for which the relation between energy and momentum is given by

$$\begin{aligned}\epsilon(p) &= \sqrt{|\mathbf{p}|^2 c^2 + m_0^2 c^4} \\ &\approx |\mathbf{p}|c.\end{aligned}$$

The expression for the density of states is modified since $p^2 \mathrm{d}p = 1/c^3 \epsilon^2 \mathrm{d}\epsilon$ and we get

$$\begin{aligned}N &= g\frac{V}{(2\pi\hbar)^3}\frac{4\pi}{c^3}\int_0^\infty \frac{\epsilon^2 \mathrm{d}\epsilon}{\mathrm{e}^{\beta(\epsilon-\mu)}+1} \\ &= B\int_0^\infty \frac{\epsilon^2 \mathrm{d}\epsilon}{\mathrm{e}^{\beta(\epsilon-\mu)}+1},\end{aligned}$$

with B a mass-independent constant. Similarly

$$\Omega^{FD} = -\frac{1}{3}B\int_0^\infty \frac{\epsilon^3 \mathrm{d}\epsilon}{\mathrm{e}^{\beta(\epsilon-\mu)}+1}.$$

We can clearly repeat the analysis for non-relativistic fermions for this situation, examining in turn, high-temperature and low-temperature limits. For the sake of future applications we determine the chemical potential for an ultra relativistic system at $T = 0$. Recall that in this limit the occupation number $n(\epsilon) = (\mathrm{e}^{\beta(\epsilon-\mu)} + 1)^{-1}$ becomes a step function, so that

$$\begin{aligned}N &= B\int_0^\mu \epsilon^2 \mathrm{d}\epsilon \\ &= \frac{1}{3}B\mu^3.\end{aligned}$$

Solving this equation for μ we find

$$\mu(0) = \left(\frac{3N}{B}\right)^{\frac{1}{3}}.$$

Note that for extremely relativistic fermions the chemical potential is mass independent at $T = 0$.

Summary

Let us summarize some of our results. For a system of non-interacting molecules which obey Fermi–Dirac statistics we found that the system continues to exert pressure even at absolute zero. Equally, the system can contain extremely relativistic particles at absolute zero. This is a dramatic example of the “strangeness” of fermions. Furthermore, both the entropy and the specific heat of such a gas vanish as the temperature approaches zero. Finally, we found that quantum effects decrease when the temperature of the system increases and when the density of the system decreases. For non-interacting Fermi–Dirac particles in the classical region we found that quantum effects led to an increase in the gas pressure from the expected classical result. In Chapter 9 we will return to the problem of quantum statistical mechanics for systems with interactions. We will find that a natural way of analyzing such a system leads to quantum field theory.

7.4 Bose–Einstein systems

For a system of non-interacting, non-relativistic molecules (or particles) which obey Bose–Einstein statistics the partition function is expressed in terms of the geometric sum

$$Z_{\Omega}^{BE} = \sum_{n_1} \mathrm{e}^{-\beta(\epsilon_1-\mu)n_1} \sum_{n_2} \mathrm{e}^{-\beta(\epsilon_2-\mu)n_2} \cdots = \prod_i \left(1 - \mathrm{e}^{-\beta(\epsilon_i-\mu)}\right)^{-1},$$

since

$$\sum_{n=0}^{\infty} \mathrm{e}^{-\beta(\epsilon-\mu)n} = \frac{1}{1-\mathrm{e}^{-\beta(\epsilon-\mu)}},$$

which is valid if $\mu \leq 0$. Therefore,

$$\Omega^{BE} = \frac{g}{\beta} \sum_i \ln\left(1 - \mathrm{e}^{-\beta(\epsilon_i-\mu)}\right).$$

Replacing $\sum_i \to V/(2\pi\hbar)^3 \int \mathrm{d}^3 p$ using $\epsilon(p) = |\, \mathbf{p}\, |^2/2m$ and integrating by parts we then find,

$$\Omega^{BE} = -\frac{2}{3} A \int_0^{\infty} \frac{\epsilon^{\frac{3}{2}} \mathrm{d}\epsilon}{\mathrm{e}^{\beta(\epsilon-\mu)} - 1},$$

$$N = A \int_0^{\infty} \frac{\epsilon^{\frac{1}{2}} \mathrm{d}\epsilon}{\mathrm{e}^{\beta(\epsilon-\mu)} - 1}.$$

There is, however, a potential problem with these expressions. Let us examine the expression for the total number of molecules N. We have

$$N = A \int_0^\infty d\epsilon \, \epsilon^{\frac{1}{2}} \, n(\epsilon),$$

with

$$n(\epsilon) = \frac{1}{e^{\beta(\epsilon-\mu)} - 1}.$$

Let us examine $n(\epsilon)$. First of all we notice that for positive chemical potential we immediately conclude that $n(0) < 0$ which is not physically meaningful. This means that μ has to be negative for a Bose–Einstein system, or equivalently,

$$z \equiv e^{\beta\mu} < 1$$

However, there is a further subtlety in the expression for the number density. For $\epsilon \to 0$ and $\beta \to 0$, the denominator goes to zero. The reason for this singular behavior is easy to understand: for a Bose–Einstein system there is no limit to the number of molecules that can occupy the lowest energy, i.e. the $\epsilon = 0$ state. In replacing the sum-over-states expression by an integral over a continuous energy variable we implicitly assumed that the integrand is a regular function and thus failed to count the presence of the $\epsilon = 0$ state. This possibility, of a system at low temperatures having a large proportion of its molecules in the $\epsilon = 0$ state, leads to a very interesting emergent phenomenon in Bose–Einstein systems. This is known as *Bose–Einstein condensation*. It appears that phenomena like superfluidity and superconductivity may have their origins in such an eventuality. The difficulty is to understand how these properties are affected in realistic systems of Bose–Einstein particles when interactions are included. We shall examine this problem in Chapter 9. For the moment we consider the simpler model without interactions. We have already seen that the fugacity z lies in the range $0 \leq z \leq 1$. Writing the average number of molecules in the energy state ϵ as

$$n(\epsilon) = \frac{e^{-\beta\epsilon} z}{1 - e^{-\beta\epsilon} z}$$

$$n(0) = \frac{z}{1-z},$$

we observe that the number of molecules in the zero energy state becomes large as $z \to 1$. On the other hand we expect molecules to settle in the lowest energy state as $T \to 0$. Thus we should expect that $z \to 1$ (or $\mu \to 0$), as $T \to 0$. This suggests that we examine

$$N = A \int_0^\infty \frac{\epsilon^{\frac{1}{2}} \mathrm{d}\epsilon}{(\mathrm{e}^{\beta\epsilon} \frac{1}{z} - 1)},$$

as a function of the temperature. Since $0 \leq z \leq 1$, we have the inequality

$$\begin{aligned} N &\leq A \int_0^\infty \frac{\epsilon^{\frac{1}{2}} \mathrm{d}\epsilon}{(\mathrm{e}^{\beta\epsilon} - 1)} \\ &= A\left(\frac{1}{\beta}\right)^{\frac{3}{2}} \left(\int_0^\infty \frac{x^{\frac{1}{2}} \mathrm{d}x}{\mathrm{e}^x - 1}\right). \end{aligned}$$

This last integral can be expressed in terms of known mathematical functions. We have

$$\int_0^\infty \frac{x^{\frac{1}{2}} \mathrm{d}x}{\mathrm{e}^x - 1} = \zeta\left(\frac{3}{2}\right)\frac{\sqrt{\pi}}{2},$$

where

$$\zeta(x) = \sum_{n=1}^{\infty} \frac{1}{n^x}$$

is the zeta function. The above inequality now becomes

$$\left(\frac{N}{Vg}\right) \leq \frac{1}{\lambda^3} \zeta\left(\frac{3}{2}\right).$$

We then conclude that for a given (N/V) a critical temperature T_c can be defined such that $(\lambda_c = \lambda(T_c))$

$$\left(\frac{N(T_c)}{Vg}\right) = \frac{1}{\lambda_c^3} \zeta\left(\frac{3}{2}\right).$$

Indeed for $T \leq T_c$ we write

$$N(T) = \left(\frac{Vg}{\lambda^3}\right) \zeta\left(\frac{3}{2}\right).$$

What happens to the $N(T_c)$ molecules when the temperature is lowered below T_c? From our discussions we conclude that $N(T)$ of the molecules occupy energy states different from $\epsilon = 0$ while the remaining $N(T_c) - N(T)$ molecules must condense down to the $\epsilon = 0$ state and are thus not taken into account by our integral formula for N. This is an interpretation of how a Bose–Einstein system can appear at low temperatures. For the argument to work for a real system it has to be shown that even in the presence of interactions such a condensation in momentum space can occur. It is also necessary that the number of molecules present in the system be well-defined.

Let us examine some experimentally measurable consequences of Bose–Einstein condensation. In Chapter 10, we will study more carefully the suggestion that the phenomenon of superfluidity observed in liquid helium at temperature below 2.2 K is due to Bose–Einstein condensation of the molecules of the system. There we will take the interaction between helium molecules into account. For the moment we remark that T_c for liquid helium obtained from the expression

$$\frac{N}{V} = \frac{1}{\lambda_c^3}\zeta\left(\frac{3}{2}\right)$$

by setting $V/N = 2.76 \times 10^{-5}$ m^3/mol, $m = 6.65 \times 10^{-24}$ g is $T_c = 3$ K. This result is close to the observed value for the onset of superfluidity in helium and hence the possibility that Bose–Einstein condensation is indeed responsible for the phenomenon of superfluidity seems plausible.

7.4.1 Equation of state and thermal properties

Let us proceed to examine a few features of this system. We determine the specific heat as a function of the temperature and we also examine the equation of state for temperatures $T \leq T_c$. To reduce clutter in the formulas we will assume $g = 1$ in this section. First we note that

$$\Omega = -\frac{VkT}{\lambda^3}\zeta_{\frac{5}{2}}(z)$$

$$N = \frac{V}{\lambda^3}\zeta_{\frac{3}{2}}(z)$$

where $\zeta_s(z) = \sum_1^\infty z^n/n^s$ is plotted in Fig. 7.4. This is a consequence of the following mathematical result

Lemma 7.2 The integral

$$I(s,z) = \int_0^\infty \frac{x^s \mathrm{d}x}{\frac{1}{z}e^{\beta x} - 1}$$

is given by

$$I(s,z) = \frac{\Gamma(s+1)}{\beta^{s+1}}\zeta_{s+1}(z)$$

where $\Gamma(s)$ is the Euler Gamma function.

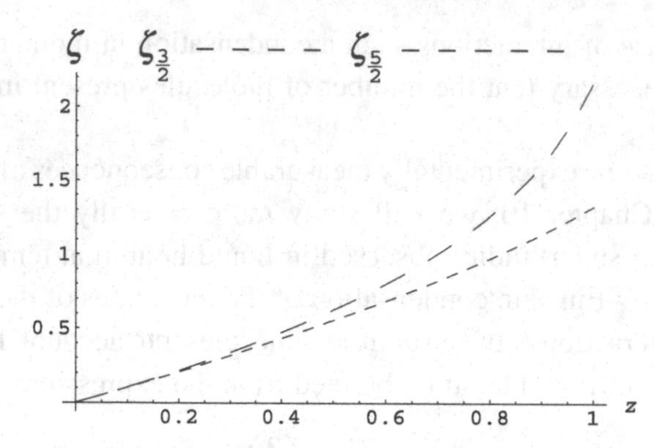

Figure 7.4 Graph for the functions $\zeta_{\frac{3}{2}}(z)$ and $\zeta_{\frac{5}{2}}(z)$.

Proof. We first write

$$I(s,z)=\int_0^{\infty}\frac{z\mathrm{e}^{-\beta x}x^s\mathrm{d}x}{1-\mathrm{e}^{-\beta x}z}$$

then using

$$\frac{1}{1-x}=\sum_{n=0}^{\infty}x^n$$

we get

$$I(s,z)=\sum_{n=0}^{\infty}z^{n+1}\left(\int_0^{\infty}\mathrm{e}^{-(n+1)\beta x}x^s\mathrm{d}x\right).$$

Now

$$\int_0^{\infty}\mathrm{e}^{-(n+1)\beta x}x^s\mathrm{d}x=\frac{1}{(\beta(n+1))^{s+1}}\int_0^{\infty}\mathrm{e}^{-y}y^s\mathrm{d}y$$

with $y=(n+1)\beta x$. Using

$$\Gamma(s+1)=\int_0^{\infty}\mathrm{e}^{-y}y^s\mathrm{d}y$$

the result follows. □

Setting $s=3/2$ and $s=1/2$ and using $\Gamma(5/2)=3/2\,\Gamma(3/2)$, the results stated for Ω and N are established. For $T\leq T_c$ we have to set z = 1 in these expressions. Using $\Omega=-PV$, we have at $T\leq T_c$

$$\begin{aligned}\frac{PV}{N(T)}&=kT\frac{\zeta_{\frac{5}{2}}(1)}{\zeta_{\frac{3}{2}}(1)}\\&=0.513(kT)\end{aligned}$$

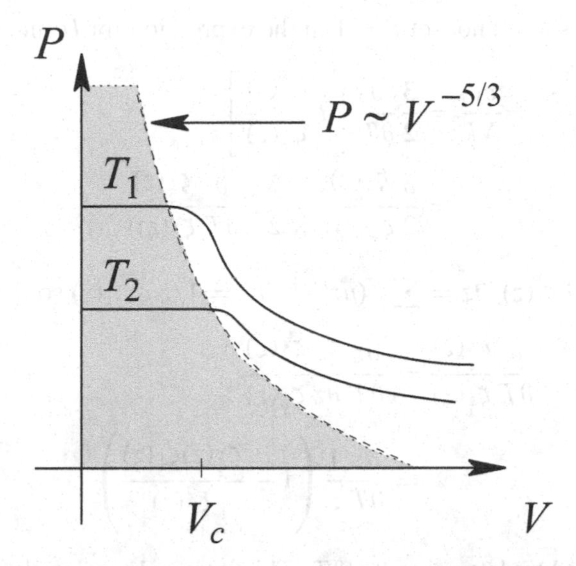

Figure 7.5 Equation of state for a non-interacting Bose–Einstein system.

where we have used $\zeta_{\frac{5}{2}}(1) = 1.341$ and $\zeta_{\frac{3}{2}}(1) = 2.612$. Note the pressure of helium molecules at $T \leq T_c$ is less than the pressure of a corresponding perfect gas with the same number of molecules. Note also that for $T \leq T_c$, $N(T)$ is no longer a constant but changes with temperature. Indeed, in this temperature range $P(T) = kT/\lambda^3\ \zeta_{\frac{5}{2}}(1)$ and is independent of the volume (see Figure 7.5).

To study the specific heat of the system we first determine the internal energy. We have seen in Section 7.3 that $U = \frac{3}{2}PV$, so that

$$U = \frac{3VkT}{2\lambda^3}\zeta_{\frac{5}{2}}(z).$$

The specific heat for temperatures $T < T_c$ is obtained by first setting $z = 1$ in this expression for U and then using

$$c_V = \left(\frac{\partial U}{\partial T}\right).$$

This gives $c_V = 15/4\ (Vk/\lambda^3)\ \zeta_{\frac{5}{2}}(1)$. Note that $c_V \to 0$ as $T \to 0$, in contrast to the classical ideal gas. Using our expression for N we can then write

$$\begin{aligned} c_V(T_c) &= \frac{15}{4}Nk\frac{\zeta_{\frac{5}{2}}(1)}{\zeta_{\frac{3}{2}}(1)} \\ &= 1.925Nk\cdot \end{aligned}$$

A result which is greater than the corresponding specific heat of a classical perfect gas with the same number of molecules namely $c_V = 3/2Nk$.

For $T > T_c$ we should not set $z = 1$ in the expression for U and we have

$$\frac{c_V}{Nk} = \frac{3}{2}\frac{\partial}{\partial T}\left[T\frac{\zeta_{\frac{5}{2}}(z)}{\zeta_{\frac{3}{2}}(z)}\right]$$
$$= \frac{3}{2}\frac{\zeta_{\frac{5}{2}}(z)}{\zeta_{\frac{3}{2}}(z))} + \frac{3}{2}T\frac{\partial}{\partial T}\frac{\zeta_{\frac{5}{2}}(z)}{\zeta_{\frac{3}{2}}(z))}.$$

Now observe that $\partial\zeta_s(z)/\partial z = \sum_1^\infty (nz^{n-1})/n^s = 1/z\,\zeta_{s-1}(z)$ so that

$$\frac{\partial}{\partial T}\frac{\zeta_{\frac{5}{2}}(z)}{\zeta_{\frac{3}{2}}(z)} = \frac{\partial z}{\partial T}\frac{\partial}{\partial z}\frac{\zeta_{\frac{5}{2}}(z)}{\zeta_{\frac{3}{2}}(z)}$$
$$= \frac{\partial z}{\partial T}\frac{1}{z}\left(1 - \frac{\zeta_{\frac{5}{2}}(z)\zeta_{\frac{1}{2}}(z)}{\zeta_{\frac{3}{2}}^2(z)}\right).$$

To proceed we need to determine $\partial z/\partial T$. This is done by using the expression

$$N = \frac{V}{\lambda^3}\zeta_{\frac{3}{2}}(z)$$

then

$$\frac{\partial}{\partial T}(\zeta_{\frac{3}{2}}(z)) = -\frac{3}{2T}\zeta_{\frac{3}{2}}(z).$$

On the other hand,

$$\frac{\partial}{\partial T}(\zeta_{\frac{3}{2}}(z)) = \frac{\partial z}{\partial T}\frac{\partial\zeta_{\frac{3}{2}}(z)}{\partial z} = \frac{1}{z}\frac{\partial z}{\partial T}\zeta_{\frac{1}{2}}(z)$$

so that

$$\frac{\partial z}{\partial T} = -\frac{3z}{2T}\frac{\zeta_{\frac{3}{2}}(z)}{\zeta_{\frac{5}{2}}(z)}.$$

Using this result we finally get

$$\frac{c_V}{Nk} = \frac{15}{4}\frac{\zeta_{\frac{5}{2}}(z)}{\zeta_{\frac{3}{2}}(z)} - \frac{9}{4}\frac{\zeta_{\frac{3}{2}}(z)}{\zeta_{\frac{1}{2}}(z)}.$$

valid for $T > T_c$. Note in the limit $T \to \infty$,

$$\frac{c_V}{Nk} \to (\frac{15}{4} - \frac{9}{4}) = \frac{3}{2}.$$

It is possible to show that $(\partial c_V/\partial T)$ is not continuous at $T = T_c$. Thus a graph of c_V as a function of temperature would look like Fig. 7.6.

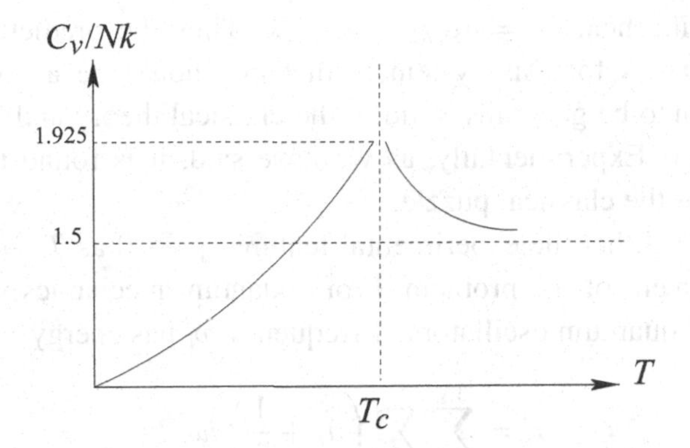

Figure 7.6 Specific heat for a Bose–Einstein gas.

7.5 Specific heat for a solid

We conclude our discussion of Bose–Einstein systems by discussing a historically important puzzle. The puzzle for statistical mechanics was to explain why the specific heat of a solid was temperature dependent. Experimentally it was found that the specific heat approached zero at low temperatures while, as we shall see in a simple example, classical statistical mechanics predicts that it should be a constant.

If we think of a solid as an assembly of molecules held in place by a short range potential then the motion of this system near its equilibrium state can be described as being equivalent to a collection of harmonic oscillators. The effective Hamiltonian for small oscillations of the system about its classical equilibrium state can be written as

$$\mathsf{H} = \sum_{i=1}^{3N} \left(\frac{p_i^2}{2m_i} + \frac{1}{2} k_i q_i^2 \right)$$

where $\omega_i = \sqrt{k_i}/m_i$ represents the frequencies of the normal modes of oscillation of the system. The canonical ensemble partition function for this system is defined by

$$Z_F = \frac{1}{(3N)!} \frac{1}{h^{3N}} \int_{-\infty}^{\infty} \mathrm{d}^{3N} p \int_{-\infty}^{\infty} \mathrm{d}^{3N} q \; \mathrm{e}^{-\beta \sum_{i=1}^{3N} (p_i^2/2m_i + 1/2 k_i q_i^2)}.$$

We simplify the model by setting $m_i = m$, $k_i = l$ for $i = 1, \ldots, 3N$. Then

$$Z_F = \frac{1}{h^{3N}} \frac{1}{(3N)!} \left(\frac{2\pi m}{\beta} \right)^{\frac{3N}{2}} \left(\frac{2\pi}{\beta l} \right)^{\frac{3N}{2}}.$$

The internal energy U for the system is then

$$U = -\frac{\partial}{\partial \beta} \ln Z_F = 3NkT.$$

Hence the specific heat $c_V = \partial U/\partial T = 3Nk$. Thus the prediction of classical statistical mechanics for this system is that c_V should be a constant. Such a result was shown to be generally valid in the classical theory and is known as the *Dulong–Petit law*. Experimentally, as we have said, it is found that $c_V \to 0$ as $T \to 0$. This was the classical puzzle.

Einstein realized that the experimental feature $c_V \to 0$ as $T \to 0$ is natural in a quantum treatment of the problem. From quantum mechanics we know that a collection of $3N$ quantum oscillators of frequency ω_i has energy

$$E = \sum_{i=1}^{3N} \sum_{n_i} \left(n_i + \frac{1}{2} \right) \hbar\omega_i ,$$

$n_i = 0, 1, \ldots, \infty$. For our simple model, we set $\omega_i = \omega,\ \forall i$. The grand canonical partition function is then given by

$$Z_\Omega = \prod_{i=1}^{3N} \sum_{n_i=0}^{\infty} \mathrm{e}^{-\beta(n_i+\frac{1}{2})\hbar\omega} .$$

This is a geometric series which is easily summed. Using $U = -\partial/\partial\beta(\ln Z_\Omega)$ it follows that

$$U = 3N\hbar\omega \left(\frac{1}{2} + \frac{1}{\mathrm{e}^{\beta\hbar\omega} - 1} \right)$$

so that

$$c_V = 3Nk(\beta\hbar\omega)^2 \frac{\mathrm{e}^{\beta\hbar\omega}}{(\mathrm{e}^{\beta\hbar\omega} - 1)^2} .$$

Note for high temperatures $c_V \to 3Nk$ in qualitative agreement with the classical calculation, while for low temperatures we have

$$c_V \to 3Nk(\beta\hbar\omega)^2 \mathrm{e}^{-\beta\hbar\omega}$$

which predicts that $c_V \to 0$ as $T \to 0$.

The assumption that all the normal mode frequencies of the system are the same is an obvious over-simplification but Einstein's example clearly demonstrates the fact that $c_V \to 0$ as $T \to 0$ is a natural feature of quantum statistical mechanics.

By removing the simplifying assumption of Einstein regarding the equality of all the normal modes, Debye was able to obtain an expression for c_V which was not just in qualitative but in excellent quantitative agreement with experimental measurements of specific heat. We will describe this approach at the end of this chapter.

7.6 Photons

As we emphasized, the phenomena of Bose–Einstein condensation can only occur if the number of molecules present in the system is well-defined. We shall see that for an important class of Bose–Einstein particles, namely photons, the number of particles is not well-defined and hence the kind of Bose–Einstein condensation described cannot occur. Let us explain: photons, or the quanta associated with light, can be described as massless particles which have energy $E = \hbar\omega$, where ω is the circular frequency of the light. Photons have two states of polarization. Furthermore for a system of photons in a cavity thermal equilibrium is achieved by photons constantly being absorbed and emitted by the walls of the container. Consequently, the number of photons in the cavity is not a well-defined concept. Indeed we should regard the number as a variable which is to be adjusted to minimize a thermodynamic function like the Gibbs function G, i.e. we require

$$\left(\frac{\partial G}{\partial N}\right)_{N=N_0,T,P} = 0.$$

But

$$\left(\frac{\partial G}{\partial N}\right)_{T,P} = \mu$$

which is the chemical potential. We thus must set the chemical potential associated with a system of photons equal to zero. We also learn, from this example, that for a thermodynamical system where the number of particles N is well-defined, the chemical potential can be determined as a function of N from the equation

$$N = \left(\frac{\partial \Omega}{\partial \mu}\right)$$

while if the number of particles varies, as for instance for photons, then $\mu = 0$. Writing $\epsilon(p) = \hbar\omega = \hbar|\mathbf{k}|c$ and setting $\mu = 0$, the thermodynamical properties for a collection of photons can be determined from:

$$\Omega = -\frac{1}{3}B\int \frac{\omega^3 \mathrm{d}\omega}{(\mathrm{e}^{\beta\hbar\omega} - 1)}$$

where

$$B = \frac{V\hbar}{\pi^2 c^3}$$

Recall that $PV = \frac{1}{3}U$ for such a system. Hence

$$\frac{U}{V} = \int_0^\infty \mathrm{d}\omega\, u(\omega, T),$$

where

$$u(\omega, T) = \frac{\hbar}{\pi^2 c^3} \frac{\omega^3}{e^{\beta\hbar\omega} - 1}$$

is Planck's celebrated radiation formula for the energy density which led to quantum theory. Carrying out the integrals using

$$\int_0^\infty \frac{x^{2n-1}\mathrm{d}x}{e^x - 1} = \frac{(2\pi)^{2n} B_n}{4n}$$

where $B_1 = \frac{1}{6}$, $B_2 = \frac{1}{30}$, $B_3 = \frac{1}{42}$, $B_4 = \frac{1}{30}, \cdots$, B_n are Bernoulli numbers we establish that

$$\frac{U}{V} = \frac{\pi^2}{15} \frac{(kT)^4}{(\hbar c)^3}$$

$$c_V = \frac{4\pi^2 k^4 T^3}{15\,(\hbar c)^3}$$

$$S = \frac{4\pi^2}{45} \frac{k^4}{(\hbar c)^3} T^3.$$

Note again that both c_V and $S \to 0$ as $T \to 0$. Hence these expressions are consistent with the third law of thermodynamics. These results when first obtained using quantum theory were found to be consistent with earlier experimental measurements. For instance, the internal energy density for photons had earlier been shown to have the form

$$\frac{U}{V} = \left(\frac{4\sigma}{c}\right) T^4$$

(with the coefficient σ determined experimentally) and was known as the Stefan–Boltzmann law. The T^4 dependence can in fact be inferred from simple dimensional analysis. The detailed quantum calculation is however necessary in order to predict the correct value of σ.

7.7 Phonons

As an improvement on Einstein's model for a solid, Debye proposed to replace the $3N$ harmonic oscillators in Einstein's approach by phonons. Phonons are quanta of sound waves in a material body. In a crystal they correspond to the normal modes of lattice oscillations. There are two types of normal modes in a lattice. These are the compression modes (sound waves) which satisfy the dispersion relation $\omega = c_s|\mathbf{k}|$, where c_s is the velocity of sound and the shear modes with $\omega = c_t|\mathbf{k}|$. For a 3-dimensional lattice consisting of N sites Debye suggested to take the $3N$

lowest-lying normal modes with frequencies $\omega_1, \ldots, \omega_{3N}$. This is in contrast to Einstein's model where all frequencies are assumed to be equal.

To simplify the calculation in this model one replaces the crystal as a continuum of volume V. We then proceed in the same way as for photons but with two modifications. First, instead of two, the phonon has three polarizations; two transverse and one longitudinal with velocities c_t and c_s respectively. We thus replace $2/c^3$ in the coefficient B appearing in the grand canonical potential for photons by

$$\frac{3}{c_{eff}^3} \equiv \left(\frac{2}{c_t^3} + \frac{1}{c_s^3}\right).$$

The second modification is to introduce a finite cutoff for the frequency ω, that is, we write

$$U = V \int_0^{\omega_D} \mathrm{d}\omega\, u(\omega, T),$$

where ω_D is chosen such that the total number of modes equals $3N$, i.e.

$$3N = \frac{3V}{(2\pi)^3} 4\pi \int_0^{k_D} k^2\, \mathrm{d}k\,.$$

This then leads to

$$k_D = \left(6\pi^2 \frac{N}{V}\right)^{\frac{1}{3}}, \qquad \omega_D = c_{eff} k_D.$$

For small T, i.e. $\hbar\omega_D\beta \gg 1$, we can ignore this cutoff due to the exponential suppression in the integrand. We then have

$$U \simeq V \int_0^{\infty} \mathrm{d}\omega\, u(\omega, T) = \frac{3\pi^4 NkT}{5}\left(\frac{T}{\theta_D}\right)^3,$$

where $\theta_D = \hbar c_{eff} k_D / k$. The specific heat is then readily evaluated to give

$$\frac{c_V}{N} = \frac{12\pi^4}{5} k \left(\frac{T}{\theta_D}\right)^3, \quad T \ll \theta_D\,.$$

Thus the specific heat vanishes like T^3 for low temperatures. This prediction is in excellent agreement with experiment and was an early triumph of quantum theory. Examples of experimental values for θ_D are

$$\theta_D = \begin{cases} 88\,\mathrm{K} & \text{lead} \\ 1860\,\mathrm{K} & \text{diamond.} \end{cases}$$

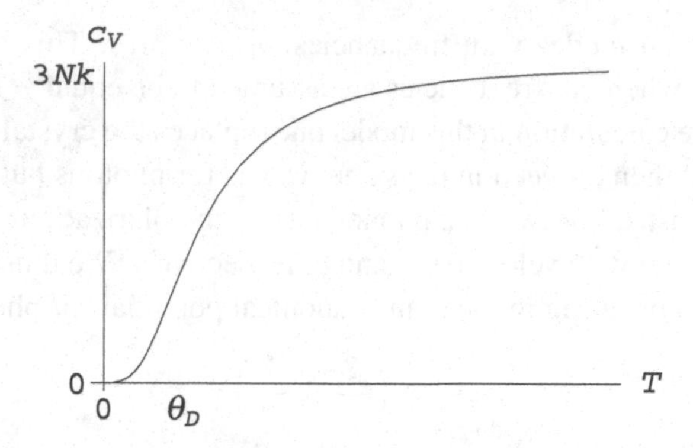

Figure 7.7 Specific heat as a function of T in the Debye model.

At high temperatures, $T >> \theta_D$ we can approximate $e^{\beta\hbar\omega} - 1 \simeq \beta\hbar\omega$, so that the internal energy becomes

$$U(T) \simeq \frac{V}{2\pi^2\beta}\left(\frac{\omega_D}{\theta_D}\right)^3 = 3NkT \, .$$

Thus $c_V = 3Nk$ in agreement with Einstein's model.

Summary

Let us summarize our main findings. A Bose–Einstein system with conserved particle number has the possibility of condensing into the lowest energy ($\epsilon = 0$) state at low temperatures. This is a quantum effect. The temperature dependence of the specific heat for a Bose–Einstein system at low temperatures is proportional to T^3 for photons and, as shown by Debye, for solids as well. For Fermi–Dirac systems the specific heat is proportional to T. At high temperatures the non-interacting Bose–Einstein and Fermi–Dirac systems both approach the perfect gas law. For Bose–Einstein particles, the deviations from the perfect gas law due to quantum effects lead to a lowering of the pressure, while for Fermi–Dirac particles the deviation from the perfect gas law due to quantum effects leads to an increase of the pressure. We shall return to discuss Bose–Einstein systems with interactions in Chapter 9.

7.8 Density matrix

In our treatment of the quantum statistical partition function we have chosen to work in the grand canonical ensemble. However, just like in classical statistical mechanics, other ensembles are possible. In quantum statistical mechanics the different ensembles are best described in terms of the *density matrix* ρ.

Let us first recall that an isolated quantum mechanical system is described by a state vector $|\psi(t)\rangle$. When expanded in terms of an eigenbasis $\{|\psi_i\rangle\}$ of the Hamilton operator the expectation value of an observable A in the state $|\psi\rangle$ takes the form

$$\langle\psi(t)|\mathsf{A}|\psi(t)\rangle = \frac{1}{\sum_n |c_n|^2}\sum_{n,m} c_n^*(t)c_m(t)\langle\psi_n|\mathsf{A}|\psi_m\rangle$$

where $|\psi(t)\rangle = \sum_m c_m(t)|\psi_m\rangle$. This provides a deterministic description of the observable A for an isolated system. The basic idea underlying equilibrium quantum statistical mechanics is that, as a result of interactions with the environment (reservoir), the coherent state $|\psi(t)$ is replaced by an *incoherent mixture* of pure states. The way this is thought to occur is that the coupling to the environment leads to a randomization of the phases of the coefficient functions $c_n(t)$ so that either averaging over time or averaging over an ensemble of independent systems leads to

$$\overline{c_n^*(t)c_m(t)} = 0, \qquad \text{for } n \neq m.$$

This postulated property is an expression of the *postulate of random phases*. This postulate formalizes the transition from a pure state to a mixture of states suggested above. What about $\overline{c_n^*(t)c_n(t)}$? Since the basis vectors $|\psi_n\rangle$ are eigenstates of the Hamiltonian we expect in analogy with the postulate of equal a priori probability introduced in Chapter 2 that $\overline{c_n^*(t)c_n(t)}$ only depends on the energy. Thus

$$\overline{c_n^*(t)c_n(t)} = \rho(E_n)$$

for some distribution function $\rho(E)$.

For a given distribution function $\rho(E)$ the quantum statistical system is then completely determined by the values of $|c_n|^2$. This is conveniently described in terms of the *density matrix* which is defined as follows:

Definition 7.1 The *density matrix* of a given equilibrium quantum statistical system is defined as

$$\rho = \sum_n |c_n|^2 |\psi_n\rangle\langle\psi_n|$$

where the sum is over all eigenstates of the Hamilton operator H of the system.

In terms of this density matrix the expectation value of any observable A of the system can then be written as

$$\begin{aligned}\langle \mathsf{A}\rangle &= \frac{1}{\sum_n |c_n|^2}\sum_n |c_n|^2\langle\psi_n|\mathsf{A}|\psi_n\rangle \\ &= \frac{\sum_{np}\langle\psi_p|\psi_n\rangle|c_n|^2\langle\psi_n|\mathsf{A}|\psi_p\rangle}{\sum_{np}\langle\psi_p|\psi_n\rangle|c_n|^2\langle\psi_n|\psi_p\rangle} \\ &= \frac{1}{\operatorname{Tr}\rho}\operatorname{Tr}(\rho\mathsf{A}).\end{aligned}$$

The advantage of this formulation in terms of the density matrix ρ is that the invariant of the choice of the basis is explicit.

We are now ready to discuss the different ensembles in quantum statistics. For the canonical ensemble we argued in Chapter 2 that the distribution function $\rho(E)$ is given by

$$\rho(E) = \mathrm{e}^{-\beta E}.$$

Thus we have for the density matrix

$$\begin{aligned}\rho &= \sum_n |\psi_n\rangle \mathrm{e}^{-\beta E_n} \langle\psi_n| \\ &= \mathrm{e}^{-\beta \mathsf{H}}\end{aligned}$$

where we have used that the Hamilton operator can be expanded in terms of the energy eigenbasis $\{|\psi_n\rangle\}$ as

$$\mathsf{H} = \sum_n E_n |\psi_n\rangle\langle\psi_n|.$$

Thus in the canonical ensemble the density matrix is just the exponential of $-\beta$ times the Hamilton operator.

Let us now consider the micro canonical ensemble. We recall from Chapter 4 that in this case the energy is fixed so that

$$|c_n|^2 = \begin{cases} 1, & E - \epsilon \leq E_n \leq E + \epsilon \\ 0, & \text{otherwise.} \end{cases}$$

In analogy with classical statistical mechanics we define the corresponding volume in phase by counting the number of states with energy in this interval, that is

$$\Delta\Gamma = \mathrm{Tr}_{\mathcal{H}_N}\, \rho$$

where the trace is over a basis of the N-particle Hilbert space $\mathcal{H}_N$. The corresponding entropy is then given by applying Boltzmann's formula

$$S(E) = k \ln(\Delta\Gamma(E)).$$

With the entropy function defined the thermodynamical properties of the system in question are then completely determined.

Finally we consider the grand canonical ensemble where, in addition to energy exchange, exchange of particles with the reservoir is assumed. Following the prescription given in Chapter 3 we generalize the canonical density matrix by including a chemical potential, that is

$$\rho = \mathrm{e}^{-\beta(\mathsf{H} - \mu \mathsf{N})}$$

where N is the particle number operator. Note that the operator ρ is now not defined on the N-particle Hilbert space $\mathcal{H}_N$ but on the direct sum

$$\mathcal{H} = \oplus_{N=0}^{\infty} \mathcal{H}_N.$$

Taking the trace $\mathrm{Tr}_{\mathcal{H}}\,\rho$ we get

$$\begin{aligned}\mathrm{Tr}_{\mathcal{H}}\,(\rho) &= \sum_N \sum_{E_N} C(E_N)\mathrm{e}^{-\beta(E_n-\mu N)} \\ &= Z_\Omega\end{aligned}$$

in accordance with the grand canonical partition sum defined in Section 7.2. This then completes the unified description of quantum statistical ensembles in terms of the density matrix.

Problems

Problem 7.1 Compute the entropy for ideal, non-relativistic Fermi–Dirac particles in the classical regime and show that Gibbs paradox is resolved automatically.

Problem 7.2 Determine U/V, c_V, S for an ultra relativistic Fermi–Dirac system with $\mu = 0$ using the identity

$$\int_0^\infty \frac{x^{2n-1}\mathrm{d}x}{\mathrm{e}^x + 1} = \left(\frac{2^{2n-1} - 1}{2n}\right)\pi^{2n} B_n$$

with $B_1 = 1/6$, $B_2 = 1/30$.

Problem 7.3 Semiconductor
A semiconductor is characterized by an energy gap $\epsilon_g > 0$ between the band of valence electrons and the band of conducting electrons. As a result of thermal fluctuations electrons can jump from the valence band to the conducting band leaving behind a "hole" in the valence band.

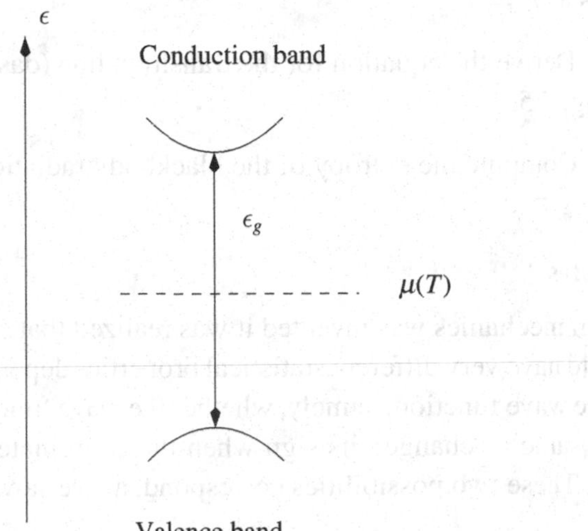

The fermion distribution functions for the electrons in the conducting band and for the holes in the valence band are

$$n_e(\epsilon) = \frac{2}{e^{\beta(\epsilon-\mu)} + 1}$$

$$n_h(\epsilon) = 2 - n_e(\epsilon) = \frac{2}{e^{\beta(\mu-\epsilon)} + 1}$$

respectively. Near the edge of the two bands the electrons and holes behave approximately like free particles with dispersion relations

$$\epsilon_h = -\frac{\mathbf{p}^2}{2m_h} \qquad \text{and} \qquad \epsilon_e = \epsilon_g + \frac{\mathbf{p}^2}{2m_e}$$

respectively, where m_h is the mass of the hole (which in realistic situations may be different from the rest mass of the electron). Let $n(T)$ be the density of conducting electrons and $p(T)$ the density of holes. We will assume that we are in the classical regime $kT \ll \epsilon_g - \mu$ and $kT \ll \mu$.

(1) Using that $n(T) = p(T)$, as a result of charge neutrality, find $\mu(T)$.
(2) For $m_h = m_e$ and $\epsilon_g = 1$ eV, compute $n(T)$ for $T = 300$ K and $T = 1000$ K.

Problem 7.4 Show that there is no Bose–Einstein condensation in two dimensions.

Problem 7.5 Show that the correction to the perfect gas law for a Bose–Einstein system at large T is of the form

$$PV = NkT\left[1 - \frac{1}{4\sqrt{2}}\left(\frac{N\lambda^3}{V}\right)\right].$$

Problem 7.6 Determine $(\partial c_V/\partial T)$ for a Bose–Einstein system and show that it is discontinuous at $T = T_c$.

Problem 7.7 Derive the equation for the transition line (dashed line) in the P–V diagram in Fig. 7.5.

Problem 7.8 Compute the entropy of the blackbody radiation as a function of T.

Historical notes

After quantum mechanics was invented it was realized that systems of N identical particles would have very different statistical properties depending on the symmetry property of the wave function, namely, whether the wave function for such a system remained the same or changed its sign when the coordinates of two particles are interchanged. These two possibilities correspond, as we saw, to the Bose–Einstein statistics and Fermi–Dirac statistics respectively. Historically the Bose–Einstein case was developed before quantum mechanics was invented. In June 1924 S. N. Bose sent Einstein a short paper in English on the derivation of Planck's radiation law asking him to arrange for its publication in *Zeitschrift für Physik*. Einstein translated the paper into German and added a note stating that he considered the new derivation an important contribution. The key new idea in Bose's derivation of Planck's law was his method of counting the states of the photon. He implicitly assumed that the photons were indistinguishable. Einstein immediately followed up Bose's approach by applying the idea of indistinguishability to an ideal gas of mono atomic molecules. He found that there was a maximum possible value for the total number N_0 of particles of non-zero energy that the system could have. Thus if the original number of particles N of the system was initially larger than this value N_0, the difference $N - N_0$ would represent molecules that have condensed into the lowest quantum state with zero energy. This was the phenomena of Bose–Einstein condensation. The idea initially was not accepted as being physically important. It was felt that the approximation of replacing a sum over states by an integration, used by Einstein, was responsible for the result. This view was based on the Ph.D. work of Uhlenbeck who was then a student of Ehrenfest. It was only much later that this view was abandoned when London, in 1938, suggested Bose–Einstein condensation as an explanation for the superfluid properties of helium at low temperatures.

The Fermi–Dirac statistics was invented separately by Fermi in a paper published in *Zeitschrift für Physik* in March 26 1926, and by Dirac in a paper presented to the Royal Society on August 26 1926. Dirac in his paper pointed out that the difference between the Bose–Einstein and Fermi–Dirac statistics corresponded to the difference between wave functions that are symmetric and antisymmetric with respect to the interchange of particles. Fermi showed that a gas of such molecules would continue to exert pressure at zero temperatures. It was immediately realized that the Pauli exclusion principle discovered by analyzing spectral lines was a consequence of Fermi–Dirac statistics.

The first physical application of Fermi–Dirac statistics was made by R. H. Fowler who in 1926 suggested that properties of white dwarf stars could be understood using these ideas.

There actually remains the possibility that any collection of Bose–Einstein molecules which are sufficiently dilute, so that they do not liquify, could settle down to their lowest energy state at low temperatures. The system would now be in a macroscopic quantum state. This is a new state of matter with properties that are actually being studied. For instance, if two condensates overlap, interference fringes result. If light travels through a cluster of Bose–Einstein condensates in a special way it can be slowed down by a factor of up to 20 (!) with the refraction index being very high. E. Cornell, C. Wieman, and W. Ketterle were awarded the physics

Nobel prize in 2001 for "Achievement of Bose–Einstein condensate in dilute gases of alkali atoms and for their studies of the properties of the condensate."

Further reading

There are many good introductory texts on quantum statistical mechanics. Here we list a few of them:

(1) W. Greiner, L. Neise, H. Stocker, *Thermodynamics and Statistical Mechanics*, Springer-Verlag (1995)
(2) L. D. Landau and E. Lifshitz, *Statistical Physics*, Pergamon Press (1959)
(3) K. Huang, *Statistical Mechanics*, J. Wiley (1987)
(4) L. E. Reichl, *A Modern Course in Statistical Physics*, Edward Arnold (1980)
(5) R. Balian, *From Microphysics to Macrophysics*, Springer-Verlag (1991)
(6) R. K. Pathria, *Statistical Mechanics*, Pergamon Press (1991)

8
Astrophysics

In this chapter we will describe how statistical mechanics can be applied to obtain some important results in astrophysics. As an application of classical statistical mechanics we will discuss the Saha ionization formula which plays a role in determining the surface temperature of a star and which will be shown to follow from an analysis of chemical reactions involving ionized particles using statistical mechanics.

We have already emphasized in the last chapter that quantum mechanics has profound implications for the equations of state and, in particular, the stability of matter. In this chapter we will illustrate this effect by considering the collapse of stellar objects. A prominent example is that of white dwarf stars which are stabilized by the Pauli exclusion principle. Understanding white dwarf stars will involve Fermi–Dirac statistics. We will also briefly discuss the fact that neutron stars contain more neutrons than protons and will show that this follows from the analysis of a particular nuclear reaction process treated as a chemical reaction.

In order to present these examples in a suitable setting we start by reviewing a few basic facts about the physics of stellar evolution and we outline the principles that are used to model these objects. This is followed by a qualitative account of stellar evolution. With this background in place the specific examples are considered. We then close this chapter with a qualitative discussion of the cosmic background radiation.

8.1 Basic problem

The Sun is part of a galaxy of stars. The typical size of a galaxy is about 30 000 parsec (1 parsec = 3.26 light years). Within a distance of 5 parsec from the Sun there are 44 known stars. Of the 44 there are 13 multiple stars. Counting multiplicity there are 59 stars in this region. It has been established by spectroscopic measurement that the main constituent of a star is hydrogen. It is also found that a typical star contains 10^{57} or more nuclei. The source of stellar energy lies in a process of fusion which initially involves hydrogen. As the hydrogen fuel is burnt, other fusion processes take over. For these to occur, the temperature of the system has to be hot enough for the appropriate nuclear reactions to take place. If the temperature is not hot enough, the process of nuclear burning stops and the star cools and dies.

It is a theoretically interesting problem to study these possibilities starting with a simple model of a star, considered as a sphere of gases of known chemical composition. The theoretical principles which one should use to study the evolution of such a system include

(1) The conditions for the hydrostatic equilibrium of the system.
(2) The conditions for thermal equilibrium of the system.
(3) The mechanisms for energy transport within the system.
(4) The opacity of the gases in the system.
(5) The mechanisms of energy generation involving nuclear processes in the star.
(6) The equation of state of the system.

To simplify matters we can take our model system to be spherical and described by variables which are dependent only on their distance r from the center of the sphere of radius R which is the star. The variables of importance are the pressure $P(r)$, the temperature $T(r)$, the luminosity $L(r)$ and the mass $M(r)$. These variables may be taken to characterize the star. As r is changed to $r + \mathrm{d}r$ the changes in these variables are to be determined using the principles listed above. Thus $\mathrm{d}P$ is determined using conditions of hydrostatic equilibrium, $\mathrm{d}T$ from the energy transport model considered, $\mathrm{d}L$ from the opacity of the gases and the method of energy generation of the star while $\mathrm{d}M$ is determined in terms of the stellar density which in turn is obtained from the equation of state of the star. There are obvious boundary conditions, e.g. $L(0) = 0$, $M(0) = 0$.

Instead of following the procedure outlined we will resort to a qualitative account of stellar evolution. This will provide a suitable framework for the applications we will be considering.

We suppose that a star contains only hydrogen atoms and is a sphere of radius R. Such a system is expected to contract due to gravitational forces. However, this process of contraction is counteracted by the random motion of the atoms. This random motion may be characterized by the temperature T of the system since the temperature is a measure of the average kinetic energy, K, of the system. We have

$$K = \sum_{i=1}^{N} \frac{1}{2} m < v_i^2 > \simeq NkT$$

for a non-relativistic system of N identical atoms of mass m with the ith atom having an average kinetic energy of $\frac{1}{2}m < v_i^2 >$. We will take this relationship between kinetic energy and temperature to be true in general.

The clumping due to gravitational forces leads to a gravitational potential energy V given by (ignoring numerical factors of order one)

$$V = \frac{GM^2}{R}$$

where G is the gravitational constant, M the mass and R the radius of the star. To continue we use the *virial theorem*. Again ignoring numerical factors, this states $K \simeq V$.

Such a result between the time average of the kinetic and potential energies can be easily proved if we assume that the motion of the system takes place in a finite region of space and the potential energy is a homogeneous function of coordinates. As this result is crucial for our discussions we give a quick proof. We take the kinetic energy K to be a quadratic function of velocities, $\mathbf{v}_i$, then by *Euler's theorem* on homogeneous functions

$$\sum_i \mathbf{v}_i \cdot \frac{\partial K}{\partial \mathbf{v}_i} = 2K .$$

Using $\partial K/\partial \mathbf{v}_i = \mathbf{p}_i$, the momentum, we have

$$2K = \sum_i \mathbf{p}_i \cdot \mathbf{v}_i = \frac{\mathrm{d}}{\mathrm{d}t} \sum_i (\mathbf{p}_i \cdot \mathbf{x}_i) - \sum_i \mathbf{x}_i \cdot \frac{\mathrm{d}\mathbf{p}_i}{\mathrm{d}t} ,$$

where $\mathbf{v}_i = \mathrm{d}\mathbf{x}_i/\mathrm{d}t$. Let us now average this equation with respect to time. Note that for any function $g = \mathrm{d}G/\mathrm{d}t$ of a bounded function $G(t)$, the time averaged mean value of g is zero. This is because

$$< g >= \lim_{T\to\infty} \frac{1}{T} \int_0^T \frac{\mathrm{d}G}{\mathrm{d}t} \mathrm{d}t = \lim_{T\to\infty} \frac{G(T) - G(0)}{T} = 0$$

Now we observe that if the system being considered executes motion in a finite region of space with finite velocities then $\sum_i \mathbf{p}_i \cdot \mathbf{x}_i$ is bounded and by the result just established the mean value of $\mathrm{d}/\mathrm{d}t(\sum_i \mathbf{p}_i \cdot \mathbf{x}_i)$ is zero. Thus

$$2K =< \sum_i \mathbf{x}_i \cdot \frac{\partial V}{\partial \mathbf{x}_i} > ,$$

where we use Newton's law in the form $\mathrm{d}\mathbf{p}_i/\mathrm{d}t = -\partial V/\partial \mathbf{x}_i$, with V the potential energy of the system.

If the potential energy is a homogeneous function of degree k in the position vector then using Euler's theorem $\sum_i \mathbf{x}_i \cdot \partial V/\partial \mathbf{x}_i = kV$. Thus

$$2K = kV ,$$

which is the virial theorem. We will take $K \simeq V$ for our qualitative discussion.

Consider now a gas of hydrogen atoms. From the virial theorem we have

$$\frac{GM^2}{R} \simeq NkT\,,$$

where N denotes the number of hydrogen atoms in the gas. This equation suggests that the radius R of the system will continue to decrease and the temperature continue to increase without limit. Inevitably this rising temperature will lead to hydrogen atoms ionizing. As R continues to decrease the separation distance between the electrons and protons will decrease. Eventually quantum effects will become important and the expression for the NkT term which represents the classical gas pressure will get modified. For a collection of electrons, which are Fermi–Dirac particles, these quantum effects will lead to an increase in the gas pressure from its classical value. In Chapter 7, an explicit expression for this modification was calculated. Here we will estimate this term qualitatively. We note that the electron and proton will both contribute to increasing the classical pressure term. However, this quantum correction term is inversely proportional to the mass of the particle and thus the proton's contribution is expected to be 2000 times smaller than that of the electron. Consequently we will neglect the proton term's contribution for the moment. To estimate the pressure increase due to quantum effects we proceed as follows. We write

$$\begin{aligned} K &= <\frac{(p_p + \Delta p_p)^2}{2m_p}> + <\frac{(p_e + \Delta p_e)^2}{2m_e}> \\ &\simeq <\frac{(p_p)^2}{2m_p}> + <\frac{(p_e)^2}{2m_e}> + <\frac{(\Delta p_e)^2}{2m_e}>\,. \end{aligned}$$

The first two terms represent the proton and electron's classical kinetic energy while the third represents the electron's quantum fluctuation energy. We next estimate the quantum energy. Let us assume that the average separation distance between electrons is r then from the uncertainty principle it follows that $r\Delta p \simeq \hbar$. Thus $< \Delta p^2 >/2m_e = \hbar^2/2m_e r^2$.

We next relate R, the radius of the star, to r, the average separation distance between electrons. We note that the average volume per constituent of the gas is proportional to r^3. Thus the volume of the gas is proportional to Nr^3, but the gas volume is proportional to R^3 therefore R^3 is proportional to Nr^3. Hence we set $R \simeq N^{\frac{1}{3}} r$. We also set M, the mass of the star, equal to Nm_p where m_p is the proton mass. Thus the virial theorem including the quantum term gives

$$\frac{GM^2}{R} \simeq NkT + \frac{N\hbar^2}{2m_e r^2}\,.$$

Rewriting this using $M \simeq Nm_p$, $R \simeq N^{\frac{1}{3}}r$ we get

$$kT \simeq \frac{A}{r} - \frac{B}{r^2},$$

where $A = GN^{\frac{2}{3}}m_p^2$ and $B = \hbar^2/2m_e$. This equation has several interesting consequences. Without the quantum term the virial theorem expression gave a relationship between the radius R and temperature T that did not lead to a stable value for R. It predicted that R would decrease and T increase without limit. The new relationship is very different. It predicts that T would increase to a maximum value:

$$kT_{max} = \frac{A^2}{4B}.$$

An interesting consequence of this result is the prediction that stars must have a minimum mass. Indeed, for nuclear reactions to occur the temperature of the star must be greater than a minimum value. This can conveniently be taken to be $m_e c^2/k$. Thus for a gas system to be a star, that is a system in which nuclear reactions take place, we must have

$$kT_{max} > m_e c^2,$$

which gives $N > 10^{57}$. Putting in the value for the mass of the proton we get $M = Nm_p > 10^{33} g$ which is approximately the mass of the Sun! It is truly remarkable that the mass of a gigantic object like the Sun is fixed not by classical but by quantum considerations.

A second consequence of the modified virial theorem is that the system, after reaching a maximum temperature, will cool, eventually reaching $T = 0$ and $R \simeq N^{\frac{1}{3}}B/A$. This final state of a star represents a white dwarf. In the next section we will present a quantitative analysis of this system.

In fact, our qualitative approach has even more to say. Suppose the mass of the original system was such that by the time quantum effects become important the electrons were highly relativistic particles with velocities close to the velocity of light. Then the quantum term would have to be modified to take this into account. We still have

$$\begin{aligned} K &= E_{proton} + E_{electron} + \langle \Delta E \rangle_{electron} \\ &= NkT + \langle \Delta E \rangle_{electron} \end{aligned}$$

but with $\Delta E \simeq N\langle \Delta p \rangle c$, since $E = \sqrt{p^2c^2 + m_e^2c^4} \simeq |\, p\,| c$ for an extremely relativistic particle. Then using $\langle \Delta p \rangle >\simeq \hbar/r$ we get $\langle \Delta E \rangle \simeq Nc\hbar/r$. The virial theorem now gives

$$\frac{GN^2m_p^2}{N^{\frac{1}{3}}r} \simeq NkT + \frac{Nc\hbar}{r}.$$

If $GN^{\frac{2}{3}}m_p^2 > \hbar c$ in this equation then no quantum force due to electronic repulsion is strong enough to prevent the collapse of the system down to $R = 0$. Thus $GN^{\frac{2}{3}}m_p^2 = \hbar c$ represents a critical mass value for the star to reach a stable $T = 0$ configuration. This critical mass value is known as the *Chandrasekhar limit*. It is, in this crude approximation, of the order of one solar mass. Thus the qualitative analysis suggests that stars much larger than the Sun cannot evolve into white dwarfs; but what is their fate?

Let us examine such a star a bit more closely. Even though the electron might be an extremely relativistic object, the proton being 2000 times more massive might still continue to be a non-relativistic object. Introducing the previously neglected proton contribution to the virial equation gives:

$$\frac{GN^2m_p^2}{N^{\frac{1}{3}}r} \simeq NkT + \frac{Nc\hbar}{r} + \frac{N\hbar^2}{2m_pr^2}.$$

The system will now reach a stable $T = 0$ configuration with a size determined by the proton quantum term. Such a system corresponds to a *neutron star*. We will understand why it is called a neutron star once we analyze the system using the statistical mechanics of chemical reactions. If the star is so massive that even the proton is extremely relativistic when quantum effects become important then this simple analysis suggests that the process of collapse does not stop and the star evolves into a *black hole*. The limiting mass value, in this crude approximation, is twice the Chandrasekhar limit. More refined calculations taking general relativistic effects into account lead to a similar prediction of an upper limit on the mass of a star.

This concludes our qualitative account of stellar evolution. In the next section we present a more quantitative account of the stability of white dwarfs and then a review of chemical reactions in statistical mechanics and their application to astrophysics.

8.2 White dwarf stars

We briefly consider the way a white dwarf star can be modeled. We will assume that the star is spherical and consists of nuclei and electrons. For simplicity we further assume that the system is at zero temperature. We will see that this simplification is in fact justified due to the very high density in the star. The gravitational collapse of the system is prevented by the quantum Fermi pressure of the electrons. Consider

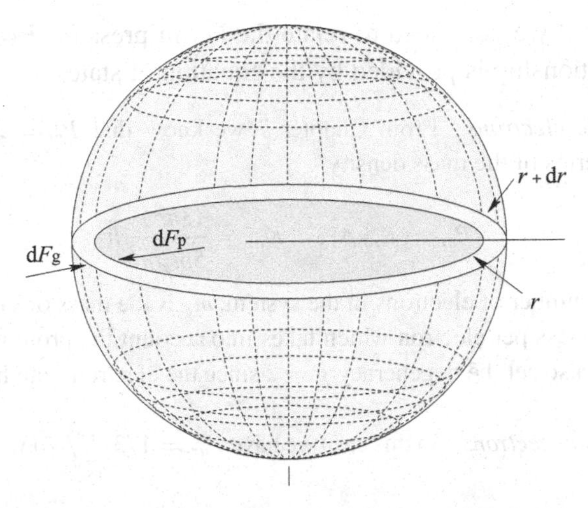

Figure 8.1 Model of a white dwarf: the Fermi pressure $\mathrm{d}F_p$ opposes gravity $\mathrm{d}F_g$.

now a thin shell of the star between radius r and $r + \mathrm{d}r$. Hydrostatic equilibrium of this shell means that the Fermi pressure term must balance the gravitational attraction term. The gravitational attraction term gives rise to a force on the shell directed inwards. The difference in the directed force $\mathrm{d}F$ between r and $r + \mathrm{d}r$ is thus given by (Figure 8.1)

$$\mathrm{d}F_g = \frac{GM(r)}{r^2} 4\pi r^2 \rho(r) \mathrm{d}r ,$$

where

$$M(r) = \int_0^r 4\pi y^2 \rho(y) \mathrm{d}y ,$$

with $\rho(y)$ the density of the star and G the gravitational constant. The difference in the force due to the pressure at r and $r + \mathrm{d}r$ in turn is given by

$$\mathrm{d}F_p = -4\pi r^2 \mathrm{d}P ,$$

where $\mathrm{d}P = P(r + \mathrm{d}r) - P(r)$. In order to have a local equilibrium for all r we then must have

$$\mathrm{d}F_p - \mathrm{d}F = 0 .$$

This gives

$$\frac{\mathrm{d}}{\mathrm{d}r}\left(\frac{r^2}{\rho(r)} \frac{\mathrm{d}P}{\mathrm{d}r}\right) = -G \frac{\mathrm{d}M}{\mathrm{d}r} = -G 4\pi r^2 \rho(r).$$

In order to proceed we now need to relate the Fermi pressure $P(r)$ to the density $\rho(r)$. Such a relationship is provided by the equation of state.

(1) *Non-relativistic electrons* From Chapter 7 we know that $P_{nr} = 2/5\,(N/V)\epsilon_F$, or expressed in terms of the mass density

$$P_{nr} = K_{nr}\rho^{\frac{5}{3}}, \quad K_{nr} = \frac{(3\pi^2)^{\frac{2}{3}}}{5m_e m^{\frac{5}{3}}}\hbar^2 .$$

Here N is the number of electrons in the system, m_e is the mass of the electron, and m is the average mass per electron which takes into account the protons and neutrons as well. We have also set the degeneracy $g = 2$ since the electrons can have up and down spin.

(2) *Ultra-relativistic electrons* In this case we have $P_{ur} = 1/3\,(U/V)$ with $U = 3/4\,N\epsilon_F$. Thus

$$P_{ur} = K_{ur}\rho^{\frac{4}{3}}, \quad K_{ur} = \frac{c\hbar}{m^{\frac{4}{3}}}\left(\frac{3\pi^2}{4}\right)^{\frac{1}{3}} .$$

To summarize, in both cases the equation of state of the system has the form

$$P = K_\gamma \rho^\gamma$$

with $\gamma = 5/3$ for the non-relativistic and $\gamma = 4/3$ for the extremely relativistic case. This equation is valid in each shell when the system is in equilibrium so that we can write

$$P(r) = K_\gamma[\rho(r)]^\gamma .$$

Substituting this expression in the equation for hydrostatic equilibrium we get

$$K_{nr}\frac{\mathrm{d}}{\mathrm{d}r}\left(r^2\frac{5}{3}\frac{1}{\rho^{\frac{1}{3}}}\frac{\mathrm{d}\rho}{\mathrm{d}r}\right) + G4\pi r^2\rho = 0 \quad \text{and}$$

$$K_{ur}\frac{\mathrm{d}}{\mathrm{d}r}\left(r^2\frac{4}{3}\frac{1}{\rho^{\frac{2}{3}}}\frac{\mathrm{d}\rho}{\mathrm{d}r}\right) + G4\pi r^2\rho = 0$$

respectively. The boundary conditions for ρ are $\rho(R) = 0$, where R is the radius of the star and ρ is regular at the origin. The latter condition implies $\rho'(0) = 0$. As in Chapter 5 we then introduce dimensionless quantities for numerical integration of these equations. The convenient choice is

$$\rho(r) = \rho(0)\Theta^{\frac{1}{\gamma-1}}(\zeta)$$

where ζ is related to r by

$$r = \left[\frac{K_\gamma\gamma}{4\pi G(\gamma-1)}\right]^{\frac{1}{2}}[\rho(0)]^{\frac{\gamma-2}{2}}\zeta$$

so that $\Theta(\zeta)$ and ζ are both dimensionless variables. We then get

$$\frac{1}{\zeta^2}\frac{\mathrm{d}}{\mathrm{d}\zeta}\left(\zeta^2\frac{\mathrm{d}\Theta}{\mathrm{d}\zeta}\right) + [\Theta]^{\frac{1}{\gamma-1}} = 0$$

with boundary conditions $\Theta(0) = 1$, $\mathrm{d}\Theta/\mathrm{d}\zeta(0) = 0$. This is known as the *Lane–Emden equation*. The numerical solution of this equation has a zero for some positive value of $\zeta = \zeta_1$. The value of ζ_1 thus determines the radius of the star. The solution for ρ is then integrated again to determine the total mass of the star with the result

$$GMR^3 = \text{const} \quad ; \quad \text{non-relativistic}$$
$$GM = \text{const} \equiv GM^* \quad ; \quad \text{ultra-relativistic}$$

where $M^* \simeq 1.5M_{Sun}$. The interpretation of this result is clear: as long as not all of the electrons are ultra-relativistic the star has a stable equilibrium with a radius that shrinks as the mass M of the star increases with increasing density ρ. As $\rho \to \rho_c$, a critical density, the electrons in the system become relativistic with M approaching M^* a value which is independent of the density for $\rho > \rho_c$. This limiting mass value M^* is known as the Chandrasekhar limit. It represents the maximum mass that a stable star supported by the Fermi pressure of electrons can have. Our brief look at white dwarfs illustrates the important role the equation of state played in the analysis.

8.3 Chemical reactions

In the first chapter we found that if a gas undergoes a thermodynamic transformation at constant temperature and pressure then the Gibbs function, G, for the system decreased. The equilibrium configuration for such a process in stellar evolution then corresponds to a minimum value for G, that is

$$\mathrm{d}G = 0\,.$$

Let us generalize our system of interest to a gas consisting of different types of molecule in which chemical reactions take place. These reactions can convert one type of molecule in the gas to a different type of molecule. For example, in a process involving hydrogen and chlorine the reaction

$$H_2 + Cl_2 \rightleftharpoons 2HCl$$

occurs. Here one hydrogen and one chlorine molecule combine to form two molecules of hydrochloric acid. The reverse process also occurs. In order to deal with such situations the first law of thermodynamics has to be generalized. We have

$$dU = \delta Q - pdV + \sum_i \mu_i dN_i$$

where dN_i represents the change in the number and μ_i the chemical potential of the ith molecule. From the definition of the Gibbs function, $G = U - TS + PV$ and using the first and second law of thermodynamics it follows that

$$dG = -SdT + VdP + \sum_i \mu_i dN_i ,$$

with $S = -(\partial G/\partial T)$; $V = (\partial G/\partial P)$; $\mu_i = (\partial G/\partial N_i)$. Let us consider the equilibrium configuration of this system under conditions of constant temperature and pressure. In this situation the average number of molecules in each species is fixed at some equilibrium value. Hence the Gibbs functions for each constituent molecule type has reached its equilibrium minimum value. We therefore expect $dG = 0$ for the entire system. This gives

$$\sum_i \mu_i dN_i = 0 ,$$

which is the key equation for analyzing chemical reactions. Note that μ_i is independent of N_i when regarded as a function of temperature and pressure. This can be verified to be true for the explicit expressions for the chemical potential we have calculated in earlier chapters and can be understood to be a consequence of the fact that the chemical potential represents the energy required to produce one extra molecule, which we take to depend on the molecule itself, and not on the number of molecules present. Then

$$G = \sum_i \mu_i N_i .$$

For a single molecular species $G = \mu N$, a result we have already used in Chapter 1.

Suppose now we have a reaction represented by the equation

$$\sum_j \nu_j A_j = 0 ,$$

where ν_j represents the number of molecules A_j which participate in the chemical reaction. The numbers ν_j can be positive or negative. For example, in the reaction involving hydrogen and chlorine to form hydrochloric acid, the ν values will be $+1, +1, -2$ respectively. Taking into account that the variation in the number of particle of type i is

$$dN_i = \nu_i dN$$

the equilibrium condition $\mathrm{d}G = 0$ becomes

$$\sum_j \nu_j \mu_j = 0\,.$$

In order to determine the equilibrium condition in chemical reactions it is then necessary to determine μ_i. This can be done in the framework of the grand canonical ensemble discussed in Chapter 3. There we found that

$$\mu_j = kT \ln\left(\frac{N_j \lambda_j^3}{V}\right),$$

where $\lambda_j = h/\sqrt{2\pi m_j kT}$ is the thermal wavelength of a particle with mass m_j and N_j represents the number of the molecules of the jth type present in volume V. Using $P = NkT/V$, we can now write

$$\mu_j = kT \ln\left(\frac{c_j P \lambda_j^3}{kT}\right),$$

where $c_j = (N_j/N)$ is the concentration of molecules of the jth type. Substituting this expression for μ_j we end up with

$$\prod_j c_j^{\nu_j} = \mathrm{e}^{-\sum_j \nu_j \ln\left(P\lambda_j^3/kT\right)} \equiv K(P,T)\,,$$

where $K(P,T)$ is known as the equilibrium constant for the reaction. We thus have the result that, for a given temperature and pressure value, the product $\prod_j c_j^{\nu_j}$ is a constant. This is known as the *law of mass action*. This is a key result in the statistical mechanical treatment of chemical reactions.

8.4 Saha ionization formula

Let us now apply the ideas of chemical equilibrium to determine the surface temperature of a star. The equilibrium properties of a star depend on the thermodynamic parameters such as the pressure P, temperature T, and density ρ of the gases in the star.

Since stars are hot, thermal ionization is expected to occur. These effects will be important when kT is comparable to the binding energy of the electrons in the star. For hydrogen, this happens when $T \simeq 10^5$ K, while for helium it happens above temperatures of $4 \cdot 10^5$ K. At low temperatures the ionization will be incomplete and the gas will be a mixture of atoms in their ground state, atoms in excited states and ionized atoms. The processes which determine the specific mixture present for a given temperature and pressure of the star are the ones involving the atom ionizing and recombining. These processes can be regarded as chemical reactions and hence analyzed using the methods of statistical mechanics. It is then possible to determine precisely the relative distribution of atoms in their ground state configurations compared to their ionized state configurations as a function of the temperature and pressure of the system. The presence of ionized atoms can be detected using spectroscopy, hence from such measurements it is possible to measure the fraction of atoms that are ionized in the star. With these results the temperature of the star can then be determined.

The chemical reaction to be considered is

$$A \rightleftharpoons A^+ + e + \gamma$$

where A denotes an atom, A^+ the ionized states of the atoms, e an electron, and γ one or more photons. Clearly A^+ can similarly be involved in further reactions which can also be analyzed as chemical processes.

The ionization reaction reaches an equilibrium state determined by the condition

$$\mu_A = \mu_{A^+} + \mu_e + \mu_\gamma$$

where μ_A, μ_{A^+}, and μ_e are the chemical potentials of the atom, the ionized atom, and the electron respectively. The chemical potential of the photon vanishes as we have seen in Chapter 7. If B is the binding energy of the electron we write the energy of the atom as

$$E = \frac{|\mathbf{p}|^2}{2m_A} - B.$$

Treating A, A^+, and e as a classical gas of non-interacting particles, the chemical potentials can be read off from our results in Chapter 3, i.e.

$$\begin{aligned}
\mu_\gamma &= 0 \\
\mu_A &= -\frac{1}{\beta}\ln\left(\frac{V g_A}{\lambda_A^3 N_A}\right) + B \\
\mu_{A^+} &= -\frac{1}{\beta}\ln\left(\frac{V g_{A^+}}{\lambda_A^3 N_{A^+}}\right) \\
\mu_e &= -\frac{1}{\beta}\ln\left(\frac{V g_e}{\lambda_e^3 N_e}\right)
\end{aligned}$$

where

$$\begin{aligned}
\lambda_A &= \frac{h}{\sqrt{2\pi m_A kT}} \\
\lambda_e &= \frac{h}{\sqrt{2\pi m_e kT}}
\end{aligned}$$

and $g_A = (2S_A + 1)$, $g_{A^+} = (2S_{A^+} + 1)$, $g_e = (2S_e + 1)$. These give

$$\frac{N_e N_{A^+}}{N_A} = \frac{g_e g_{A^+}}{g_A} \frac{(2\pi m_e kT)^{\frac{3}{2}}}{h^3} e^{-\frac{B}{kT}}.$$

This is the Saha ionization formula. Using $P_e = N_e kT$ where P_e is the electron pressure we can write this as

$$\frac{N_{A^+}}{N_A} = \frac{g_e g_{A^+}}{g_A} \frac{(2\pi m_e)^{\frac{3}{2}}}{h^3} \frac{(kT)^{\frac{5}{2}}}{P_e} e^{-\frac{B}{kT}}.$$

If P_e is known, this equation can be used to determine the temperature of the star by measuring N_{A^+}/N_A by spectroscopic means.

In this calculation we have simplified the problem by considering only the atom and the ionized atom neglecting all internal excitations of these systems. These excitations can be incorporated by replacing the factors g_A by $Z_A = \sum d_r e^{-\beta \hat{E}_r}$ where $\hat{E}_r = B - E_r$, E_r is the binding energy of the rth excited state, and d_r is its degeneracy. The difficulties with this approach are discussed in Chapter 12.

8.5 Neutron stars

In our qualitative account of stellar evolution, we concluded that if a star had mass bigger than the Chandrasekhar limit then the final state of the star was not determined by the electron but by the proton Fermi pressure. Furthermore, the mass of such an object at $T = 0$ also had a limiting value. We would like to understand the chemical composition of a star when its mass is close to this limiting value. We will see that a simple analysis of the relevant chemical reactions will lead to the result that the ratio of the number of neutrons to protons in such a system is 8 : 1. The star is a neutron star.

To establish this result we realize that the electron and proton for the system close to its limiting mass value are both extremely relativistic objects. The composition of the star will be determined by analyzing the chemical reaction in which the electron and proton interact to form neutrons and neutrinos. The reaction can be written as

$$e + p \rightleftharpoons n + \nu_e.$$

The neutrino is an extremely weakly interacting particle and leaves the system after it is formed. We now recall from Chapter 7 that the chemical potential of an extremely relativistic particle at $T = 0$ is

$$\mu_i = K(s) N_i^{\frac{1}{3}},$$

where K is a mass-independent constant which depends on the spin s of the ith particle, and N_i is the number of particles of type i present in the system. We are now ready to analyze the chemical reaction involving the electron and the proton. We have

$$\mu_e + \mu_p = \mu_n + \mu_{\nu_e}.$$

As explained above the density of neutrinos is small so that μ_{ν_e} can be neglected. There is an additional constraint present in the system. Namely the total charge of the system has to be zero. This is because we started with a collection of neutral hydrogen atoms and as charge conservation is respected throughout the evolution of this system we must impose the constraint that the total charge of the system is zero. For the system containing electrons, protons, neutrons, and neutrinos this means that we must have

$$N_e = N_p.$$

Now, the electron and proton have the same spin. Hence $N_e = N_p$ implies for extremely relativistic electrons and protons that

$$\mu_e = \mu_p = K(s)N_p^{\frac{1}{3}}.$$

The neutron also has the same spin as the electron and the proton so that we get

$$2K(s)N_p^{\frac{1}{3}} = K(s)N_n^{\frac{1}{3}}, \quad \text{or}$$
$$\frac{N_n}{N_p} = 8.$$

A final comment is necessary. It is known that the neutron is an unstable particle so that, even though the nuclear process just analyzed gives $N_n/N_p = 8$, these neutrons might be expected to ultimately decay and convert to protons and electrons. This does not happen. The reason is that the decay of a neutron can only produce an electron with $E_e \leq 1\,\text{MeV}$. This is a consequence of the energy available for producing an electron which depends on the mass difference between the neutron and the proton. However, the Fermi energy for the extremely relativistic system we have considered leads to $\mu_e > 1\,\text{MeV}$. So that as a result of the Pauli exclusion principle the neutron cannot decay. Energy states of electrons which could be produced by the decay of the neutron are already occupied. The environment in such a collapsed star thus makes the neutrons stable.

8.6 Blackbody spectrum of the Universe

In 1965, A. Penzias and R. Wilson accidentally discovered that there was radiation coming from all directions in the sky and that it had a temperature of approximately 2.7 K. Subsequent measurements have established that this radiation has a

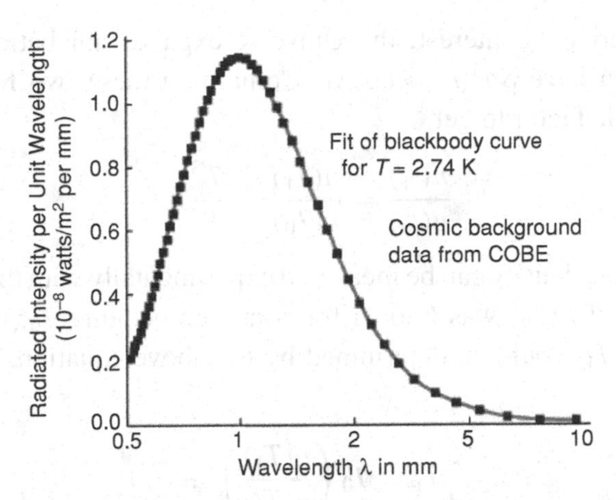

Figure 8.2 Measured intensity of the background radiation by the COBE satellite experiment. This data was adapted from J. C. Mather *et al.*, *Astro. Jour.* **354**, L37 (1990).

blackbody spectrum and that it is isotropic. Indeed recent satellite experiments gave a curve which fit the blackbody spectrum with a temperature, $T = (2.725 \pm 0.001)\,\mathrm{K}$ to an accuracy which is better than any blackbody spectrum measured in a laboratory (see Figure 8.2). The isotropic nature of this microwave radiation is better than one part in ten thousand. These experimental observations indicate that the Universe started with a *big bang*.

It turns out that a background temperature today of the order of 10 K can be motivated using a simple argument duc to Gamov in 1948. This argument goes as follows: at present the Universe consists mostly of helium which is formed at very early stages of the evolution. After the temperature has dropped below $kT \simeq 0.1$ MeV, the number of helium molecules is conserved during the expansion, that is

$$n(t)a^3(t) = \text{constant}$$

where $n(t)$ is the density of helium atoms and $a(t)$ is the scale factor, measuring the relative size of the Universe at different times.

The entropy, on the other hand, is dominated by the contribution from the light particles. These are the photons and the light neutrinos which are ultra relativistic. Recalling our treatment of photons in Chapter 7 we thus have

$$S = \int \frac{\mathrm{d}E}{T} \simeq \kappa \frac{16\sigma}{3} T^3,$$

where σ is the Stefan–Boltzmann constant introduced in Section 7.6 and κ is a coefficient of order one which depends on the number of light degrees of freedom.

If, during the period of interest, the Universe expands adiabatically and quasi-statically we then have $Sa^3(t) = const$. Combining these two formulas we can eliminate the scale factor to get

$$\frac{n(t_1)}{n(t_0)} = \frac{n(T_1)}{n(T_0)} = \frac{T_1^3}{T_0^3}.$$

The current particle density can be measured experimentally with the result $n(T_1) \simeq 10^{-6}\text{cm}^{-3}$. Thus if $n(T_0)$ was known for some temperature T_0, then the present day temperature T_1 would be determined by the above equation. Indeed we then have

$$T_1 = T_0 \left(\frac{n(T_1)}{n(T_0)}\right)^{\frac{1}{3}}.$$

Gamov suggested to fix $n(T_0)$ by the following argument. Since helium, like all matter, is built up by successive nuclear reactions, we first need to produce a sufficient density of deuterium which is obtained by combining a neutron and a proton through the reaction

$$n + p \rightleftharpoons d + \gamma.$$

However, deuterium has a small binding energy of approximately 2 MeV. Consequently, the dissociation process of deuterium by energetic photons is very efficient at $kT \geq 1$ MeV. Below $kT \simeq 0.1$ MeV, or $T \simeq 10^9$ K, on the other hand, the production of deuterium becomes inefficient. Thus, in order to explain the dominance of helium in the Universe today we must assume that the fraction of protons that have been transformed into deuterium by the time the temperature falls below $T_0 = 10^9$ K is of order one, that is (see also Chapter 9)

$$\sigma_{tot} v\, n(t_0)\, t_0 \simeq 1,$$

where $\sigma_{tot} v \simeq 10^{-20}\text{cm}^3/\text{s}$ is the total cross-section for the above reaction times v, the relative velocity between the protons and neutrons which we take to be the thermal velocity $v = \sqrt{2kT/m}$. Next, $n(t)$ is the number density of protons which, at $t = t_0$, is of the same order of magnitude as the number density of deuterium atoms. Finally, t_0 is the time when the Universe has cooled down to 10^9 K.

We can determine t_0 with the help of the following simple, non-relativistic physics argument. Consider a particle of mass μ and velocity $\vec{v}$ at the edge of the universe, which is assumed to have a mass M. Using non-relativistic energy convervation we have

$$E(r) = \frac{1}{2}\mu v^2 - \frac{G\mu M}{r} = k.$$

There are three possibilities: $k > 0$, representing an expanding universe, $k < 0$, a universe which after expansion, slows down and then collapses, and finally $k = 0$ which just manages to expand to infinity. This is the case we consider. We also assume that the mass of the Universe is dominated by rediation, that is

$$M = \frac{4\pi\rho_o r^3}{3} = \frac{4\pi r^3}{3}\sigma T^4 c^2.$$

The factor c^2 is required to convert the energy density of radiation to its equivalent mass. Also, since the entropy is constant we have $T_r = \textit{constant}$. If we then write the energy conservation law as

$$\frac{1}{2}\dot{r}^2 = G\frac{4}{3}\pi\frac{\sigma T^4}{c^2}\frac{r^3}{r},$$

we get

$$r^2\dot{r}^2 = \frac{8G\pi\sigma T^4 r^4}{3c^2} = \alpha^2,$$

where α is a constant as a consequence of the conservation of the entropy. Taking the square root of both sides and integrating ($\int r\mathrm{d}r = \alpha \int \mathrm{d}t$) we then end up with

$$t = \frac{1}{T^2}\sqrt{\frac{3c^2}{32\pi\sigma G}}.$$

If we include the numerical values and set $T = 0.1$ MeV, we find $t_0 \simeq 10^2$ s as an order of magnitude estimate. With t_0 determined $n(t_0)$ follows and we therefore have all ingredients. Substituting numbers we end up with $T_1 \simeq 10$ K, again as an order of magnitude estimate.

Problems

Problem 8.1 Determine the equilibrium distribution of electrons and positrons at very high temperatures ($kT >> m_e c^2$) where m_e is the mass of the electron.

Problem 8.2 Show that the equilibrium ratio of α particles and neutrons, n, to iron nuclei (Fe) in the reaction

$$\gamma + \mathrm{Fe} \rightleftharpoons 13\alpha + 4n$$

is given by

$$\frac{n_\alpha^{13} n_n^4}{n_{\mathrm{Fe}}} = \frac{2^{43}}{(56)^{\frac{3}{2}}(1.4)}\left(\frac{m_u kT}{2\pi\hbar^2}\right)^2 4\mathrm{e}^{-\frac{Q}{kT}}$$

where $Q = c^2(13m_\alpha + 4m_n - m_{\text{Fe}}) = 124.4$ MeV and the mass of a nucleus of atomic weight A is written as Am_u.

Problem 8.3 Background radiation:

(1) Assuming that the energy of the Universe is dominated by radiation at all times, estimate the present temperature of the background radiation given that the age of the Universe today (determined by experimental value of the Hubble constant) is $t_1 \simeq 10^{18}$ s.
(2) In reality, the measured matter energy density of the Universe at present is found to be about 10^4 times that of the radiation so that our simplifying assumption about radiation domination is not justified. In a matter dominated Universe the energy $u(T)$ is proportional to T^3. Assuming that matter domination has set in after 10^{12} s, estimate the error made in the estimate for t_1 in the approximation at 1.

Historical notes

S. Chandrasekhar derived the existence of a limiting mass for a white dwarf star while traveling by boat from Madras to Venice in 1930 to take up a Government scholarship in Cambridge where he hoped to work for his doctoral degree under the supervision of Fowler. With him he had a paper he had completed just before his departure from Madras. In it he had developed Fowler's theory of white dwarfs further. Fowler had demystified the long-standing puzzle regarding the high densities encountered in white dwarfs making use of the recently discovered Fermi–Dirac statistics. Combining Fowler's ideas with Eddington's polytropic considerations for a star, Chandrasekhar had found that in the center of a white dwarf the density could be as high as one ton per cubic inch! It suddenly occurred to him that relativistic effects could become important for such a high density.

According to Pauli's exclusion principle no two electrons could be in the same quantum state. Because of the high density one would thus invariably find electrons with energies higher than their rest mass, so that relativistic effects would become important. Chandrasekhar started his calculation on the boat journey expecting to find a neat, relativistic generalization of Fowler's theory. The calculation was finished by the time the boat arrived in Venice. Much to his surprise he found that the relativistic generalization led to a completely different result. Contrary to Fowler's theory, if the mass of the star exceeded a limiting mass for which essentially all electrons had become relativistic, then the star could not evolve into a white dwarf!

Further reading

A good text on the physics of stars accessible at undergraduate level is R. Bowers and T. Deeming, *Astrophysics 1: Stars*, Jones and Bartlett (1984). A more advanced, but very readable account on astrophysics can be found in S. L. Shapiro and S. A. Teukolsky, *Black Holes, White Dwarfs and Neutron Stars*, John Wiley (1983). A self-contained text in which the properties of model stars are derived starting from Carathedory's approach to thermodynamics, with many interesting historical comments and notes, is S. Chandrasekhar, *Introduction to the Study of Stellar Structure*, Chicago University Press (1939). A detailed discussion of the Saha ionization formula is contained in S. Brush, *Statistical Physics and the Atomic Theory of Matter*, Princeton University Press (1983) as well as L. D. Landau and E. Lifshitz, *Statistical Physics*, Pergamon Press (1959). Finally, for a detailed discussion of big bang cosmology see P. J. E. Peebles, *Principles of Physical Cosmology*, Princeton (1993). A concise but very accessible popular book on cosmology is S. Weinberg's *The First Three Minutes*, Basic Books (1977).

9

Non-relativistic quantum field theory

In this rather technical chapter we develop an efficient tool to include interactions in quantum statistical mechanics. We will provide the basic framework of quantum field theory and perturbation theory. This will be formalized in terms of Wick's theorem and Feynman rules. We furthermore adapt these rules to include finite temperature and density, in order to compute the grand canonical ensemble partition function Z_Ω for an interacting quantum system of identical molecules. This will include, in particular, a discussion of temperature-dependent Green functions. The ideas described in this chapter will play an essential role in our discussion of the superfluidity properties of helium in Chapter 10.

9.1 The quantum field theory formulation

Let us begin by repeating the calculation of the quantum mechanical partition function for non-interacting particles in a notation which is adapted to the quantum field theory formalism. In Chapter 7 we introduced the time-independent N-particle wave function $\psi(\mathbf{x}_1, \ldots, \mathbf{x}_N)$ for a system of N-identical particles. This wave function will be symmetric or antisymmetric when exchanging two coordinates $\mathbf{x}_i$ and $\mathbf{x}_j$ depending whether we describe Bose–Einstein particles or Fermi–Dirac particles respectively. Now, just as we write the 1-particle wave function $\psi(\mathbf{x})$ as

$$\psi(\mathbf{x}) = \langle \mathbf{x} | \psi \rangle$$

where $|\psi\rangle$ is the corresponding state vector, we can write

$$\psi(\mathbf{x}_1, \ldots, \mathbf{x}_N) = \langle \mathbf{x}_1, \ldots, \mathbf{x}_N | \mathbf{k}_1, \ldots, \mathbf{k}_N \rangle$$

where $|\mathbf{k}_1, \ldots, \mathbf{k}_N\rangle$ is the state vector for an N-particle system with momenta $\mathbf{k}_1, \ldots, \mathbf{k}_N$, correctly symmetrized. However, instead of specifying the momentum of each particle, we could specify the state vector by simply stating how many particles have momentum $\mathbf{k}_i$, for all allowed momenta $\mathbf{k}_i$. Since, in quantum mechanics we cannot distinguish between identical particles with the same quantum numbers, these two ways of specifying a state vector are equivalent. We then write

$$|\psi\rangle = |n_1, \ldots, n_i, \ldots\rangle$$

where n_i is the *occupation number* of the state with momentum $\mathbf{k}_i$. If there are further quantum numbers, such as the spin, then the set $\{n_\alpha\}$ is enlarged such as to specify the occupation number of each possible state with momentum $\mathbf{k}_i$ and spin j. Now, in the absence of interactions between the different particles, the total energy of the state $|\psi\rangle$ is $E = \sum_i \epsilon_i n_i$. Furthermore, for non-interacting particles the occupation numbers n_i do not change in time and we therefore conclude that the state $|\psi\rangle$ is an eigenstate of the Hamilton operator H and that

$$\mathsf{H}|n_1, \ldots, n_i, \ldots\rangle = \sum_i n_i \epsilon_i |n_1, \ldots, n_i, \ldots\rangle.$$

Similarly, we can introduce a *number operator*, N, with the property

$$\mathsf{N}|n_1, \ldots, n_i, \ldots\rangle = \sum_i n_i |n_1, \ldots, n_i, \ldots\rangle.$$

With the use of these operators we can now write the grand canonical partition sum Z_Ω, introduced in Chapter 7 simply as a trace over the Hilbert space of states, i.e.

$$\begin{aligned} Z_\Omega &= \sum_{n_1, n_2, \ldots} \mathrm{e}^{-\beta \sum_i (\epsilon_i - \mu) n_i} \\ &= \mathrm{Tr}\left(\mathrm{e}^{-\beta(\mathsf{H} - \mu \mathsf{N})}\right). \end{aligned}$$

The problem we consider below is whether for an interacting system a description of this kind is meaningful given that the energy of an N-particle system is no longer just the sum of 1-particle energies. In such a description the notion of "occupation number" in some form should be present and the energy eigenvalue E_N should, for the N-particle system, reduce to $E_N = \sum \epsilon_i n_i$ when interactions are neglected. When interactions are included we would require that E_N be an eigenvalue corresponding to the N-body Schrödinger equation

$$\left[\sum_{i=1}^{N}\left(-\frac{\hbar^2}{2m}\nabla_i^2\right) + \sum_{i<j} V(|\,\mathbf{x}_i - \mathbf{x}_j\,|)\right]\psi(\mathbf{x}_1, \ldots, \mathbf{x}_N) = E_N \psi(\mathbf{x}_1, \ldots, \mathbf{x}_N)$$

where $V(|\mathbf{x}_i - \mathbf{x}_j|)$ is the potential energy representing the interaction between the molecule at $\mathbf{x}_i$ and the one at $\mathbf{x}_j$, while the N-body wave function $\psi(\mathbf{x}_1, \ldots, \mathbf{x}_N)$ is totally symmetric in the location of the N molecules for Bose–Einstein particles and is totally antisymmetric for Fermi–Dirac particles.

We will now show that such a description does indeed exist. To see how this works we first give an explicit construction of the Hamilton and particle number operators defined above for non-interacting particles and subsequently include interactions in the formalism. For this we have to distinguish between Fermi–Dirac and Bose–Einstein systems.

9.1.1 Non-interacting bosons

As we will now argue, for bosons the Hamilton operator H and the particle number operator N can be expressed in terms of operators known from the quantum mechanical harmonic oscillator. Concretely, we write

$$\mathsf{H} = \sum_i \epsilon_i \mathsf{a}_i^\dagger \mathsf{a}_i$$
$$\mathsf{N} = \sum_i \mathsf{a}_i^\dagger \mathsf{a}_i \,,$$

where the $\mathsf{a}_i^\dagger$, and a_i represent a collection of harmonic oscillator raising and lowering operators which satisfy the commutation relations

$$[\mathsf{a}_i, \mathsf{a}_j^\dagger] = \delta_{ij}$$
$$[\mathsf{a}_i, \mathsf{a}_j] = 0$$
$$[\mathsf{a}_i^\dagger, \mathsf{a}_j^\dagger] = 0 \,.$$

We assume, furthermore, that the *vacuum*, or zero-particle state is annihilated by all lowering operators a_i

$$\mathsf{a}_i |0, \ldots, 0, \ldots\rangle = 0 \,.$$

States with positive occupation numbers n_i are then obtained by acting successively with the raising operators $\mathsf{a}_i^\dagger$ on the vacuum, i.e.

$$\mathsf{a}_i^\dagger \mid \ldots, n_i, \ldots\rangle = \sqrt{n_i + 1} \mid \ldots, n_i + 1, \ldots\rangle$$

and similarly

$$\mathsf{a}_i \mid \ldots, n_i, \ldots\rangle = \sqrt{n_i} \mid \ldots, n_i - 1, \ldots\rangle \,.$$

Furthermore, applying the above commutation rules to the composite operator $\mathsf{a}_i^\dagger \mathsf{a}_i$ we readily obtain the relations

$$[\mathsf{a}_i^\dagger \mathsf{a}_i, \mathsf{a}_i] = -\mathsf{a}_i \qquad \text{and} \qquad [\mathsf{a}_i^\dagger \mathsf{a}_i, \mathsf{a}_i^\dagger] = \mathsf{a}_i^\dagger \,,$$

which in turn implies

$$\mathsf{a}_i^\dagger \mathsf{a}_i \mid \ldots, n_i, \ldots\rangle = n_i \mid \ldots, n_i, \ldots\rangle \,.$$

Thus for each label i the operators a_i acting on an eigenstate of the operator $\mathsf{a}_i^\dagger \mathsf{a}_i$ with eigenvalue n_i lowers the eigenvalue by 1 to $n_i - 1$, while the operator $\mathsf{a}_i^\dagger$ acting on such an eigenstate increases n_i by one. The operators a_i^+ and a_i thus create and destroy a particle of energy ϵ_i.

To complete the proof that the Hamilton operator defined above is the correct N-particle Hamiltonian we need to show that H reproduces the N-particle Schrödinger equation

$$\sum_{i=1}^{N}\left(-\frac{\hbar^2}{2m}\nabla_i^2\right)\psi_{E_N,N}(x_1,\ldots,x_N) = E_N\psi_{E_N,N}(x_1,\ldots,x_N)\,.$$

We note that for free particles

$$E_N = \sum_{\mathbf{k}}\frac{\hbar^2|\mathbf{k}|^2}{2m}n(|\mathbf{k}|),$$

and

$$\psi_{E_N,N}(\mathbf{x}_1,\ldots,\mathbf{x}_n) = \frac{1}{\sqrt{N!}}\sum_{\sigma}\frac{\mathrm{e}^{\mathrm{i}\mathbf{k}_1\cdot\mathbf{x}_{\sigma(1)}}}{\sqrt{V}}\cdots\frac{\mathrm{e}^{\mathrm{i}\mathbf{k}_N\cdot\mathbf{x}_{\sigma(N)}}}{\sqrt{V}},$$

where $k_i = 2\pi n_i/L$, $i = 1, 2, 3$ when the N identical particles are in a cubic box of side L and volume $V = L^3$. The sum is over the $N!$ permutations of $\mathbf{x}_1,\ldots,\mathbf{x}_N$. The labels $i_1,\ldots,i_N$ range over $1,\ldots,N$. Comparing this with the form of our proposed Hamiltonian we find

$$\mathsf{H} = \sum_{\mathbf{k}}\frac{\hbar^2|\mathbf{k}|^2}{2m}\mathsf{a}_{\mathbf{k}}^\dagger\mathsf{a}_{\mathbf{k}}\,,$$

and similarly for the particle number operator

$$\mathsf{N} = \sum_{\mathbf{k}}\mathsf{a}_{\mathbf{k}}^\dagger\mathsf{a}_{\mathbf{k}}$$

where the raising and lowering operators are now labeled by the momentum $\mathbf{k}$ so that

$$[\mathsf{a}_{\mathbf{k}}, \mathsf{a}_{\mathbf{k}'}^\dagger] = \delta_{\mathbf{k},\mathbf{k}'}$$

with other commutators vanishing. To continue we introduce position-dependent operators, $\Psi(\mathbf{x})$ and $\Psi^\dagger(\mathbf{x})$, by means of the Fourier transform

$$\Psi(\mathbf{x}) = \sum_{\mathbf{k}}\mathsf{a}_{\mathbf{k}}\frac{\mathrm{e}^{\mathrm{i}\mathbf{k}\cdot\mathbf{x}}}{\sqrt{V}}$$

$$\Psi^\dagger(\mathbf{x}) = \sum_{\mathbf{k}}\mathsf{a}_{\mathbf{k}}^\dagger\frac{\mathrm{e}^{-\mathrm{i}\mathbf{k}\cdot\mathbf{x}}}{\sqrt{V}}\,.$$

It is then not hard to see that these operators satisfy the commutation relations

$$[\Psi(\mathbf{x}), \Psi^\dagger(\mathbf{y})] = \delta^{(3)}(\mathbf{x}-\mathbf{y})$$
$$[\Psi(\mathbf{x}), \Psi(\mathbf{y})] = 0$$
$$[\Psi^\dagger(\mathbf{x}), \Psi^\dagger(\mathbf{y})] = 0\,.$$

We now write H and N in terms of these new operators as

$$\mathsf{H} = \int d^3x\, \Psi^\dagger(\mathbf{x}) \left(-\frac{\hbar^2}{2m}\nabla^2_{\mathbf{x}}\right) \Psi(\mathbf{x})$$
$$\mathsf{N} = \int d^3x\, \Psi^\dagger(\mathbf{x})\Psi(\mathbf{x})\,.$$

Note that H is expressed in terms of operators $\Psi^\dagger(\mathbf{x})$ and $\Psi(\mathbf{x})$ which carry the continuum label $\mathbf{x}$, and are thus quantum field operators. We are thus dealing, in this formulation, with a *quantum field theory*. The expressions for H and N have a simple intuitive meaning. The operator $\Psi^\dagger(\mathbf{x})\Psi(\mathbf{x})$ simply represents the "number density" operator at position $\mathbf{x}$. The integral of this density is N, the particle number operator. H is related to the product of the energy operator at $\mathbf{x}$ which is given by $-\hbar^2/2m\nabla^2$ with the number density operator at $\mathbf{x}$.

After this preparation we are now in position to make the connection with the N-particle Schrödinger equation precise. Concretely, we will reproduce the N-particle Schrödinger equation for the wave function

$$\psi(\mathbf{x}_1, \ldots, \mathbf{x}_N) = \frac{1}{\sqrt{N!}}\langle 0 \mid \Psi(\mathbf{x}_1)\ldots\Psi(\mathbf{x}_N) \mid E_N, N\rangle\,,$$

where the vacuum state $\mid 0\rangle$ is defined so that $\Psi \mid 0\rangle = \mathsf{H} \mid 0\rangle = \mathsf{N} \mid 0\rangle = 0$. For this we first note that

$$\frac{1}{\sqrt{N!}}\langle 0 \mid \Psi(\mathbf{x}_1)\ldots\Psi(\mathbf{x}_N)\mathsf{H} \mid E_N, N\rangle = E_N\Psi(\mathbf{x}_1, \ldots, \mathbf{x}_N)\,.$$

In fact, since $\langle 0 \mid \mathsf{H} = 0$ we also have

$$\frac{1}{\sqrt{N!}}\langle 0 \mid \Psi(\mathbf{x}_1)\ldots\Psi(\mathbf{x}_N)\mathsf{H} \mid E_N, N\rangle$$
$$= \frac{1}{\sqrt{N!}}\langle 0 \mid [\Psi(\mathbf{x}_1)\ldots\Psi(\mathbf{x}_N), \mathsf{H}] \mid E_N, N\rangle\,.$$

On the other hand the commutator above equals

$$[\Psi(\mathbf{x}_1)\ldots\Psi(\mathbf{x}_N), \mathsf{H}] = \sum_i \Psi(\mathbf{x}_1)\ldots\Psi(\mathbf{x}_{i-1})[\Psi(\mathbf{x}_i), \mathsf{H}]\Psi(\mathbf{x}_{i+1})\ldots\Psi(\mathbf{x}_N)\,.$$

This can be established by a simple inductive argument. Namely we observe that

$$[\Psi(\mathbf{x}_1)\Psi(\mathbf{x}_2), \mathsf{H}] = [\Psi(\mathbf{x}_1), \mathsf{H}]\Psi(\mathbf{x}_2) + \Psi(\mathbf{x}_1)[\Psi(\mathbf{x}_2), \mathsf{H}]\,,$$

as can be checked from the definition of the commutator. Next we suppose this result is true for n $\Psi(\mathbf{x}_i)$'s and we show that it must then be true for $(n+1)$ $\Psi(\mathbf{x}_i)$'s.

Let us now evaluate each commutator separately. Using the commutation relations between $\Psi(\mathbf{x})$ and $\Psi^\dagger(\mathbf{x})$ as well as the definition of H we have

$$\begin{aligned}
[\Psi(\mathbf{x}_i), \mathsf{H}] &= \int d^3y \left[\Psi(\mathbf{x}_i), \Psi^\dagger(\mathbf{y})\left(-\frac{\hbar^2}{2m}\nabla^2_{\mathbf{y}}\right)\Psi(\mathbf{y})\right] \\
&= \int d^3y \left[\Psi(\mathbf{x}_i), \Psi^\dagger(\mathbf{y})\right]\left(-\frac{\hbar^2}{2m}\nabla^2_{\mathbf{y}}\right)\Psi(\mathbf{y}) \\
&= \left(-\frac{\hbar^2}{2m}\nabla^2_{\mathbf{x}}\right)\Psi(\mathbf{x}_i).
\end{aligned}$$

To continue we substitute this result into the equation for $\Psi(\mathbf{x}_1)\ldots\Psi(\mathbf{x}_N)$ which then gives

$$\begin{aligned}
&\frac{1}{\sqrt{N!}}\langle 0 \mid [\Psi(\mathbf{x}_1)\ldots\Psi(\mathbf{x}_N), \mathsf{H}] \mid E_N, N\rangle \\
&= \sum_i^N \left(-\frac{\hbar^2}{2m}\nabla^2_1\right)\frac{1}{\sqrt{N!}}\langle 0 \mid \Psi(\mathbf{x}_1)\ldots\Psi(\mathbf{x}_N) \mid E_N, N\rangle \\
&= \sum_i^N \left(-\frac{\hbar^2}{2m}\nabla^2_i\right)\Psi(\mathbf{x}_1\ldots\mathbf{x}_N),
\end{aligned}$$

showing that $\psi(\mathbf{x}_1,\ldots,\mathbf{x}_N)$ satisfies the N-body Schrödinger equation in the case where the N particles do not interact. A pleasant feature of this approach is that the total symmetry of the wave function $\psi(\mathbf{x}_1,\ldots,\mathbf{x}_N)$ in the coordinate labels $\mathbf{x}_1,\ldots,\mathbf{x}_N$ is manifest. This is because switching $\mathbf{x}_i$ with $\mathbf{x}_j$ is equivalent to interchanging $\Psi(\mathbf{x}_i)$ with $\Psi(\mathbf{x}_j)$ but since these operators commute for arbitrary labels i and j this interchange symmetry holds.

9.1.2 Interacting bosons

We now turn to the problem of introducing interactions in this framework. From our discussions it should be clear how this is to be done. Interactions represent contributions to the energy due to pairs of particles, while for non-interacting particles the energy is due to each particle separately. For the case of non-interacting particles we found that

$$\mathsf{H} = \int d^3y\, \Psi^\dagger(\mathbf{y})\left(-\frac{\hbar^2}{2m}\nabla^2_{\mathbf{y}}\right)\Psi(\mathbf{y}),$$

which we interpreted as an integration over the energy of a particle at $\mathbf{y}$, represented by the operator $(-\hbar^2/2m\nabla^2_{\mathbf{y}})$, and "multiplied" by the number density operator

$\Psi^\dagger(\mathbf{y})\Psi(\mathbf{y})$ at $\mathbf{y}$. We thus expect the interaction energy term to involve the density operator at $\mathbf{y}_1$, the density operator at $\mathbf{y}_2$, multiplied by the potential energy $V(|\mathbf{y}_1 - \mathbf{y}_2|)$ associated with this pair of particles. This suggests the ansatz

$$\mathsf{V} = \frac{1}{2}\int d^3y_1 d^3y_2\, \Psi^\dagger(\mathbf{y}_1)\Psi^\dagger(\mathbf{y}_2)V(|\mathbf{y}_1 - \mathbf{y}_2|)\Psi(\mathbf{y}_1)\Psi(\mathbf{y}_2)\,.$$

The factor $1/2$ takes into account that we are dealing with identical molecules. It should be clear how these ideas generalize. If energy was associated to collections of three molecules which was different from the energy due to interactions $V(|\mathbf{y}_1 - \mathbf{y}_2|)$ between pairs of molecules, i.e. if a new $V(\mathbf{y}_i, \mathbf{y}_j, \mathbf{y}_k)$ term or a "three-body force" was present, then the corresponding V would include the number density operators at $\mathbf{y}_1$, $\mathbf{y}_2$, and $\mathbf{y}_3$ and would have the form

$$\mathsf{V}^{(3)} = \frac{1}{3!}\int d^3y_1 d^3y_2 d^3y_3\, \Psi^\dagger(\mathbf{y}_1)\Psi^\dagger(\mathbf{y}_2)\Psi^\dagger(\mathbf{y}_3)V(\mathbf{y}_1, \mathbf{y}_2, \mathbf{y}_3)\Psi(\mathbf{y}_1)\Psi(\mathbf{y}_2)\Psi(\mathbf{y}_3)\,.$$

We will now show that our expectations are indeed justified. We formulate this result as a theorem

Theorem 9.1 If the Hamiltonian and number operator are defined as

$$\begin{aligned}\mathsf{H} &= \int d^3y\, \Psi^\dagger(\mathbf{y})\left(-\frac{\hbar^2}{2m}\nabla^2_{\mathbf{y}}\right)\Psi(\mathbf{y}) \\ &\quad + \frac{1}{2!}\int d^3y_1 d^3y_2\, \Psi^\dagger(\mathbf{y}_1)\Psi^\dagger(\mathbf{y}_2)V(|\mathbf{y}_1 - \mathbf{y}_2|)\Psi(\mathbf{y}_1)\Psi(\mathbf{y}_2) \\ \mathsf{N} &= \int d^3y\, \Psi^\dagger(\mathbf{y})\Psi(\mathbf{y})\end{aligned}$$

respectively, and if $|\, E_N, N\rangle$ satisfies

$$\begin{aligned}\mathsf{H}\,|\, E_N, N\rangle &= E_N\,|\, E_N, N\rangle \\ \mathsf{N}\,|\, E_N, N\rangle &= N\,|\, E_N, N\rangle\,,\end{aligned}$$

with $\mathsf{H}\,|\,0\rangle = \mathsf{N}\,|\,0\rangle = \Psi\,|\,0\rangle = 0$, then the wave function

$$\Psi(\mathbf{x}_1, \ldots, \mathbf{x}_N) = \frac{1}{\sqrt{N!}}\langle 0\,|\,\Psi(\mathbf{x}_1)\ldots\Psi(\mathbf{x}_N)\,|\,E_N, N\rangle$$

satisfies the interacting N-particle Schrödinger equation

$$\left(\sum_{i=1}^{N}\left(-\frac{\hbar^2}{2m}\nabla_i^2\right) + \sum_{i<j} V(|\mathbf{x}_i - \mathbf{x}_j|)\right)\Psi(\mathbf{x}_1, \ldots, \mathbf{x}_N) = E_N\Psi(\mathbf{x}_1, \ldots, \mathbf{x}_N)\,.$$

Before we proceed to the proof of this theorem a comment concerning the order of the operators Ψ and $\Psi^\dagger$ is appropriate. Instead of the order chosen above we could have written

$$\mathsf{V}' = \frac{1}{2} \int \mathrm{d}^3 y_1 \mathrm{d}^3 y_2 \Psi^\dagger(\mathbf{y}_1)\Psi(\mathbf{y}_1) V(|\mathbf{y}_1 - \mathbf{y}_2|)\Psi^\dagger(\mathbf{y}_2)\Psi(\mathbf{y}_2)$$

for example. How would this affect the result? This question is no sooner asked than answered. Using the commutation relation of $\Psi^\dagger$ and Ψ it follows immediately that

$$\mathsf{V}' = \mathsf{V} + \frac{1}{2} \int \mathrm{d}^3 y \, V(0)\Psi^\dagger(\mathbf{y})\Psi(\mathbf{y}) \,.$$

Thus the alternative ordering is equivalent adding the constant $1/2V(0)$ to the kinetic energy. However, for most potentials arising in practical applications, such as for example the Coulomb potential, $V(0)$ is infinite. It is thus preferable to choose the order where all $\Psi^\dagger$'s stand to the left of the Ψ's. This is a special instance of the so-called *normal ordering*. Normal ordering is the order of operators where all creation operators are to the left of the annihilation operators. Physically, this means that no particles are created out of the vacuum by the Hamiltonian and that the energy of the vacuum vanishes, which is what one intuitively expects.

Proof. It is convenient to define

$$\begin{aligned} \mathsf{H}_0 &= \int \mathrm{d}^3 y \, \Psi^\dagger(\mathbf{y}) \left(-\frac{\hbar^2}{2m} \nabla^2_{\mathbf{y}} \right) \Psi(\mathbf{y}) \\ \mathsf{H}_I &= \frac{1}{2} \int \mathrm{d}^3 y_1 \mathrm{d}^3 y_2 \, \Psi^\dagger(\mathbf{y}_1)\Psi^\dagger(\mathbf{y}_2) V(|\mathbf{y}_1 - \mathbf{y}_2|)\Psi(\mathbf{y}_1)\Psi(\mathbf{y}_2) \,. \end{aligned}$$

We then observe that

$$\begin{aligned} & E_N \Psi(\mathbf{x}_1, \ldots, \mathbf{x}_N) \\ & \quad = \frac{1}{\sqrt{N!}} \langle 0 \mid \Psi(\mathbf{x}_1) \ldots \Psi(\mathbf{x}_N) \mathsf{H} \mid E_N, N \rangle \\ & \quad = \frac{1}{\sqrt{N!}} \langle 0 \mid [\Psi(\mathbf{x}_1) \ldots \Psi(\mathbf{x}_N), \mathsf{H}] \mid E_N, N \rangle \end{aligned}$$

where we have used that $\mathsf{H} \mid 0 \rangle = 0$. To continue we then consider

$$\begin{aligned} & [\Psi(\mathbf{x}_1) \ldots \Psi(\mathbf{x}_N), \mathsf{H}] \\ & \quad = \sum_{i=1}^{N} \Psi(\mathbf{x}_1) \ldots \Psi(\mathbf{x}_{i-1}) [\Psi(\mathbf{x}_i), \mathsf{H}] \Psi(\mathbf{x}_{i+1}) \ldots \Psi(\mathbf{x}_N). \end{aligned}$$

We now decompose the Hamiltonian into free and interaction parts

$$[\Psi(\mathbf{x}), \mathsf{H}] = [\Psi(\mathbf{x}), \mathsf{H}_0] + [\Psi(\mathbf{x}), \mathsf{H}_I].$$

We have already seen that

$$[\Psi(\mathbf{x}_i), \mathsf{H}_0] = \left(-\frac{\hbar^2}{2m}\nabla_i^2\right)\Psi(\mathbf{x}_i),$$

so we need only consider $[\Psi(\mathbf{x}), \mathsf{H}_I]$ for which we have

$$\begin{aligned}
&[\Psi(x_i), \mathsf{H}_I] \\
&= \frac{1}{2}\int \mathrm{d}^3y_1\mathrm{d}^3y_2\, V(|\mathbf{y}_1 - \mathbf{y}_2|)[\Psi(\mathbf{x}_i), \Psi^\dagger(\mathbf{y}_1)\Psi^\dagger(\mathbf{y}_2)\Psi(\mathbf{y}_1)\Psi(\mathbf{y}_2)] \\
&= \frac{1}{2}\int \mathrm{d}^3y_1\mathrm{d}^3y_2\, V(|\mathbf{y}_1 - \mathbf{y}_2|)\delta^3(\mathbf{x}_i - \mathbf{y}_1)\Psi^\dagger(\mathbf{y}_2)\Psi(\mathbf{y}_1)\Psi(\mathbf{y}_2) \\
&\quad + \frac{1}{2}\int \mathrm{d}^3y_1\mathrm{d}^3y_2\, V(|\mathbf{y}_1 - \mathbf{y}_2|)\delta^3(\mathbf{x}_i - \mathbf{y}_2)\Psi^\dagger(\mathbf{y}_1)\Psi(\mathbf{y}_1)\Psi(\mathbf{y}_2) \\
&= \int \mathrm{d}^3y\, V(|\mathbf{x}_i - \mathbf{y}|)\Psi^\dagger(\mathbf{y})\Psi(\mathbf{y})\Psi(\mathbf{x}_i)\,.
\end{aligned}$$

Thus we have

$$\begin{aligned}
E_N\Psi(\mathbf{x}_1,\ldots,\mathbf{x}_N) = {}& \sum_{i=1}^{N}\left(-\frac{\hbar^2}{2m}\nabla_i^2\right)\Psi(\mathbf{x}_1,\ldots,\mathbf{x}_N) \\
&+ \sum_{i=1}^{N}\frac{1}{\sqrt{N!}}\int \mathrm{d}^3y V(|\mathbf{x}_i - \mathbf{y}|) \\
&\times\langle 0 \mid \Psi(\mathbf{x}_1)\ldots\Psi^\dagger(\mathbf{y})\Psi(\mathbf{y})\Psi(\mathbf{x}_i)\ldots\Psi(\mathbf{x}_N) \mid E_N, N\rangle\,.
\end{aligned}$$

The first term on the right-hand side of this equation comes from H_0, the second from H_I. Let us examine the second term

$$\begin{aligned}
&\sum_{i=1}^{N}\frac{1}{\sqrt{N!}}\int \mathrm{d}^3y V(|\mathbf{x}_i - \mathbf{y}|) \\
&\quad\times\langle 0 \mid \Psi(\mathbf{x}_1)\ldots\Psi(\mathbf{x}_{i-1})\Psi^\dagger(\mathbf{y})\Psi(\mathbf{y})\Psi(\mathbf{x}_i)\ldots\Psi(\mathbf{x}_N) \mid E_N, N\rangle.
\end{aligned}$$

Define

$$\mathbf{K} = \Psi^\dagger(\mathbf{y})\Psi(\mathbf{y})\Psi(\mathbf{x}_i)\ldots\Psi(\mathbf{x}_N)$$

then this term is

$$\sum_{i=1}^{N}\frac{1}{\sqrt{N!}}\int \mathrm{d}^3y V(|\mathbf{x}_i - \mathbf{y}|)\langle 0 \mid \Psi(\mathbf{x}_1)\ldots\Psi(\mathbf{x}_{i-1})K \mid E_N, N\rangle.$$

Now, since $\langle 0 \mid \mathsf{K} = 0$ we can write the sum equivalently as

$$\sum_{i=1}^{N} \frac{1}{\sqrt{N!}} \int \mathrm{d}^3 y V(|\mathbf{x}_i - \mathbf{y}|) \langle 0 \mid [\Psi(\mathbf{x}_1) \ldots \Psi(\mathbf{x}_{i-1}), K] \mid E_N, N\rangle .$$

We are thus left with the evaluation of the commutator

$$\begin{aligned}
&[\Psi(\mathbf{x}_1) \ldots \Psi(\mathbf{x}_{i-1}), \mathsf{K}] \\
&\quad = \sum_{j=1}^{i-1} \Psi(\mathbf{x}_1) \ldots \Psi(\mathbf{x}_{j-1}) [\Psi(\mathbf{x}_j), \mathsf{K}] \Psi(\mathbf{x}_{j+1}) \ldots \Psi(\mathbf{x}_{i-1})
\end{aligned}$$

with

$$\begin{aligned}
[\Psi(\mathbf{x}_j), \mathsf{K}] &= [\Psi(\mathbf{x}_j), \Psi^\dagger(\mathbf{y}) \Psi(\mathbf{y}) \Psi(\mathbf{x}_{i+1}) \ldots \Psi(\mathbf{x}_N)] \\
&= \delta^3(\mathbf{x}_j - \mathbf{y}) \Psi(\mathbf{y}) \Psi(\mathbf{x}_{i+1}) \ldots \Psi(\mathbf{x}_N) .
\end{aligned}$$

Therefore

$$\begin{aligned}
[\Psi(\mathbf{x}_1) \ldots \Psi(\mathbf{x}_{i-1}), \mathsf{K}] = \sum_{j=1}^{i-1} & \Psi(\mathbf{x}_1) \ldots \Psi(\mathbf{x}_{j-1}) \\
& \times \delta^3(\mathbf{x}_j - \mathbf{y}) \Psi(\mathbf{y}) \\
& \Psi(\mathbf{x}_{j+1}) \ldots \Psi(\mathbf{x}_N)
\end{aligned}$$

so that

$$\begin{aligned}
&\sum_{i=1}^{N} \frac{1}{\sqrt{N!}} \int \mathrm{d}^3 y \langle 0 \mid [\Psi(\mathbf{x}_1) \ldots \Psi(\mathbf{x}_{i-1}), \mathsf{K}] \mid E_N, N\rangle \\
&\quad = \sum_{i=1}^{N} \sum_{j=1}^{i-1} V(|\mathbf{x}_i - \mathbf{x}_j|) \frac{1}{\sqrt{N!}} \langle 0 \mid \Psi(\mathbf{x}_1) \ldots \Psi(\mathbf{x}_N) \mid E_N, N\rangle \\
&\quad = \sum_{j<i} V(|\mathbf{x}_i - \mathbf{x}_j|) \Psi(\mathbf{x}_1 \ldots \mathbf{x}_N) ,
\end{aligned}$$

which establishes the theorem. □

We have succeeded in setting up a formulation for N interacting identical bosons in which the different states of the system are characterized by occupation numbers that range over all integer values. In the process of doing so we have generalized the quantum mechanical framework by introducing operator-valued fields, or quantum fields, which individually create and annihilate particles in the system. Nevertheless, the Hamilton operator we have constructed is such that the total particle number is conserved, although a particle with a given quantum number may be replaced by a particle with a different quantum number due to interactions. The identical nature of the particles is manifest in the symmetry of the N-body wave function

$$\psi(\mathbf{x}_1 \dots \mathbf{x}_N) = \frac{1}{\sqrt{N!}} \langle 0 \mid \Psi(\mathbf{x}_1) \dots \Psi(\mathbf{x}_N) \mid E_N, N \rangle .$$

9.1.3 Non-interacting fermions

In this section we describe the modifications that arise when dealing with Fermi–Dirac particles instead of bosons. As we have seen already the crucial difference with Fermi–Dirac particles is that the occupation number can only take the values zero or one. The question is how this can be built into the present framework? It turns out that all we have to do is to change the commutation relations of the creation and annihilation operators. That is, we replace a_i and $\mathsf{a}_i^\dagger$ by operators, b_i and $\mathsf{b}_i^\dagger$ with commutation relations

$$\begin{aligned} \left\{\mathsf{b}_i, \mathsf{b}_j^\dagger\right\} &\equiv \mathsf{b}_i \mathsf{b}_j^\dagger + \mathsf{b}_j^\dagger \mathsf{b}_i = \delta_{i,j} \\ \left\{\mathsf{b}_i, \mathsf{b}_j\right\} &\equiv \mathsf{b}_i \mathsf{b}_j + \mathsf{b}_j \mathsf{b}_i = 0 \\ \left\{\mathsf{b}_i^\dagger, \mathsf{b}_j^\dagger\right\} &\equiv \mathsf{b}_i^\dagger \mathsf{b}_j^\dagger + \mathsf{b}_j^\dagger \mathsf{b}_i^\dagger = 0 . \end{aligned}$$

As before we define the number operator for the state i as

$$\mathsf{n}_i \equiv \mathsf{b}_i^\dagger \mathsf{b}_i .$$

Making use of the associativity of the operator product as well as the commutation relations above it is not hard to see that N_i is a projection operator, i.e.

$$\mathsf{n}_i^2 = (\mathsf{b}_i^\dagger \mathsf{b}_i)(\mathsf{b}_i^\dagger \mathsf{b}_i) = \mathsf{b}_i^\dagger (\mathsf{b}_i \mathsf{b}_i^\dagger) \mathsf{b}_i = \mathsf{b}_i^\dagger (1 - \mathsf{b}_i^\dagger \mathsf{b}_i) \mathsf{b}_i = \mathsf{b}_i^\dagger \mathsf{b}_i = \mathsf{n}_i .$$

The eigenvalues of N_i are thus 0, 1. More precisely

$$\mathsf{n}_i \mid \dots n_i \dots \rangle = n_i \mid \dots n_i \dots \rangle$$

with $n_i = 0, 1$. The vectors $\mid \dots, n_i, \dots \rangle$ form a complete set of basis vectors in the Hilbert space with Fermi–Dirac statistics.

Repeating the steps for the Bose–Einstein case we then introduce operators $\Psi(\mathbf{x})$ and $\Psi^\dagger(\mathbf{x})$ and write

$$\mathsf{H} = \int \mathrm{d}^3 y \Psi^\dagger(\mathbf{y}) \left(-\frac{\hbar^2}{2m} \nabla_y^2 \right) \Psi(\mathbf{y}) ,$$

where we now have

$$\begin{aligned} \{\Psi(\mathbf{y}_1), \Psi^\dagger(\mathbf{y}_2)\} &= \delta^3(\mathbf{y}_1 - \mathbf{y}_2) \\ \{\Psi(\mathbf{y}_1), \Psi(\mathbf{y}_2)\} &= 0 \\ \{\Psi^\dagger(\mathbf{y}_1), \Psi^\dagger(\mathbf{y}_2)\} &= 0 \end{aligned}$$

i.e. the previous commutation relations are now replaced by anticommutation relations. It remains to show that if

$$\mathsf{H} \mid E_N, N\rangle = E_N \mid E_N, N\rangle$$

with $\mathsf{H} \mid 0\rangle = \Psi \mid 0\rangle = 0$, then the corresponding wave function satisfies the Schrödinger equation, i.e.

$$\begin{aligned} E_N \Psi(\mathbf{x}_1, \ldots, \mathbf{x}_N) &= E_N \frac{1}{\sqrt{N!}} \langle 0 \mid \Psi(\mathbf{x}_1) \ldots \Psi(\mathbf{x}_N) \mid E_N, N\rangle \\ &= \sum_{i=1}^{N} \left(-\frac{\hbar^2}{2m} \nabla_i^2 \right) \Psi(\mathbf{x}_1, \ldots, \mathbf{x}_N) . \end{aligned}$$

The proof proceeds as follows. We observe, as before, that

$$\begin{aligned} & E_N \Psi(\mathbf{x}_1, \ldots, \mathbf{x}_N) \\ &= \frac{1}{\sqrt{N!}} \langle 0 \mid \Psi(\mathbf{x}_1) \ldots \Psi(\mathbf{x}_N) \mathsf{H} \mid E_N, N\rangle \\ &= \frac{1}{\sqrt{N!}} \langle 0 \mid [\Psi(\mathbf{x}_1) \ldots \Psi(\mathbf{x}_N), \mathsf{H}] \mid E_N, N\rangle \\ &= \frac{1}{\sqrt{N!}} \langle 0 \mid \Psi(\mathbf{x}_1) \ldots \Psi(\mathbf{x}_{i-1}) \\ & \times [\Psi(\mathbf{x}_i), \mathsf{H}] \Psi(\mathbf{x}_{i+1}) \ldots \Psi(\mathbf{x}_N) \mid E_N, N\rangle . \end{aligned}$$

We now consider

$$[\Psi(\mathbf{x}_i), \mathsf{H}] = \int d^3y [\Psi(\mathbf{x}), \Psi^\dagger(\mathbf{y}) \left(-\frac{\hbar^2}{2m} \nabla_y^2 \right) \Psi(\mathbf{y})].$$

It is at this stage that we have to take into account the fact that the anticommutators rather than the commutators of the fields are known. To do so we use the following identity, valid for any three operators A, B, C,

$$[A, BC] = \{A, B\} C - B \{A, C\} .$$

Using this identity we can write

$$\begin{aligned} \left[\Psi(\mathbf{x}_i), \Psi^\dagger(\mathbf{y}) \left(-\frac{\hbar^2}{2m} \nabla_y^2 \right) \Psi(\mathbf{y}) \right] &= \{\Psi(\mathbf{x}), \Psi^\dagger(\mathbf{y})\} \left(-\frac{\hbar^2}{2m} \nabla_y^2 \right) \Psi(\mathbf{y}) \\ & \quad - \Psi^\dagger(\mathbf{y}) \left\{ \Psi(\mathbf{x}_i), \left(-\frac{\hbar^2}{2m} \nabla_y^2 \right) \Psi(\mathbf{y}) \right\} \\ &= \delta^3(\mathbf{x}_i - \mathbf{y}) \left(-\frac{\hbar^2}{2m} \nabla_y^2 \right) \Psi(\mathbf{y}) . \end{aligned}$$

Therefore

$$[\Psi(\mathbf{x}_i), \mathsf{H}] = \left(-\frac{\hbar^2}{2m}\nabla^2_{\mathbf{x}_i}\right)\Psi(\mathbf{x}_i)$$

so that

$$E_N\Psi(\mathbf{x}_1, \ldots, \mathbf{x}_N) = \sum_{i=1}^{N}\left(-\frac{\hbar^2}{2m}\nabla^2_i\right)\Psi(\mathbf{x}_1, \ldots, \mathbf{x}_N),$$

and we again have Schrödinger's equation.

The inclusion of interactions is then a straightforward (although instructive) exercise which we will leave to the reader.

9.2 Perturbation theory

So far we have considered the time-independent Schrödinger equation $\mathsf{H} \mid \Psi_E\rangle = E \mid \Psi_E\rangle$. However, we will also need to understand the time evolution of our system. Therefore we now turn to the problem of determining the changes in a state vector due to interactions. We recall that a state vector evolves with time according to Schrödinger's equations, namely

$$\mathrm{i}\hbar\frac{\partial}{\partial t} \mid \Psi(t)\rangle_S = \mathsf{H} \mid \Psi(t)\rangle_S\,.$$

In this description the operator H is time independent. This is the Schrödinger picture of quantum mechanics. We can integrate this equation to write

$$\mid \Psi(t)\rangle_S = \exp\left(-\mathrm{i}\frac{\mathsf{H}t}{\hbar}\right) \mid \Psi(0)\rangle\,.$$

The subscript S in both these equations draws attention to the fact that we are in the Schrödinger picture. Of course, unless the molecules are non-interacting we can not expect to be able to solve the evolution equation analytically. We will therefore have to come up with some suitable perturbative method. For this we introduce $\mid \Psi(t)\rangle_I$, a state vector in the *interaction picture*. We define this vector to change with time only if interactions are present. Writing $\mathsf{H} = \mathsf{H}_0 + \mathsf{H}_1$, where H_1 is the interaction part of the Hamiltonian, we define

$$\mid \Psi(t)\rangle_I = \exp\left(\mathrm{i}\frac{\mathsf{H}_0 t}{\hbar}\right)\exp\left(-\mathrm{i}\frac{\mathsf{H}t}{\hbar}\right) \mid \Psi(0)\rangle.$$

Observe that if $\mathsf{H} = \mathsf{H}_0$, then $\mid \Psi(t)\rangle_I = \mid \Psi(0)\rangle$, i.e. $\mid \Psi\rangle_I$ is time independent in the absence of interactions. On the other hand, differentiating the above equation with respect to t gives

$$i\hbar\frac{\partial}{\partial t}\mid\Psi(t)\rangle_I = \exp\left(i\frac{\mathsf{H}_0 t}{\hbar}\right)\mathsf{H}_1\exp\left(-i\frac{\mathsf{H}_0 t}{\hbar}\right)\mid\Psi(t)\rangle_I$$
$$\equiv \mathsf{H}_I(t)\mid\Psi(t)\rangle_I\,.$$

The operator H_I just defined is the *interaction Hamiltonian*. These equations give the time evolution of state vectors and operators in the *interaction picture*.

Alternatively we can write the time evolution of $\mid\Psi\rangle_I$ in the form

$$\mid\Psi(t_2)\rangle_I = \mathsf{U}(t_2,t_1)\mid\Psi(t_1)\rangle_I$$

where the evolution operator $\mathsf{U}(t_2,t_1)$ is the interaction picture analogue on the operator $\exp\left(-i\mathsf{H}(t_2-t_1)/\hbar\right)$ in the Schrödinger picture. It is easy to see that

$$i\hbar\frac{\partial}{\partial t_2}\mid\Psi(t_2)\rangle_I = \mathsf{H}_I(t_2)\mid\Psi(t_2)\rangle_I$$
$$= \mathsf{H}_I(t_2)\mathsf{U}(t_2,t_1)\mid\Psi(t_1)\rangle_I\,.$$

This equation is valid for a complete set of vectors, $\mid\Psi(t_2)\rangle_I$. Hence the operator $\mathsf{U}(t_2,t_1)$ must satisfy the equation

$$i\hbar\frac{\partial}{\partial t}\mathsf{U}(t_2,t_1) = \mathsf{H}_I(t_2)\mathsf{U}(t_2,t_1)$$

with the boundary condition $\mathsf{U}(t_1,t_1) = I$. This evolution equation for $U(t_2,t_1)$ is equivalent to the integral equation

$$\mathsf{U}(t_2,t_1) = I - \frac{i}{\hbar}\int_{t_1}^{t_2} d\tau\,\mathsf{H}_I(\tau)\mathsf{U}(\tau,t_1)\,.$$

Once the operator $\mathsf{U}(t_2,t_1)$ is determined by solving this integral equation then the time evolution of the interaction picture state $\mid\Psi\rangle_I$ is determined. We now claim that

$$\mathsf{U}(t_2,t_1) = \sum_{n=0}^{\infty}\left(-\frac{i}{\hbar}\right)^n\frac{1}{n!}\int_{t_1}^{t_2} d\tau_1\ldots\int_{t_1}^{t_2} d\tau_n\,T(\mathsf{H}_I(\tau_1)\ldots\mathsf{H}_I(\tau_n))\,,$$

where T is the *time-ordering operator*. For two bosonic operators $\mathsf{A}_1(t_1)$, $\mathsf{A}_1(t_2)$ time ordering is defined as follows

$$T(\mathsf{A}_1(t_1)\mathsf{A}_2(t_2)) = \theta(t_1-t_2)\mathsf{A}_1(t_1)\mathsf{A}_2(t_2) + \theta(t_2-t_1)\mathsf{A}_2(t_2)\mathsf{A}_1(t_1)$$

where

$$\theta(t) = \begin{Bmatrix} 0, & t<0 \\ 1, & t>0 \end{Bmatrix}.$$

For a monomial of n bosonic operators $\mathsf{A}_1(\tau_1)\ldots\mathsf{A}_n(\tau_n)$, the time-ordered product $T(\mathsf{A}_1(\tau_1)\ldots\mathsf{A}_n(\tau_n))$ consists of $n!$ terms. Each term corresponds to an ordering of the operators reflecting the ordering of the distinct time variables $\tau_1\ldots\tau_n$. For example

$$T(\mathsf{A}_1(\tau_1)\ldots\mathsf{A}_n(\tau_n)) = \mathsf{A}_1(\tau_1)\ldots\mathsf{A}_n(\tau_n) \quad \text{if} \quad \tau_1 > \tau_2 > \cdots > \tau_n.$$

For two fermionic operators the corresponding time ordered product is given by $\mathsf{B}_1(t_1)$, $\mathsf{B}_1(t_2)$:

$$T(\mathsf{B}_1(t_1)\mathsf{B}_2(t_2)) = \theta(t_1 - t_2)\mathsf{B}_1(t_1)\mathsf{B}_2(t_2) - \theta(t_2 - t_1)\mathsf{B}_2(t_2)\mathsf{B}_1(t_1).$$

The fermionic expression can be derived from the bosonic one if an additional rule is introduced, namely, whenever operator orderings are changed a factor $(-1)^\pi$ has to be introduced, where π represents the number of permutations. In the example, the second term involved one permutation of the operators relative to the first term. This leads to the factor (-1). For a product of n fermionic operators, there will similarly be $n!$ terms. These can be obtained from the bosonic expression supplemented by the rule just stated.

The proof of the claim which we provide is formal and does not consider questions of the convergence. Regarding the integral equation for U formally as a *Volterra integral equation*, the lowest order approximation to $\mathsf{U}(t_2, t_1)$ is obtained by substituting I for $\mathsf{U}(\tau, t_1)$, i.e.

$$\mathsf{U}(t_2, t_1) \approx I - \frac{\mathrm{i}}{\hbar}\int_{t_1}^{t_2} d\tau\, \mathsf{H}_I(\tau).$$

Repeating this procedure we get

$$\begin{aligned}\mathsf{U}(t_2, t_1) \approx I &- \frac{\mathrm{i}}{\hbar}\int_{t_1}^{t_2} d\tau\, \mathsf{H}_I(\tau)\\ &+ \left(-\frac{\mathrm{i}}{\hbar}\right)^2 \int_{t_1}^{t_2} d\tau_1\, \mathsf{H}_I(\tau_1)\int_{t_1}^{\tau_1} d\tau_2\, \mathsf{H}_I(\tau_2).\end{aligned}$$

Assuming this formal expansion in powers of the operator $\mathsf{H}_I(\tau)$ converges to $\mathsf{U}(t_2, t_1)$ we have

$$\mathsf{U}(t_2, t_1) = \sum_{n=0}^{\infty}\left(-\frac{\mathrm{i}}{\hbar}\right)^n \int_{t_1}^{t_2} d\tau_1 \int_{t_1}^{\tau_1} d\tau_2 \cdots \int_{t_1}^{\tau_{n-1}} d\tau_n\, \mathsf{H}_I(\tau_1)\cdots\mathsf{H}_I(\tau_n).$$

Consider, as an example, the term

$$\int_{t_1}^{t_2} d\tau_1 \int_{t_1}^{\tau_1} d\tau_2\, \mathsf{H}_I(\tau_1)\mathsf{H}_I(\tau_2)$$

and observe that this term equals

$$\frac{1}{2!}\int_{t_1}^{t_2} d\tau_1 \int_{t_1}^{t_2} d\tau_2\, T(\mathsf{H}_I(\tau_1)\mathsf{H}_I(\tau_2)).$$

This is easily seen to be true since

$$\begin{aligned}
&\frac{1}{2!}\int_{t_1}^{t_2} d\tau_1 \int_{t_1}^{t_2} d\tau_2\, T(\mathsf{H}_I(\tau_1)\mathsf{H}_I(\tau_2)) \\
&= \frac{1}{2!}\int_{t_1}^{t_2} d\tau_1 \int_{t_1}^{t_2} d\tau_2 [\theta(\tau_1-\tau_2)\mathsf{H}_I(\tau_1)\mathsf{H}_I(\tau_2) + \theta(\tau_2-\tau_1)\mathsf{H}_I(\tau_2)\mathsf{H}_I(\tau_1)] \\
&= \frac{1}{2!}\int_{t_1}^{t_2} d\tau_2 \int_{t_1}^{t_1} d\tau_1\, \mathsf{H}_I(\tau_1)\mathsf{H}_I(\tau_2) \\
&\quad + \frac{1}{2!}\int_{t_1}^{t_2} d\tau_2 \int_{t_1}^{t_2} d\tau_1 \mathsf{H}_I(\tau_2)\mathsf{H}_I(\tau_1) \\
&= \int_{t_1}^{t_2} d\tau_1 \int_{t_1}^{t_1} d\tau_2\, \mathsf{H}_I(\tau_1)\mathsf{H}_I(\tau_2).
\end{aligned}$$

Similarly

$$\begin{aligned}
&\int_{t_1}^{t_2} d\tau_1 \int_{t_1}^{t_1} d\tau_2 \ldots \int_{t_1}^{t_{n-1}} d\tau_n\, \mathsf{H}_I(\tau_1)\cdots \mathsf{H}_I(\tau_n) \\
&= \frac{1}{n!}\int_{t_1}^{t_2} d\tau_1 \ldots \int_{t_1}^{t_2} d\tau_n T(\mathsf{H}_I(\tau_1)\cdots \mathsf{H}_I(\tau_n)).
\end{aligned}$$

This then establishes our claim.

9.3 Wick's theorem

From our result in the last section it is clear that, in order to determine the evolution operator $\mathsf{U}(t_2, t_1)$ to any given order in perturbation theory, we must learn how to evaluate vacuum expectation values of the form

$$\langle 0|T(\mathsf{A}_1(t_1)\ldots \mathsf{A}_n(t_n))|0\rangle$$

where $\mathsf{A}_1(t_1)\ldots \mathsf{A}_n(t_n)$ are operators. Wick's theorem provides an efficient algorithm to compute such products to any order in perturbation theory. It does so by establishing a relation between time-ordered products and normal-ordered products which vanish when evaluated in the vacuum.

Theorem 9.2 If $\mathsf{A}_i(t_i) = \mathsf{A}_i^+(t_i) + \mathsf{A}_i^-(t_i)$, $i = 1, \ldots, n$ are bosonic operators such that $\mathsf{A}_i^-(t_i) \mid 0\rangle = \langle 0 \mid \mathsf{A}_i^+(t_i) = 0$ and such that $[\mathsf{A}_i(t_i), \mathsf{A}_j(t_j)]$ is a complex number valued function (or distribution), then

$$\langle 0|T(\mathsf{A}_1(t_1)\dots\mathsf{A}_n(t_n))|0\rangle = \begin{cases} 0 & n \text{ odd} \\ \sum_\sigma \prod_{i=1,3,\cdots}^{n-1} \langle 0 \mid T\left(\mathsf{A}_{\sigma(i)}(t_{\sigma(i)})A_{\sigma(i+1)}(t_{\sigma(i+1)})\right) \mid 0\rangle & n \text{ even} \end{cases}$$

where the sum is over all permutations of $1, \dots, n$ which do not lead to identical expressions. Identical expressions are obtained if the arguments within the same time-ordered product are interchanged, i.e.

$$\langle 0 \mid T(\mathsf{A}_1(t_1)\mathsf{A}_2(t_2)) \mid 0\rangle = \langle 0 \mid T(\mathsf{A}_2(t_2)\mathsf{A}_1(t_1)) \mid 0\rangle .$$

In other words, the sum on the right-hand side is over all possible pairings of the integers $(i, j) \in \{1, \dots, n\}$.

Proof. For $n = 1$ there is nothing to prove. Before considering the case $n > 2$ we note that the time-ordered product $T(\mathsf{A}_i(t_i)\mathsf{A}_j(t_j))$ can alternatively be written as a commutator. Indeed let us suppose $t_i > t_j$ then

$$\begin{aligned} T(\mathsf{A}_i\mathsf{A}_j) &= \mathsf{A}_i\mathsf{A}_j \\ &= (\mathsf{A}_i^+ + \mathsf{A}_i^-)(\mathsf{A}_j^+ + \mathsf{A}_j^-) \\ &= \mathsf{A}_i^+\mathsf{A}_j^+ + \mathsf{A}_i^+\mathsf{A}_j^- + \mathsf{A}_i^-\mathsf{A}_j^+ + \mathsf{A}_i^-\mathsf{A}_j^- \\ &= \mathsf{A}_i^+\mathsf{A}_j^+ + \mathsf{A}_i^+\mathsf{A}_j^- + \mathsf{A}_j^+\mathsf{A}_i^- - \mathsf{A}_j^+\mathsf{A}_i^- + \mathsf{A}_i^-\mathsf{A}_j^+ + \mathsf{A}_i^-\mathsf{A}_j^- \\ &= :\mathsf{A}_i\mathsf{A}_j: + [\mathsf{A}_i^-, \mathsf{A}_j^+], \end{aligned}$$

where, to simplify the notation, A_i stands for $\mathsf{A}_i(t_i)$. Here : : stands for the normal-ordered product, i.e. the A^+'s are placed to the left of the A^-'s. Now, since $[\mathsf{A}_j^+, \mathsf{A}_i^-]$ is not an operator but a function we have

$$\langle 0|T(\mathsf{A}_i\mathsf{A}_j)|0\rangle = \langle 0| : \mathsf{A}_i\mathsf{A}_j : |0\rangle + [\mathsf{A}_j^+, \mathsf{A}_i^-].$$

But $\mathsf{A}_i^-|0\rangle = 0$, so that $\langle 0| : \mathsf{A}_i\mathsf{A}_j : |0\rangle = 0$ and therefore $[\mathsf{A}_i^-, \mathsf{A}_j^+] = \langle 0|T(\mathsf{A}_i\mathsf{A}_j)|0\rangle$.

Let us now assume the result for n and $n + 1$ operators and consider the time-ordered product of $n + 2$ operators,

$$T(\mathsf{A}_1 \dots \mathsf{A}_n\mathsf{A}_{n+1}\mathsf{A}(t)).$$

Now, since for the time-ordered product we have

$$T(\mathsf{A}_1 \dots \mathsf{A}_i\mathsf{A}_j \cdots \mathsf{A}_n\mathsf{A}_{n+1}\mathsf{A}(t)) = T(\mathsf{A}_1 \dots \mathsf{A}_j\mathsf{A}_i \cdots \mathsf{A}_n\mathsf{A}_{n+1}\mathsf{A}(t)),$$

let us therefore assume $t_1 > t_2 > \cdots > t_{n+1}$, that is, we assume the first $n+1$ times are already ordered so that

$$T(\mathsf{A}_1 \cdots \mathsf{A}_i \mathsf{A}_j \cdots \mathsf{A}_n \mathsf{A}_{n+1} \mathsf{A}(t)) = \begin{cases} \mathsf{A}_1 \cdots \mathsf{A}_n \mathsf{A}_{n+1} \mathsf{A}(t) & ; \quad t_{n+1} > t \\ \mathsf{A}_1 \ldots \mathsf{A}_n \mathsf{A}(t) \mathsf{A}_{n+1} & ; \quad t_n > t > t_{n+1} \\ \cdots & \\ \cdots & \\ \mathsf{A}(t) \mathsf{A}_1 \ldots \mathsf{A}_n \mathsf{A}_{n+1} & ; \quad t < t_1 \end{cases}$$

To continue we write $\mathsf{A}(t) = \mathsf{A}^+(t) + \mathsf{A}^-(t)$ and move A^+ to the left and A^- to the right respectively. Each time we move $\mathsf{A}^\pm$ past any of the A_i's we pick up a commutator so that

$$T(\mathsf{A}_1 \ldots \cdots \mathsf{A}_{n+1} \mathsf{A}(t)) = \mathsf{A}^+(t) \mathsf{A}_1 \ldots \mathsf{A}_{n+1} + \mathsf{A}_1 \ldots \mathsf{A}_{n+1} \mathsf{A}^-(t)$$

$$+ \begin{cases} \sum_{i=1}^{n+1} [\mathsf{A}_i^-, \mathsf{A}^+(t)] \mathsf{A}_1 \ldots \underline{\mathsf{A}_i} \cdots \mathsf{A}_{n+1} & ; \quad t_{n+1} > t \\ \sum_{i=1}^{n} [\mathsf{A}_i^-, \mathsf{A}^+(t)] \mathsf{A}_1 \ldots \underline{\mathsf{A}_i} \cdots \mathsf{A}_{n+1} & ; \quad t_n > t > t_{n+1} \\ \cdots & \\ \cdots & \\ [\mathsf{A}_1^-, \mathsf{A}^+(t)] \mathsf{A}_2 \cdots\cdots\cdots \mathsf{A}_{n+1} & ; \quad t_1 > t > t_2 \end{cases}$$

$$+ \begin{cases} [\mathsf{A}^-(t), \mathsf{A}_{n+1}^+] \mathsf{A}_1 \cdots\cdots\cdots \mathsf{A}_n & ; \quad t_n > t > t_{n+1} \\ \cdots & \\ \cdots & \\ \sum_{i=2}^{n+1} [\mathsf{A}^-(t), \mathsf{A}_i^+] \mathsf{A}_1 \ldots \underline{\mathsf{A}_i} \cdots \mathsf{A}_{n+1} & ; \quad t_1 > t > t_2 \\ \sum_{i=1}^{n+1} [\mathsf{A}^-(t), \mathsf{A}_i^+] \mathsf{A}_1 \cdots \underline{\mathsf{A}_i} \cdots \mathsf{A}_{n+1} & ; \quad t > t_1 \end{cases}$$

Here underlined operators do not appear in the product since they appear in the commuators with A^+ and A^- respectively. Re-introducing the time-ordering operator the above equation becomes

$$T\left(\mathsf{A}_1 \ldots \cdots \mathsf{A}_{n+1} \mathsf{A}(t)\right) = \mathsf{A}^+(t) T(\mathsf{A}_1 \ldots \mathsf{A}_{n+1}) + T(\mathsf{A}_1 \ldots \mathsf{A}_{n+1}) \mathsf{A}^-(t)$$
$$+ \sum_{i=1}^{n+1} \langle 0 \mid T(\mathsf{A}(t), \mathsf{A}_i) \mid 0 \rangle T(\mathsf{A}_1 \ldots \underline{\mathsf{A}_i} \cdots \mathsf{A}_{n+1}) .$$

To complete the proof we then note that the remaining operator products are time-ordered products of $n+1$ and n operators respectively for which the result holds by assumption. Note that $\sum_{i=1}^{n+1} \langle 0 \mid T\left(\mathsf{A}(t)\mathsf{A}_i\right) \mid 0\rangle$ sums over all non-equivalent pairings of $\mathsf{A}(t)$ with the other A_i's. This is just what is required to get the correct sum over all pairings on the right-hand side of the theorem. □

Although the theorem just proved contains the basic result, it is not directly applicable to perturbation theory in the form presented above. Indeed, the operators appearing in the perturbative expansion of the evolution operator in the last section are normal-ordered composite operators H_I. These do not have a decomposition into a sum of creation operators $\mathsf{A}^+ + \mathsf{A}^-$ as assumed in the proof. However, there is a straightforward generalization of Wick's theorem which takes care of this situation and which we state as a corollary without proof.

Corollary 9.3 Let $:\mathsf{A}_i:(t_i)$ be bosonic normal-ordered local products of operators. Then

$$\langle 0 \mid T\left(:\mathsf{A}_1:(t_1)\cdots:\mathsf{A}_n:(t_n)\right) \mid 0\rangle$$

is again given by the product of all possible pairings of the fundamental operators appearing in the normal-ordered products $:\mathsf{A}_i:(t_i)$ with the restriction that no pairings of two operators within the same normal-ordered monomial $:\mathsf{A}_i:(t_i)$ appear. We leave the proof of this corollary as an exercise to the reader.

Wick's theorem makes it clear that the correlation function $\langle 0|T(\mathsf{A}_i(t_i)\mathsf{A}_j(t_j))|0\rangle$ plays a central role in perturbation theory. In particular, for

$$\mathsf{A}_i(t_i) = \Psi(\mathbf{x}_i, t_i)\,, \qquad \text{or}$$
$$\mathsf{A}_i(t_i) = \Psi^+(\mathbf{x}_i, t_i)$$

the corresponding correlator $\langle 0|T\left(\Psi(\mathbf{x}_i, t_i)\Psi^+(\mathbf{x}_j, t_j)\right)|0\rangle$ provides the building block of perturbation theory. We will thus study this correlator in detail in the next section.

9.4 Green functions

In this section we discuss the two-point correlation function

$$G(\mathbf{x}, t_1, \mathbf{y}, t_2) = \langle 0|T\left(\Psi(\mathbf{x}, t_1)\Psi^+(\mathbf{y}, t_2)\right)|0\rangle\,.$$

We allow $\Psi(\mathbf{x}, t)$ to be a bosonic or fermionic field operator. Instead of evaluating this correlator directly we will first derive a differential equation satisfied by $G(\mathbf{x}, t_1, \mathbf{y}, t_2)$ and then discuss the solutions of this equation in turn. For this we need the time evolution of the field operator $\Psi(\mathbf{x}, t)$ which is determined by the Heisenberg equation

$$\mathrm{i}\hbar\frac{\partial \Psi}{\partial t} = -[\mathsf{H}_0, \Psi],$$

where $\mathsf{H}_0 = \int \mathrm{d}^3 y\, \Psi^+(\mathbf{y}, t)\left(-\frac{\hbar^2}{2m}\nabla_y^2\right)\Psi(\mathbf{y}, t)$ is the Hamilton operator for non-interacting particles. This means that

$$\begin{aligned}
\mathrm{i}\hbar\frac{\partial \Psi}{\partial t}(\mathbf{x}, t) &= -\int \mathrm{d}^3 y\left[\Psi^+(\mathbf{y}, t)\left(-\frac{\hbar^2}{2m}\nabla_y^2\right)\Psi(\mathbf{y}, t), \Psi(\mathbf{x}, t)\right]\\
&= \int \mathrm{d}^3 y \delta^{(3)}(\mathbf{y} - \mathbf{x})\left(-\frac{\hbar^2}{2m}\nabla_y^2\right)\Psi(\mathbf{y}, t)\\
&= -\frac{\hbar^2}{2m}\nabla_x^2\Psi(\mathbf{x}, t)
\end{aligned}$$

and similarly for $\Psi^\dagger$. On the other hand, when taking the time derivative of $G(\mathbf{x}, t_1, \mathbf{y}, t_2)$ we have to take into account the time-ordering operator, that is

$$\begin{aligned}
&\mathrm{i}\hbar\frac{\partial}{\partial t_1}\langle|T(\Psi(\mathbf{x}, t_1)\Psi^+(\mathbf{y}, t_2))|0\rangle\\
&= \mathrm{i}\hbar\frac{\partial}{\partial t_1}\{\theta(t_1 - t_2)\langle 0|\Psi(\mathbf{x}, t_1)\Psi^+(\mathbf{y}, t_2)|0\rangle\\
&\quad \pm \theta(t_2 - t_1)\langle 0|\Psi^+(\mathbf{y}, t_2)\Psi(\mathbf{x}, t_1)|0\rangle\}
\end{aligned}$$

where the $\pm$ corresponds to the definition of time ordering for bosonic and fermionic operators respectively. Using $\frac{\partial}{\partial t_1}\theta(t_1 - t_2) = \delta(t_1 - t_2)$ we get

$$\mathrm{i}\hbar\frac{\partial}{\partial t_1}G(\mathbf{x}, t_1, \mathbf{y}, t_2) = \mathrm{i}\hbar[\Psi(\mathbf{x}, t_1), \Psi^+(\mathbf{y}, t_2)]_\pm\delta(t_1 - t_2) - \left(\frac{\hbar^2}{2m}\nabla_x^2\right)G(\mathbf{x}, t_1, \mathbf{y}, t_2)$$

where $[\ ,\ ]_\pm$ stands for the commutator in the case of bosons and the anticommutator for fermions. Recalling the commutation relations for Ψ and $\Psi^\dagger$ we then get

$$\left(\mathrm{i}\hbar\frac{\partial}{\partial t_1} + \frac{\hbar^2}{2m}\nabla_x^2\right)G(\mathbf{x}, t_1, \mathbf{y}, t_2) = \mathrm{i}\hbar\,\delta^3(\mathbf{x} - \mathbf{y})\delta(t_1 - t_2).$$

This equation, which is valid for both bosonic and fermionic systems, simply states that $G(\mathbf{x}, t_1, \mathbf{y}, t_2)$ is a Green function for the Schrödinger equation. To continue, it is convenient to change to the momentum representation, which is obtained by the Fourier transform of $G(\mathbf{x}, t_1\ \mathbf{y}, t_2)$. Since the Schrödinger equation is independent of the time t and the position $\mathbf{x}$ the Green function $G(\mathbf{x}, t_1, \mathbf{y}, t_2)$ only depends on the difference $t_1 - t_2$ and $\mathbf{x} - \mathbf{y}$. We can thus write

$$G(\mathbf{x}, t_1, \mathbf{y}, t_2) = \int \frac{d^3k}{(2\pi)^3} \int_{-\infty}^{\infty} \frac{dw}{2\pi} e^{i\mathbf{k}\cdot(\mathbf{x}-\mathbf{y})} e^{-iw(t_1-t_2)} \tilde{G}(\mathbf{k}, w) .$$

Upon substitution into the differential equation above we end up with an algebraic equation for $\tilde{G}(\mathbf{k}, w)$

$$\left(w - \frac{\hbar|\mathbf{k}|^2}{2m} \right) \tilde{G}(\mathbf{k}, w) = i .$$

We then obtain the coordinate representation by substitution of $\tilde{G}(\mathbf{k}, \omega)$, that is

$$G(\mathbf{x}, t) = \int \frac{d^3k}{(2\pi)^3} \int_{-\infty}^{\infty} \frac{dw}{2\pi} e^{i\mathbf{k}\cdot\mathbf{x}} e^{-iwt} \left(\frac{i}{w - \frac{\hbar|\mathbf{k}|^2}{2m}} \right) ,$$

where we set $\mathbf{x} = \mathbf{x} - \mathbf{y}, \quad t = t_1 - t_2$. The evaluation of this integral, however, requires some care, since the integrand has a pole at $w = \hbar|\mathbf{k}|/2m$. The prescription of how to deal with this pole depends on the boundary conditions imposed on $\tilde{G}(\mathbf{k}, \omega)$. For the time-ordered Green function the appropriate prescription is

$$w - \frac{\hbar|\mathbf{k}|^2}{2m} \quad \to \quad w - \frac{\hbar|\mathbf{k}|^2}{2m} + i\epsilon, \quad \epsilon > 0,$$

i.e. we move the pole to the lower half plane. That this is the correct description follows from the fact that, for $t < 0$, we have to close the contour above the real line, leading the integral to vanish since there is no singularity present. This is as it should be since for $t < 0$ the operators are ordered as $\Psi^\dagger\Psi$ and since $\Psi \mid 0\rangle = 0$ the correlation function vanishes in this case. For $t > 0$ we close the contour below the real line so that the contour now encircles the pole and the integral is non-zero. Closing the contour in the upper half plane is not possible since $e^{-itw} \to 0$ only if $w_I < 0$. Closing the contour as shown we get

$$\int \frac{dw}{2\pi} e^{-iwt} \left(\frac{i}{w - \frac{\hbar|\mathbf{k}|^2}{2m} + i\epsilon} \right) = e^{-it\frac{\hbar|\mathbf{k}|^2}{2m}} .$$

Thus

$$G(\mathbf{x}, t) = - \int \frac{d^3k}{(2\pi)^3} e^{i\mathbf{k}\cdot\mathbf{x}} e^{-i\frac{\hbar|\mathbf{k}|^2}{2m}t} .$$

Absorbing the i in t this is the Fourier transform of a Gaussian and is thus given by

$$G(\mathbf{x}, t) = \frac{i}{(2\pi)^{\frac{3}{2}}} \left(\frac{m}{it\hbar} \right)^{\frac{3}{2}} e^{-\frac{m|\mathbf{x}|^2}{it\hbar}} .$$

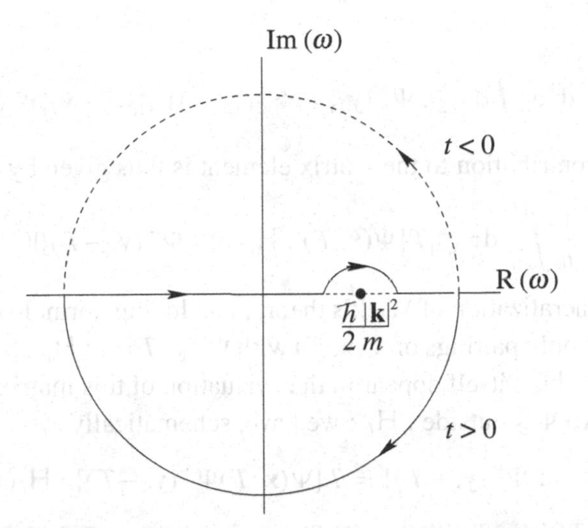

Figure 9.1 The dashed contour is chosen for $t < 0$ and the full line for $t > 0$.

Both the momentum and coordinate representation of this Green function play a key role in the systematic construction of perturbation theory as we will see in the next section.

9.5 Feynman rules

In this section we formulate a set of graphical rules which represent the perturbation series. These are the *Feynman rules*, to which we now turn.

To begin with, we consider the propagation of a single particle in the presence of an interaction term in the Hamiltonian. This is described by the matrix element

$$\langle 0|\Psi(\mathbf{x}, T)\mathsf{U}(T, -T)\Psi^{+}(\mathbf{y}, -T)|0\rangle$$

with

$$\mathsf{U}(T, -T) \approx I - \frac{\mathrm{i}}{\hbar}\int_{-T}^{T} \mathrm{d}\tau : \mathsf{H}_I(\tau) : +O\big(: \mathsf{H}_I^2 : \big)$$

where the : : means that we take H_I to be normal ordered. Here we focus on the leading correction to the free propagation, that is we neglect terms of $O(\mathsf{H}_I^2)$. The first order part of $\mathsf{U}(T, -T)$ is then

$$\mathsf{U}^{(1)}(T, -T) = -\frac{\mathrm{i}}{\hbar}\int_{-T}^{T} \mathrm{d}\tau \ : \mathsf{H}_I : (\tau) ,$$

where

$$:\mathsf{H}_I:(\tau)=\frac{1}{2}\int \mathrm{d}^3y_1\int \mathrm{d}^3y_2 :\Psi^+(\mathbf{y}_1,\tau)\Psi^+(\mathbf{y}_2,\tau)V(|\mathbf{y}_1-\mathbf{y}_2|)\Psi(\mathbf{y}_1,\tau)\Psi(\mathbf{y}_2,\tau):$$

The first-order contribution to the matrix element is thus given by

$$-\frac{\mathrm{i}}{\hbar}\int_{-T}^{T}\mathrm{d}\tau\,\langle 0|T[\Psi(\mathbf{x},T):\mathsf{H}_I(\tau):\Psi^+(\mathbf{y},-T)]|0\rangle\,.$$

Recalling the generalization of Wick's theorem including normal-ordered products we conclude that only pairings of $\Psi(\mathbf{x},T)$ with $\Psi^\dagger(\mathbf{y},T)$ or $:\mathsf{H}_I:$ but no pairings of operators within $:\mathsf{H}_I:$ itself appear in the evaluation of this matrix element. Since there are only two Ψ's outside $:\mathsf{H}_I:$ we have, schematically

$$T[\Psi(\mathbf{x},T):\mathsf{H}_I(\tau):\Psi^+(\mathbf{y},-T)]=T[\Psi(\mathbf{x},T)\Psi^+(\mathbf{y},-T)]:\mathsf{H}_I(\tau):+$$
$$\frac{1}{2}\int V(|\mathbf{y}_1-\mathbf{y}_2|)T[\Psi(\mathbf{x},T)\Psi^+(\mathbf{y}_1,\tau)]T[\Psi(\mathbf{y}_1,\tau)\Psi^+(\mathbf{y},-T)]:\Psi^+(\mathbf{y}_2,\tau)\Psi(\mathbf{y}_2,\tau):$$

plus similar terms with $\mathbf{y}_1$ and $\mathbf{y}_2$ interchanged. Here the integral is over $\mathbf{y}_1$ and $\mathbf{y}_2$. We see that there will be at least two uncontracted Ψ's left in the normal-ordered product. On the other hand, $\langle 0|:\mathcal{O}:|0\rangle$ vanishes for any operator $\mathcal{O}$. Thus, this matrix element vanishes altogether.

Let us now give a graphical representation of this calculation.

(1) A *propagator* $\langle 0|T(\Psi,(\mathbf{x},t_1)\Psi^+(\mathbf{y},t_2))|0\rangle=G(\mathbf{x}-\mathbf{y},t_1-t_2)$ is represented as a line from y to x:

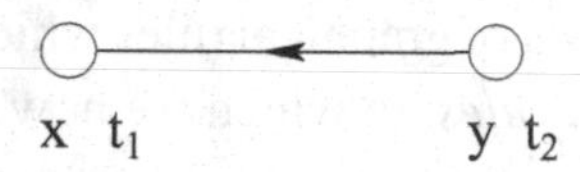

(2) An interaction term of the form

$$-\frac{\mathrm{i}}{2\hbar}\int_{-T}^{T}\mathrm{d}t\int \mathrm{d}^3u\int \mathrm{d}^3v\,V(|\mathbf{u}-\mathbf{v}|)$$

has the graphical representation as a *vertex*

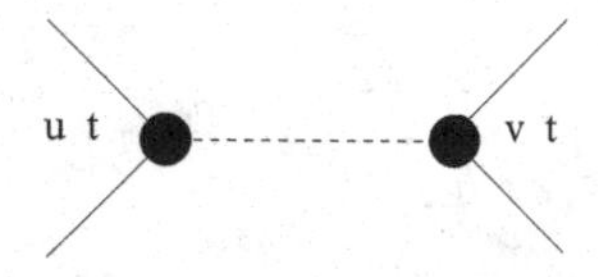

The sum over pairings in Wick's theorem is then represented graphically as sum over all graphs obtained by combining propagators and vertices. For instance, in the case at hand we could draw graphs like in Figure 9.2. Note, however, that all of these graphs correspond to pairings of operators within $:\mathsf{H}_I:$ which, according

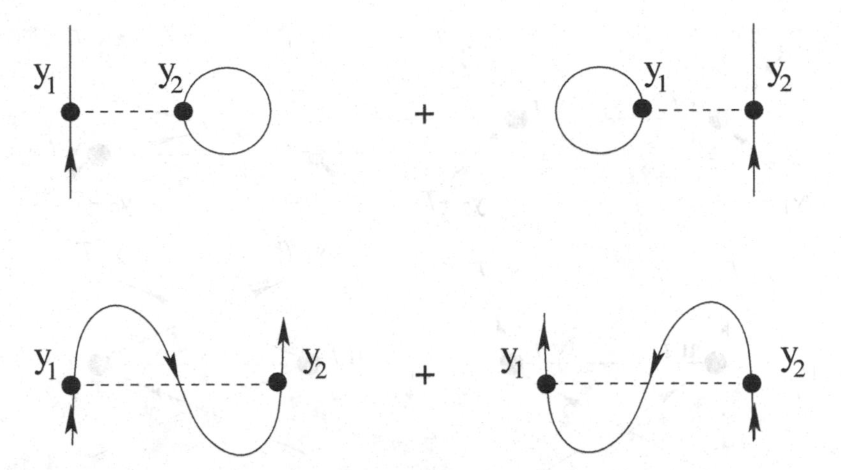

Figure 9.2 Feynman graphs for the 1-boson propagator at first order in $:\mathsf{H}_I:$

to the generalized Wick's theorem, do not appear. These are therefore precisely the graphs which have to be left out in the presence of normal-ordered operators. We will see, however, that in the finite temperature case, discussed in Section 9.8, such graphs will contribute to the result.

Let us now consider a slightly more involved example, that is the scattering of two bose particles. In this case the matrix element to first order in $:\mathsf{H}_I:$ is given by

$$
\begin{aligned}
&\langle 0|\Psi(\mathbf{x}_1,T)\Psi(\mathbf{x}_2,T)\mathsf{U}(T,-T)\Psi^+(\mathbf{y}_1,-T)\Psi^+(\mathbf{y}_2,-T)|0\rangle \\
&= -\frac{\mathrm{i}}{2\hbar}\int_{-T}^{T}\mathrm{d}t\int \mathrm{d}^3u\,\mathrm{d}^3v\,V(|\mathbf{u}-\mathbf{v}|)\,\langle 0|T[\Psi(\mathbf{x}_1,T)\Psi(\mathbf{x}_2,T) \\
&\qquad :\Psi^+(\mathbf{u},t)\Psi^+(\mathbf{v},t)\Psi(\mathbf{u},t)\Psi(\mathbf{v},t):\Psi^+(\mathbf{y}_1,-T)\Psi^+(\mathbf{y}_2,-T)]|0\rangle\,.
\end{aligned}
$$

Wick's theorem then instructs us to replace the above expression by all possible pairings of Ψ's and $\Psi^\dagger$'s outside the normal-ordered product with a $\Psi^\dagger$ and $\Psi^\dagger$ inside the normal-ordered product. Since

$$
\langle 0|T(\Psi(\mathbf{x},t_1)\Psi(\mathbf{y},t_2))|0\rangle = \langle 0|T(\Psi^+(\mathbf{x},t_1)\Psi^+(\mathbf{y},t_2))|0\rangle = 0
$$

this gives

$$
\begin{aligned}
&-\frac{\mathrm{i}}{2\hbar}\int_{-T}^{T}\mathrm{d}t\int \mathrm{d}^3u\,\mathrm{d}^3v\,V(|\mathbf{u}-\mathbf{v}|) \\
&\qquad\times\big\{G(\mathbf{x}_1,T,\mathbf{u},t)G(\mathbf{x}_2,T,\mathbf{v},t)G(\mathbf{u},t,\mathbf{y}_1,-T)G(\mathbf{v},t,\mathbf{y}_2,-T) \\
&\qquad+(\mathbf{x}_1\leftrightarrow\mathbf{x}_2)+(\mathbf{y}_1\leftrightarrow\mathbf{y}_2)+(\mathbf{x}_1\leftrightarrow\mathbf{x}_2,\mathbf{y}_1\leftrightarrow\mathbf{y}_2)\big\}.
\end{aligned}
$$

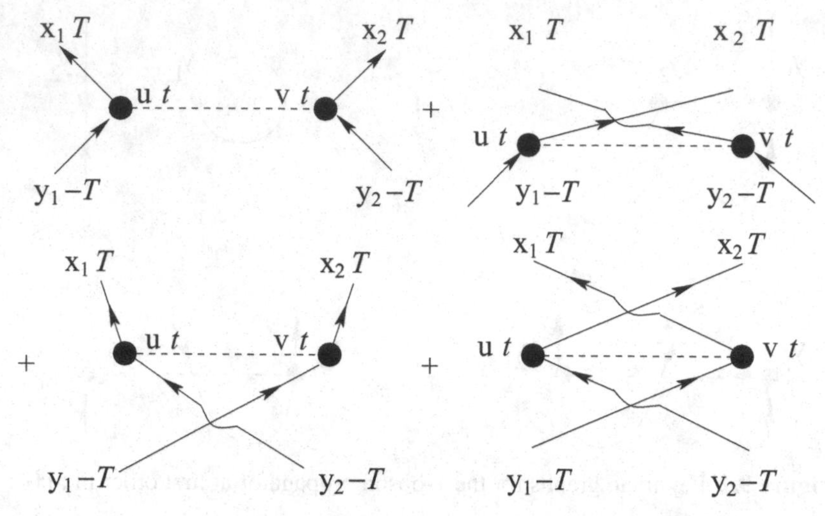

Figure 9.3 Feynman graphs for the 2-boson scattering at first order in : H_I :

This has the graphical representation Figure 9.3. Let us now state the Feynman rules for this graphical representation:

(1) Sum over all topologically non-equivalent graphs. In order to account for all equivalent pairings appearing in Wick's theorem we have to multiply each graph with a symmetry factor which counts the number of permutations of *internal lines* that do not change the topology of the graph. Internal lines are the lines that do not connect with any of the coordinates $\mathbf{x}_i, \mathbf{y}_i$.

(2) In the case of fermions, each fermion loop comes with a factor (-1).

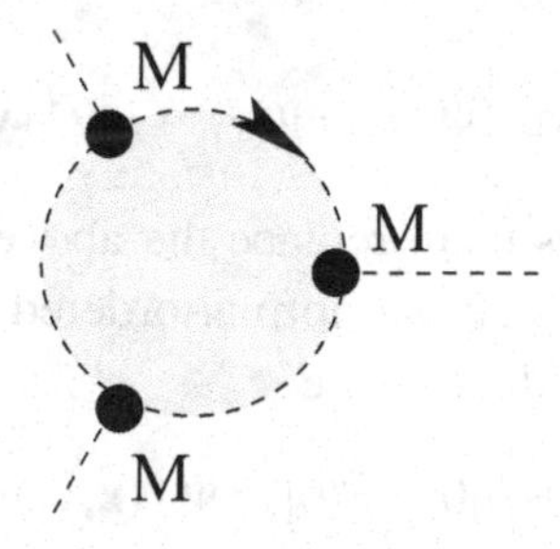

This is because in order to pair all fermion operators in a loop in the order $\Psi^\dagger\Psi$ the operator $\Psi_{\alpha_1}(x_1)$ has to be moved through an odd number of fermionic operators thus picking up a factor (-1). i.e.

$$\begin{aligned}&\langle 0|\Psi_{\alpha_1}(x_1)M^{\alpha_1\beta_1}\Psi^\dagger_{\beta_1}(x_1)\Psi_{\alpha_2}(x_2)\cdots M^{\alpha_n\beta_n}\Psi^\dagger_{\beta_n}(x_n)|0\rangle\\&=(-1)\langle 0|\,\mathrm{Tr}(M\Psi(x_1)\Psi^\dagger(x_2)M\ldots M\Psi(x_n)\Psi^\dagger(x_1))\rangle\,.\end{aligned}$$

(3) A global minus sign arises for each permutation of external fermions.

9.5.1 Feynman rules in momentum space

It is often preferable to perform the computation of transition amplitudes in momentum space. This is the case, for instance in a scattering experiment, where the asymptotic momenta of the particles, rather than their positions, are fixed. Let us now derive the momentum space Feynman rules.

To begin with we introduce a convenient shorthand notation for space and time coordinates borrowed from the theory of special relativity. That is we write

$$x \equiv (\mathbf{x}, t)\,, \qquad \mathrm{d}^4x \equiv \mathrm{d}^3x\,\mathrm{d}t$$
$$k \equiv (\mathbf{k}, \omega)\,, \qquad \mathrm{d}^4k \equiv \mathrm{d}^3k\,\mathrm{d}\omega\,,$$

where ω is the frequency corresponding to the energy $E = \hbar\omega$. Also $k \cdot x \equiv \mathbf{k}\cdot\mathbf{x} - \omega t$. Finally we replace the instantaneous interaction potential $V(|\mathbf{u} - \mathbf{u}'|)$ by

$$U(u, u') \equiv V(|\mathbf{u} - \mathbf{u}'|)\delta(t - t').$$

Let us now consider the interaction vertex which, in the new notation reads $-\mathrm{i}/2\hbar \int \mathrm{d}^4u\mathrm{d}^4v\; U(u - v)$. We focus on the left corner of the diagram for the vertex with the two legs paired with external particles

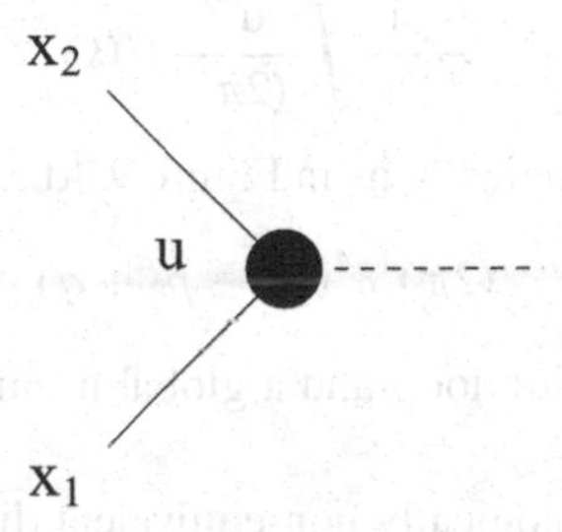

Taking into account the integral over u only (u' goes with the r.h.s.) we have

$$\begin{aligned}
&\int \mathrm{d}^4u\; G(x_2, u)U(u, v)G(u, x_1) \\
&= \int \mathrm{d}^4u \frac{\mathrm{d}^4k}{(2\pi)^4}\mathrm{e}^{\mathrm{i}k\cdot(x_2-u)}\tilde{G}(k)\frac{\mathrm{d}^4q}{(2\pi)^4}\mathrm{e}^{\mathrm{i}q\cdot(u-v)}U(q)\frac{\mathrm{d}^4p}{(2\pi)^4}\mathrm{e}^{\mathrm{i}p\cdot(u-x_1)}\tilde{G}(p) \\
&= \frac{1}{(2\pi)^8}\int \mathrm{d}^4k\,\mathrm{d}^4q\,\mathrm{d}^4p\,\delta^4(p-q-k)\,\tilde{G}(k)U(q)\tilde{G}(p)\mathrm{e}^{\mathrm{i}(k\cdot x_2 - p\cdot x_1 - q\cdot v)}\,.
\end{aligned}$$

To complete the transition into momentum space we Fourier transform with respect to the coordinates of the external particles x_1, x_2, by the operation

$$\int \mathrm{d}^4x_1\mathrm{e}^{\mathrm{i}p_1\cdot x_1}\int \mathrm{d}^4x_2\mathrm{e}^{\mathrm{i}p_2\cdot x_2}.$$

This fixes k and p in terms of p_1 and p_2. The remaining exponential $e^{-iq\cdot v}$ will be absorbed by transforming the right-hand side of the interaction vertex. The left-hand side of the vertex then has the momentum space representation

$$\int \frac{d^4q}{(2\pi)^4} \tilde{G}(p_2)U(q)\tilde{G}(p_1)(2\pi)^4\delta^4(p_1 - p_2 + q)\,.$$

The interpretation of the delta function is that it ensures the energy-momentum conservation at each vertex. We are now ready to formulate the Feynman rules in momentum space.

(1) External lines: each external line contributes a factor

$$\left(\omega - \frac{\hbar|k|^2}{2m}\right)\tilde{G}(p) = \mathrm{i}$$

and is represented graphically as in Figure 9.4a.

(2) Internal lines are represented graphically as in Figure 9.4b and give a contribution

$$\int \frac{d^4q}{(2\pi)^4}\tilde{G}(q)\,.$$

(3) Interaction potentials are represented by a dashed line as in Figure 9.4c and contribute

$$-\frac{\mathrm{i}}{2\hbar}\int \frac{d^4k}{(2\pi)^4}U(k)\,.$$

(4) Vertices are represented graphically as in Figure 9.4d and give a contribution

$$(2\pi)^4\delta^4(p_1 - p_2 + q)\,.$$

(5) A minus sign for each fermion loop and a global minus sign for each permutation of external fermions.

(6) Sum over all connected topologically non-equivalent diagrams.

9.6 Scattering cross-section for two helium atoms

As an application of the momentum space perturbation theory we compute the scattering cross-section for two helium atoms to the lowest order in the interaction. Let us consider two atoms whose initial state can be described by the initial energy and momentum of each atom which we take to be $(E_1, \mathbf{p}_1)$ and $(E_2, \mathbf{p}_2)$ respectively. We want to determine the probability that the final state of the two atoms after they scatter is $(E_1', \mathbf{q}_1)$ and $(E_2', \mathbf{q}_2)$. where these represent the energy and momentum values of the atoms respectively.

We will write $\mid \Psi(-T)\rangle$ as the initial state, i.e.

$$\mid \Psi(-T)\rangle = \mid E_1, \mathbf{p}_1; E_2, \mathbf{p}_2\rangle\,.$$

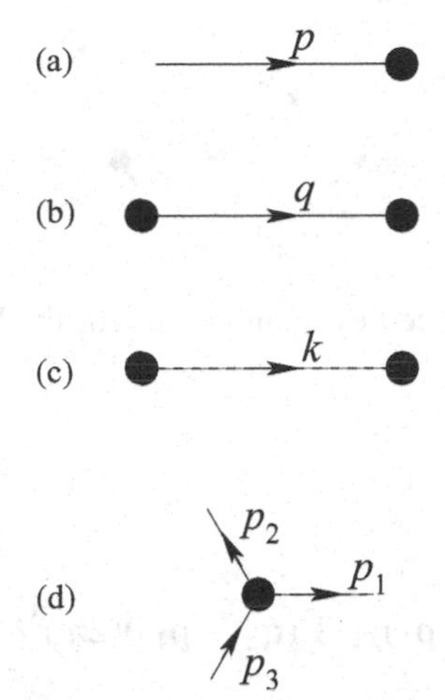

Figure 9.4 Feynman rules in momentum space.

In the interaction picture this state evolves from time $-T$ to T as

$$| \Psi(T)\rangle = \mathsf{U}(T, -T) \mid \Psi(-T)\rangle ,$$

where $\mathsf{U}(T, -T)$ is the evolution operator in the interaction picture. The probability for $| \Psi(T)\rangle$ with the state $| E_1', \mathbf{q}_1; E_2', \mathbf{q}_2\rangle$ is given by $| \langle E_1', \mathbf{q}_1; E_2', \mathbf{q}_2 \mid \Psi(T)\rangle |^2$.

We will take the limit $T \to \infty$ and assume that for large values of T the states of the particles can be described as eigenstates of the non-interacting part of the Hamiltonian which describes the dynamics of the system, i.e. if

$$\mathsf{H} = \mathsf{H}_0 + \mathsf{H}_I$$

then $\lim | \Psi(T) >$, $T \to \pm\infty$ are eigenstates of H_0.

This assumption corresponds to the physical picture of the scattering process which we have used. Namely that the initial and final states of the system can be specified by the energy and momentum values of each of the atoms separately. To first order in perturbation theory the scattering amplitude is given by

$$-\frac{\mathrm{i}}{\hbar}\int_{-T}^{T} \mathrm{d}t \langle E_1', \mathbf{q}_1 E_2', \mathbf{q}_2 \mid : \mathsf{H}_I(t) : \mid E_1, \mathbf{p}_1, E_2, \mathbf{p}_2\rangle .$$

There are in total four Feynman diagrams that contribute to this amplitude (see Figure 9.3) but only two of them give different contributions. In momentum space these are the diagrams in Figure 9.5. Using rules (1) to (4) from Section 9.5.1, these

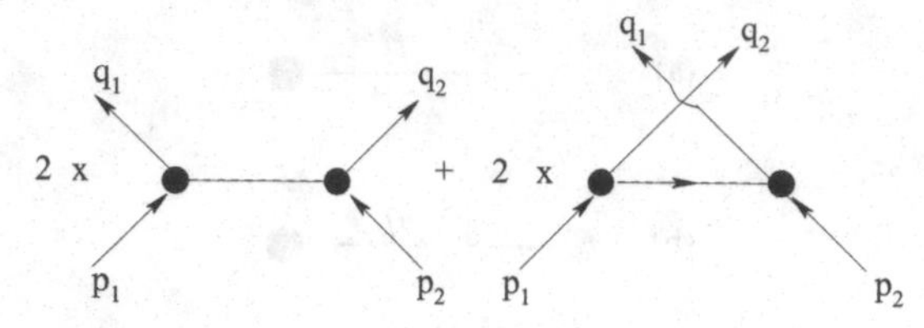

Figure 9.5 Momentum space Feynman graphs for the helium–helium scattering at first order in : H_I :

two graphs add up to

$$\mathcal{A} = 2\left(\frac{-\mathrm{i}}{2\hbar}\right)(\tilde{V}(\mathbf{q}_1 - \mathbf{p}_1) + \tilde{V}(\mathbf{q}_2 - \mathbf{p}_1))(2\pi)^3\delta^3(\mathbf{p}_1 + \mathbf{p}_2 - \mathbf{q}_1 - \mathbf{q}_2)$$
$$\cdot (2\pi)\delta(E_1 + E_2 - E_1' - E_2') .$$

From this result it is clear that the field theory description of this process automatically takes the identical nature of the atoms into account. If we had considered the problem of scattering involving two identical Fermi–Dirac particles then the second diagram would come with a minus sign since it involves the permutation of two external lines. As a result the scattering amplitude would get modified to

$$\mathcal{A} = 2\left(\frac{-\mathrm{i}}{2\hbar}\right)(\tilde{V}(\mathbf{q}_1 - \mathbf{p}_1) - \tilde{V}(\mathbf{q}_2 - \mathbf{p}_1))(2\pi)^3\delta^3(\mathbf{p}_1 + \mathbf{p}_2 - \mathbf{q}_1 - \mathbf{q}_2)$$
$$\cdot (2\pi)\delta(E_1 + E_2 - E_1' - E_2') .$$

To compare this result with experiment we need to compute the *scattering cross-section* which is is a measure of the probability of detecting a scattering particle within a given range of momentum values. The probability for this process to occur is then obtained by taking the absolute value square of the probability amplitude, $| \mathcal{A} |^2$. It is useful to define the transition rate $R_{f \leftarrow i}$ as

$$R_{f \leftarrow i} = \frac{\text{Probability of a transition from state } i \text{ to state } f}{(\text{ Time })(\text{ Volume })} .$$

If we interpret the square of the delta function as

$$(2\pi)^2 \, |\delta(E_i - E_f)|^2 = (2\pi)\delta(E_i - E_f) \int_{-T/2}^{T/2} \frac{\mathrm{d}t}{\hbar} \mathrm{e}^{-\mathrm{i}t(E_i - E_f)/\hbar}$$
$$= (2\pi)\delta(E_i - E_f)(\frac{T}{\hbar}) , \qquad (T \text{ large})$$

and similarly

$$(2\pi)^6 \,|\, \delta^3(\mathbf{p}_1+\mathbf{p}_2-\mathbf{q}_1-\mathbf{q}_2)\,|^2 = (2\pi)^3\delta^3(\mathbf{p}_1+\mathbf{p}_2-\mathbf{q}_1-\mathbf{q}_2)\int_V \mathrm{d}^3x\, \mathrm{e}^{-\mathrm{i}\mathbf{x}\cdot(\mathbf{p}_1+\mathbf{p}_2-\mathbf{q}_1-\mathbf{q}_2)}$$

$$= (2\pi)^3\delta^3(\mathbf{p}_1+\mathbf{p}_2-\mathbf{q}_1-\mathbf{q}_2)V, \qquad (V \text{ large})$$

then no squares of the delta functions appear in the transition rate. Let us temporarily go to the coordinate frame where one of the particles is at rest, say $\mathbf{p}_1 = 0$. This particle then plays the role of the target onto which the second particle is infalling. The scattering cross-section σ_{fi}, is then defined as

$$\sigma_{fi} = \frac{R_{f\leftarrow i}}{[\text{ flux of incoming particles }]} \cdot [\text{ phase space volume of final states }] .$$

In a general coordinate system, the flux of incoming particles is simply the relative velocity $|\mathbf{v}_{12}|$. The differential cross-section thus takes the form

$$\mathrm{d}\sigma_{fi} = \frac{R_{i\mapsto f}}{|\,\mathbf{v}_{12}\,|} \frac{\mathrm{d}^3q_1}{(2\pi\hbar)^3} \frac{\mathrm{d}^3q_2}{(2\pi\hbar)^3} ,$$

where $\mathrm{d}^3q_i/(2\pi\hbar)^3$ is the phase space volume for outgoing particles with momenta between $\mathbf{q}$ and $\mathbf{q}+\mathrm{d}\mathbf{q}$. Substituting the transition probability and using our definition of the square of the delta function we then find

$$\mathrm{d}\sigma_{fi} = \frac{2\pi}{\hbar}\delta(E_1'+E_2'-E_1+E_2)\,|\,\tilde{V}(\mathbf{q}_1-\mathbf{p}_1)+\tilde{V}(\mathbf{q}_2-\mathbf{p}_1)\,|^2$$

$$\cdot \frac{m}{|\,\mathbf{p}_{12}\,|} \frac{\mathrm{d}^3q_1}{(2\pi\hbar)^3} \frac{\mathrm{d}^3q_2}{(2\pi\hbar)^3} \delta^3(\mathbf{p}_1+\mathbf{p}_2-\mathbf{q}_1-\mathbf{q}_2)(2\pi\hbar)^3 .$$

If we detect a particle with momentum $\mathbf{q}_1$ in the center of mass coordinate system $(\mathbf{p}_1+\mathbf{p}_2 = 0)$ we have

$$\mathrm{d}\sigma_{fi} = \frac{2\pi}{\hbar}\delta(E_{tot}'-E_{tot})\,|\,\tilde{V}(\mathbf{q}-\mathbf{p})+\tilde{V}(\mathbf{q}+\mathbf{p})\,|^2 \frac{m}{|\,\mathbf{p}\,|} \frac{\mathrm{d}^3q_1\mathrm{d}^3q_2}{(2\pi\hbar)^3}\delta^3(\mathbf{q}_1+\mathbf{q}_2)$$

where we set $\mathbf{p}_1 = \mathbf{p}$ and $\mathbf{q}_1 = \mathbf{q}$. Integrating this expression with respect to d^3q_2 then removes the delta function for the momenta, i.e.

$$\mathrm{d}\sigma_{fi} = \frac{2\pi}{\hbar}\delta(E'-E)\,|\,\tilde{V}(\mathbf{q}-\mathbf{p})+\tilde{V}(\mathbf{q}+\mathbf{p})\,|^2 \frac{1}{|\,\mathbf{p}\,|} \frac{m}{(2\pi\hbar)^3} q^2\mathrm{d}q\mathrm{d}\Omega_q .$$

Finally, the differential cross-section for finding an outgoing particle in a given direction is then obtained by integrating over q, that is

$$\frac{d\sigma_{fi}}{d\Omega_q} = \frac{2\pi}{\hbar}\frac{m}{|\mathbf{p}|}\int \delta\left(\frac{\mathbf{p}^2}{2m} - \frac{\mathbf{q}^2}{2m}\right) | \tilde{V}(\mathbf{q}-\mathbf{p}) + \tilde{V}(\mathbf{q}+\mathbf{p}) |^2 \frac{q^2 dq}{(2\pi\hbar)^3}.$$

This is the result we were aiming for.

9.7 Fermions at finite density

In this section we extend the quantum field theory description to include interactions in a degenerate Fermi gas, that is, we consider fermions at finite density but zero temperature. As we have seen in Chapter 7 the Pauli exclusion principle implies that all energy levels up to the Fermi energy ϵ_F are occupied even at zero temperature. We expect this fact to have consequences for the way interactions are described within quantum field theory since now the ground state of the system is no longer the vacuum state $| 0\rangle$ but some state which we denote by $| \Omega\rangle$ and in which all energy levels up to ϵ_F are occupied.

Indeed, an immediate consequence is that $| \Omega\rangle$ is no longer annihilated by all destruction operators $b_{\mathbf{k}}$. In particular, it is therefore not annihilated by Ψ, i.e. $\Psi | \Omega\rangle \neq 0$. How will this affect the time-ordered Green function $G(\mathbf{x}_1, t_1, \mathbf{x}_2, t_2)$? At finite density a convenient definition of $G(\mathbf{x}_1, t_1, \mathbf{x}_2, t_2)$ is given by

$$\begin{aligned} G(\mathbf{x}_1, t_1, \mathbf{x}_2, t_2) = \frac{1}{\langle\Omega \| \Omega\rangle} & \big\{\theta(t_1 - t_2)\langle\Omega | \Psi^\dagger(\mathbf{x}_1, t_1)\Psi(\mathbf{x}_2, t_2) | \Omega\rangle \\ & - \theta(t_2 - t_1)\langle\Omega | \Psi(\mathbf{x}_2, t_2)\Psi^\dagger(\mathbf{x}_1, t_1) | \Omega\rangle\big\}. \end{aligned}$$

The numerator is introduced as a convenient normalization of the correlation function. At zero density, we have $\langle\Omega \| \Omega\rangle = \langle 0 \| 0\rangle = 1$. In order to determine the momentum representation of G we substitute the momentum representation of the field operators Ψ and $\Psi^\dagger$ into the above expectation value. Now, since

$$\langle\mathbf{k}', \alpha' \| \mathbf{k}, \alpha\rangle = \delta_{\alpha'\alpha}(2\pi)^3\delta^3(\mathbf{k}' - \mathbf{k})$$

where $\alpha = 1, 2$ is the spinor index, only $\mathsf{b}_\alpha(\mathbf{k})\mathsf{b}^\dagger_\alpha(\mathbf{k})$, or $\mathsf{b}^\dagger_\alpha(\mathbf{k})\mathsf{b}_\alpha(\mathbf{k})$, but not $\mathsf{b}^\dagger_\alpha(\mathbf{k}')\mathsf{b}_\alpha(\mathbf{k})$ with $\mathbf{k} \neq \mathbf{k}'$ have non-vanishing expectation values. More precisely,

$$\frac{\langle\Omega \mid \mathsf{b}_\alpha(\mathbf{k})\mathsf{b}^\dagger_{\alpha'}(\mathbf{k}) \mid \Omega\rangle}{\langle\Omega \mid\mid \Omega\rangle} = \begin{cases}\delta_{\alpha'\alpha} & ; \quad k > k_F \\ 0 & ; \quad k < k_F\end{cases}$$

$$\frac{\langle\Omega \mid \mathsf{b}^\dagger_\alpha(\mathbf{k})\mathsf{b}_{\alpha'}(\mathbf{k}) \mid \Omega\rangle}{\langle\Omega \mid\mid \Omega\rangle} = \begin{cases}0 & ; \quad k > k_F \\ \delta_{\alpha'\alpha} & ; \quad k < k_F\end{cases}$$

If we then substitute these expressions into the momentum space formula of G we end up with

$$\tilde{G}_{\alpha\beta}(\omega, \mathbf{k}) = \mathrm{i}\delta_{\alpha\beta}\left[\frac{\theta(k-k_F)}{\omega - \frac{\hbar|\mathbf{k}|^2}{2m} + \mathrm{i}\epsilon} + \frac{\theta(k_F-k)}{\omega - \frac{\hbar|\mathbf{k}|^2}{2m} - \mathrm{i}\epsilon}\right].$$

The ϵ-prescription for the poles is again dictated by the choice of the contour. The difference with the zero density situation is that now the correlation function is non-vanishing even for $t < 0$ due to the contributions of the *Fermi sea*, i.e. the states with energy $E < \epsilon_F$ and that are occupied at zero temperature.

Particles and holes There is an alternative interpretation of the Fermi sea which consists of introducing a new type of particle which we call a *hole*. The idea is this:

- For $k > k_F$ we say that $\mathsf{b}^\dagger_\alpha(\mathbf{k})$ creates a fermion with momentum $\mathbf{k}$ out of $\mid \Omega\rangle$ while $\mathsf{b}_\alpha(\mathbf{k}) \mid \Omega\rangle = 0$.
- For $k < k_F$ we say that $\mathsf{b}_\alpha(\mathbf{k})$ annihilates a fermion with momentum $\mathbf{k}$ out of $\mid \Omega\rangle$ or, equivalently, creates a *hole* with momentum $\mathbf{k}$.

Therefore we can define creation and annihilation operators for particles and holes respectively as follows:

- Particles: $k > k_F$

$$\mathsf{c}^\dagger_\alpha(\mathbf{k}) \equiv \mathsf{b}^\dagger_\alpha(\mathbf{k})$$
$$\mathsf{c}_\alpha(\mathbf{k}) \equiv \mathsf{b}_\alpha(\mathbf{k})$$

- Holes: $k < k_F$

$$\mathsf{d}^\dagger_\alpha(\mathbf{k}) \equiv \mathsf{b}_\alpha(\mathbf{k})$$
$$\mathsf{d}_\alpha(\mathbf{k}) \equiv \mathsf{b}^\dagger_\alpha(\mathbf{k})$$

with $\mathsf{c}_\alpha(\mathbf{k}) \mid \Omega\rangle = \mathsf{d}_\alpha(\mathbf{k}) \mid \Omega\rangle = 0$ so that formally the structure is the same as in the zero density case, in that $\mid \Omega\rangle$ is annihilated by all annihilation operators. This is the advantage of introducing holes as "particles".

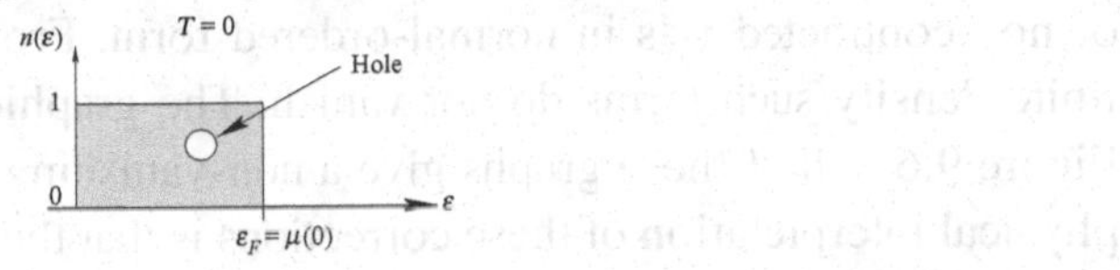

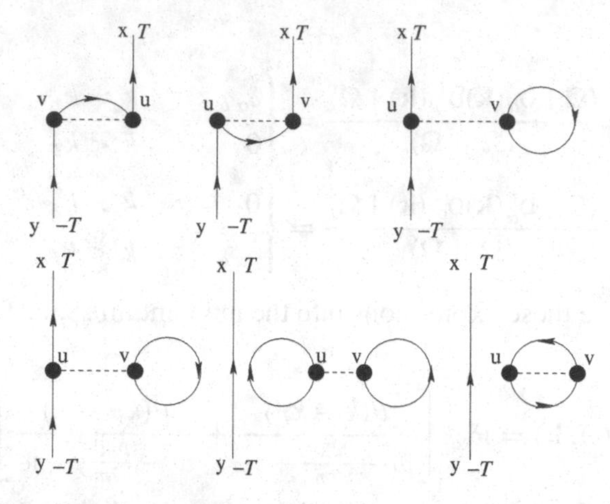

Figure 9.6 Feynman diagrams for the first order correction to the finite density propagator.

We can similarly expand the field operators

$$\Psi_\alpha(\mathbf{x}) = \sum_{|\mathbf{k}|\leq k_F} \mathsf{d}^\dagger_\alpha(\mathbf{k})\, \mathrm{e}^{\mathbf{ik\cdot x}} + \sum_{|\mathbf{k}|>k_F} \mathsf{c}_\alpha(\mathbf{k})\, \mathrm{e}^{\mathbf{ik\cdot x}}$$
$$\equiv \Psi_+(\mathbf{x}) + \Psi_-(\mathbf{x})$$
$$\Psi^\dagger_\alpha(\mathbf{x}) = \Psi^\dagger_-(\mathbf{x}) + \Psi^\dagger_+(\mathbf{x})$$

with $\Psi_-(\mathbf{x}) \mid \Omega\rangle = \Psi^\dagger_-(\mathbf{x}) \mid \Omega\rangle = 0$.

It is not hard to see that Wick's theorem applies without modification to particles as well as holes since we have not assumed any specific interpretation of the operators A and $\mathsf{A}^\dagger$ in the theorem. Note however that if $: \mathsf{H}_I :$ is normal-ordered with respect to $\mid 0\rangle$ it will not be normal-ordered with respect to $\mid \Omega\rangle$. Consequently, unlike in the zero-density case, pairings, or contractions within normal-ordered products, will now appear in the expansion of time-ordered products. As an example we consider the time-ordered fermion two-point function at first order in H_I.

$$-\frac{\mathrm{i}}{\hbar}\int_{-T}^{T} \mathrm{d}t\, \langle\Omega|T[\Psi(\mathbf{x}, T) : \mathsf{H}_I(t) : \Psi^+(\mathbf{y}, -T)]|\Omega\rangle.$$

This is the same example as that at the beginning of the last section. There we concluded that the first-order correction vanishes as a consequence of the presence of non-contracted ψ's in normal-ordered form. From what we have just said, at finite density such terms do not vanish. The graphical representation is given in Figure 9.6. All of these graphs give a non-vanishing contribution for $k_F > 0$. The physical interpretation of these corrections is that they represent the interactions of the particle with the Fermi sea. Note, however that not all of graphs in Figure 9.6 contribute to the normalized Green function $G_{\alpha\beta}(\mathbf{x}, T, \mathbf{y}, -T)$. Indeed the *vacuum graphs* such as

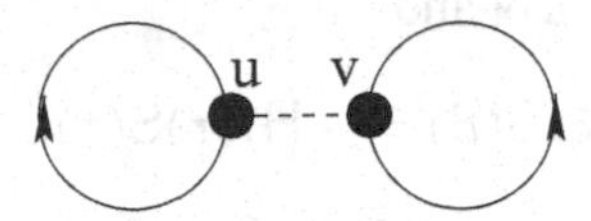

equally appear in the numerator $\langle\Omega \mid\mid \Omega\rangle$. They get thus divided out. Concretely we can write

$$\frac{\langle\Omega|T[\Psi(\mathbf{x}, T) : \mathsf{H}_I(t) : \Psi^+(\mathbf{y}, -T)]|\Omega\rangle}{\langle\Omega \mid\mid \Omega\rangle} = \langle\Omega|T[\Psi(\mathbf{x}, T) : \mathsf{H}_I(t) : \Psi^+(\mathbf{y}, -T)]|\Omega\rangle^c$$

where c indicated that only connected graphs are taken into account. Thus only the connected graphs appear in the expansion of $G_{\alpha\beta}(\mathbf{x}_2, t_2, \mathbf{x}_1, t_1)$.

9.8 Finite temperature perturbation theory

In this section we develop a perturbative formalism for the computation of the grand canonical partition sum Z_Ω. This proceeds in close analogy with the perturbative evaluation of the evolution operator $\mathsf{U}(T, -T)$. Indeed, we recall from the beginning of this chapter that the grand canonical partition sum Z_Ω can be written as a trace of an operator

$$Z_\Omega = \mathrm{Tr}\, \mathrm{e}^{-\beta(\mathsf{H}-\mu\mathsf{N})} \equiv \mathrm{Tr}(\rho(\beta))$$

where $\rho(\beta) = \mathrm{e}^{-\beta(\mathsf{H}-\mu\mathsf{N})}$ is called the *density operator* and $\mathsf{H} = \mathsf{H}_0 + \mathsf{H}_1$ is the Hamilton operator with interaction H_1. We will make use of the formal similarity between the density operator and the evolution operator $\mathsf{U}(T, 0)$ introduced in Section 9.2. Indeed, if we define the evolution operator in the Schrödinger picture as

$$\mathsf{U}_S(T, 0) \equiv \exp\left(-\frac{\mathrm{i}T}{\hbar}\mathsf{H}\right),$$

so that $\mid \Psi(T)\rangle_S = \mathsf{U}_S(T, 0) \mid \Psi(0)\rangle$, then we see that $\mathsf{U}_S(T, 0)$ and ρ are related to each other by the substitution $T \to -\mathrm{i}\beta\hbar$, or

$$\rho(\beta) = \mathsf{U}_S(-\mathrm{i}\beta\hbar, 0).$$

We can change from the Schrödinger picture to the interaction picture by writing $\mathsf{U}_S(T, 0) = \mathsf{U}_0(T, 0)\mathsf{U}(T, 0)$, so that $\mathsf{U}(T, 0)$ satisfies the differential equation

$$\mathrm{i}\hbar\partial_t \mathsf{U}(t, 0) = \mathsf{H}_I \mathsf{U}(t, 0).$$

Repeating these steps for ρ we get

$$\rho(\beta) = \rho_0(\beta)\mathsf{S}(\beta)\,,$$

with $\rho_0 = \mathrm{e}^{-\beta \mathsf{K}_0}$, $\mathsf{K}_0 = \mathsf{H}_0 - \mu\mathsf{N}$ and

$$\partial_\tau \mathsf{S}(\beta) = -\mathsf{H}_I(\tau)\mathsf{S}(\tau)\,,$$

where $0 \leq \tau \leq \beta$ and $\mathsf{H}_I(\tau) = \mathrm{e}^{\tau \mathsf{K}_0}\mathsf{H}_1\mathrm{e}^{-\tau\mathsf{K}_0}$. We then solve this equation by the formal series expansion

$$\mathsf{S}(\beta) = \sum_{n\geq 0}\frac{(-1)^n}{n!}\int_0^\beta \mathrm{d}\tau_1 \ldots \int_0^\beta \mathrm{d}\tau_n T[\mathsf{H}_I(\tau_1)\cdots \mathsf{H}_I(\tau_n)].$$

Note that T is now the ordering with respect to τ (or imaginary time). Finally, upon substitution of this last expression into the grand canonical partition sum

$$Z_\Omega = \mathrm{Tr}\left(\mathrm{e}^{-\beta\mathsf{K}_0}\mathsf{S}(\beta)\right)$$

we obtain a perturbative expansion of the partition sum in analogy with that of the evolution operator U. Before we can apply this formalism to the computation of Z_Ω we need to analyze the finite temperature versions of the time-ordered Green functions and Wick's theorem.

Finite temperature Green functions and Wick's theorem In analogy with the finite density Green function we define the finite temperature version of $G(\mathbf{x}_1, t_1, \mathbf{x}_2, t_2)$ as

$$G^\beta(\mathbf{x}_1, t_1, \mathbf{x}_2, t_2) = \frac{\mathrm{Tr}(\rho_0 T[\Psi^+(\mathbf{x}_1, t_2)\Psi(\mathbf{x}_2, t_1)])}{\mathrm{Tr}(\rho_0)}$$

or, in momentum space

$$\tilde{G}_{\mathbf{k}_1,\mathbf{k}_2}(\tau_1, \tau_2, \beta) = \frac{\mathrm{Tr}[\rho_0(\beta)T[\mathsf{a}^+_{\mathbf{k}_1}(\tau_1)\mathsf{a}_{\mathbf{k}_2}(\tau_2)]]}{\mathrm{Tr}\ \rho_0(\beta)}.$$

In analogy with the finite density case we have that the only non-vanishing matrix elements are between states of the same quantum numbers, i.e.

$$\mathrm{Tr}[\rho_o(\beta)\mathsf{a}^+_{\mathbf{k}}\mathsf{a}_{\mathbf{k}'}] = \mathrm{Tr}[\rho_o(\beta)\mathsf{a}_{\mathbf{k}}\mathsf{a}^+_{\mathbf{k}'}] = 0$$

unless $\mathbf{k} = \mathbf{k}'$. On the other hand $\mathsf{a}^+_{\mathbf{k}}\mathsf{a}_{\mathbf{k}} = \mathsf{n}(\mathbf{k})$ is the number operator at momentum $\mathbf{k}$. Thus

$$\tilde{G}_{\mathbf{k}_1,\mathbf{k}_2}(\tau_1, \tau_2; \beta) = (2\pi)^3\delta(\mathbf{k}_1 - \mathbf{k}_2)\tilde{G}_{\mathbf{k}}(\tau; \beta)$$

where $\mathbf{k} = \mathbf{k}_1$ and

$$\tilde{G}_{\mathbf{k}}(\tau;\beta) = e^{+\tau(\epsilon(\mathbf{k})-\mu n(\mathbf{k}))}[\theta(\tau)n_\beta(\mathbf{k}) + \theta(-\tau)(1+n_\beta(\mathbf{k}))]$$

$$n_\beta(\mathbf{k}) = \left(\frac{1}{e^{\beta(\epsilon(\mathbf{k})-\mu n(\mathbf{k})}-1}\right), \quad -\beta \le \tau = \tau_1 - \tau_2 \le \beta.$$

Finally we need to reconsider Wick's theorem at finite temperature. It turns out that there is no finite temperature version of Wick's theorem for (imaginary) time-ordered products. However, there is a weaker result which is sufficient for our purpose.

Theorem 9.4 If $\mathsf{A}_i(\tau_i) \in \{\mathsf{a}_{\mathbf{k}_i}(\tau_i), \mathsf{a}^+_{\mathbf{k}_i}(\tau_i), \mathsf{b}_{\mathbf{k}_i}(\tau_i), \mathsf{b}^+_{\mathbf{k}_i}(\tau_i)\}$, then

$$\frac{1}{Z_0}\mathrm{Tr}(\rho_0 T(\mathsf{A}_1(\tau_1)\cdots \mathsf{A}_n(\tau_n))) = \text{ sum over all completely contracted terms,}$$

where each contraction is replaced by a temperature-dependent Green function.

Proof. The proof of this theorem proceeds along the same lines as that given for the zero temperature version in Section 9.3. we will therefore leave it to the interested reader. □

Let us now illustrate our formalism by computation of the first-order correction to the grand canonical potential in the presence of a two-body interaction potential $V(|\mathbf{x}_i - \mathbf{x}_j|)$. We write

$$Z_\Omega = Z_0 + Z_1$$

where $Z_0 = \mathrm{Tr}(\rho_0)$ and

$$Z_1 = -\int_0^\beta d\tau\, \mathrm{Tr}(\rho_0\, \mathsf{H}_I(\tau)).$$

For a two-body interaction potential $V(|\mathbf{x}_i - \mathbf{x}_j|)$ we have already determined the interaction Hamiltonian in Section 9.2. In the momentum space representation this becomes for bosons

$$\mathsf{H}_I(\tau) = \frac{1}{2}\int \frac{d^3k_1}{(2\pi)^3}\frac{d^3k_2}{(2\pi)^3}\frac{d^3q}{(2\pi)^3} \times \tilde{V}(|\mathbf{q}|)\mathsf{a}^+(\mathbf{k}_1+\mathbf{q},\tau)\mathsf{a}^+(\mathbf{k}_2-\mathbf{q},\tau)\mathsf{a}(\mathbf{k}_1,\tau)\mathsf{a}(\mathbf{k}_2,\tau).$$

The two Feynman diagrams contributing to this trace are given in Figure 9.7. Applying the Feynman rules we then end up with

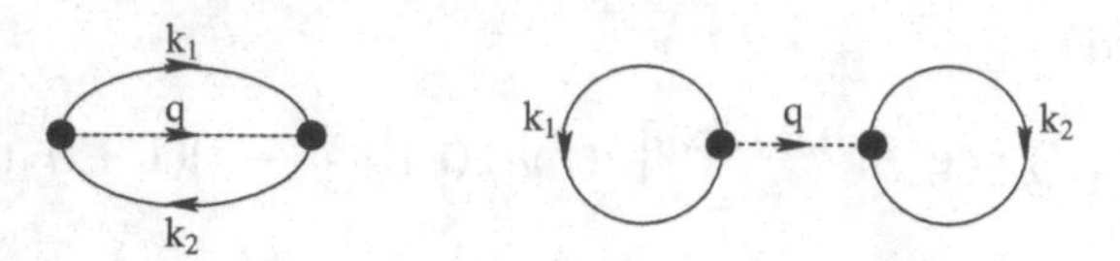

Figure 9.7 Feynman diagrams for the first-order correction to the grand canonical potential.

$$
\begin{aligned}
Z_1 = -Z_0\beta \int & \frac{d^3k_1}{(2\pi)^3}\frac{d^3k_2}{(2\pi)^3}\frac{d^3q}{(2\pi)^3}\tilde{V}(|\mathbf{q}|) \\
& \times \left\{(2\pi)^3\delta^3(\mathbf{k}_1+\mathbf{q}-\mathbf{k}_2)(2\pi)^3\delta^3(\mathbf{k}_2-\mathbf{q}-\mathbf{k}_1)n(\mathbf{k}_2)n(\mathbf{k}_1)\right. \\
& \left. + (2\pi)^3\delta^3(\mathbf{q})(2\pi)^3\delta^3(-\mathbf{q})n(\mathbf{k}_1)n(\mathbf{k}_2)\right\}.
\end{aligned}
$$

Using the integral representation to regularize the square of the delta functions we end up with

$$
Z_1 = -Z_0(\beta V)\int \frac{d^3k_1}{(2\pi)^3}\frac{d^3k_2}{(2\pi)^3}\{\tilde{V}(|\mathbf{k}_2-\mathbf{k}_1|)+\tilde{V}(0)\}n(\mathbf{k}_2)n(\mathbf{k}_1)\,.
$$

For Fermi–Dirac particles the only difference is that the sign in front of $\tilde{V}(0)$ changes. In particular, for a delta function potential ($\tilde{V} = const.$) the first-order correction vanishes identically. The physical interpretation of this sign is clear: due to the exclusion principle two spin particles cannot sit on top of each other. Therefore this configuration does not contribute to the partition sum.

In closing this section we should recall that the perturbation expansion described here is meaningful only if the interaction potential $V(|\mathbf{y}_i - \mathbf{y}_j|)$ is small. If this condition is not met any truncation of the infinite powers series expansion in H_I will in general not produce a reliable prediction. Furthermore, even when all terms are summed up the power series expansion may not converge. In this case one has to invent alternative approximation schemes. For example, if the potential is short-ranged, then a quantum cluster expansion (see Chapter 4) may be applicable.

9.9 Relativistic effects

We have seen in Chapter 7 that in the statistical mechanical description of massless particles such as photons one has to replace the non-relativistic dynamics by relativistic dynamics. Furthermore, massive fermions have to be treated relativistically, even at low temperature, if the density is sufficiently high, as a result of Pauli's exclusion principle. This occurs, for example, for sufficiently massive white dwarf stars as we have seen in Chapter 8.

In this section we briefly discuss how this affects the quantum field theoretic description of the system. Concretely this means that we have to replace the non-relativistic dispersion relation $\epsilon(\mathbf{p}) = \mathbf{p}^2/2m$ with the relativistic dispersion relation, $\epsilon^2(\mathbf{p}) = c^2(\mathbf{p}^2 + m^2c^2)$, which, in the case of massless particles, becomes $\epsilon(\mathbf{p}) = c^2\mathbf{p}^2$. Let us consider the case of non-interacting bosons and discuss the expansion of the field operator $\Psi(\mathbf{x}, t)$. The key observation is then that due to the presence of the square root in the dispersion relation, there is an ambiguity in the sign of the energy, $\epsilon(\mathbf{p}) = \pm c\sqrt{\mathbf{p}^2 + m^2c^2}$. It turns out that both signs have a physical interpretation, the positive sign as annihilation of a state with energy ϵ and the negative sign as creation of a state with energy $-\epsilon$. This suggests the expansion of $\Psi(\mathbf{x}, t)$ in the infinite volume limit, as

$$\begin{aligned}\Psi_\alpha(\mathbf{x}, t) &= \frac{1}{(2\pi)^4}\int \frac{\mathrm{d}^3k}{2\omega(\mathbf{k})}\left\{\mathrm{c}^\dagger(\mathbf{k})\,\mathrm{e}^{ik\cdot x} + \mathrm{c}(\mathbf{k})\,\mathrm{e}^{-ik\cdot x}\right\}\\ &\equiv \Psi_+(\mathbf{x}, t) + \Psi_-(\mathbf{x}, t)\\ \Psi^\dagger(\mathbf{x}, t) &= \Psi(\mathbf{x}, t)\end{aligned}$$

where $k \cdot x \equiv \mathbf{k}\cdot\mathbf{x} - \omega t$ as in Section 9.5.1 and the factor $2\omega(\mathbf{k})$ in the integral measure is to ensure relativistic covariance. Note the formal analogy with the expansion of the field operator at finite density in terms of particles and holes. Indeed if we set $\mid\Omega\rangle = \mid 0\rangle$, with $\Psi_-(\mathbf{x}, t)\mid 0\rangle = 0$, the vacuum for a relativistic quantum field theory can be thought of as a "Fermi sea" in which all negative energy states are filled. An immediate consequence is then that the finite density version of Wick's theorem stated in Section 9.7 applies without modification to the relativistic case.

Another important modification in a relativistic quantum field theory is that instantaneous two-body interactions described in terms of a potential $V(|\mathbf{y}_i - \mathbf{y}_j|)$ are not consistent with relativistic covariance, since the notion of simultaneity does not exist for spacially separated points. Thus, in any relativistic quantum field theory all interactions have to take place locally at a single point of interaction.

To close this chapter let us make a comment about higher-order effects in quantum field theory. In generic calculations in which interactions are included perturbatively one encounters Feynman diagrams involving closed loops (see e.g. Figure 9.6). Such loops involve integrals over momenta which are typically divergent. Mathematically there is a well-defined procedure to deal with such divergencies in which the infinities are absorbed in a redefinition of the fields and coupling constants parametrizing the interactions. This is known as *renormalization*. The physical phenomenon behind renormalization is that interactions "polarize" the vacuum in a similar fashion as is known, for instance, in the classical theory of electromagnetism in a dielectric medium where the fields are "shielded" as a result of polarization. In analogy the observed couplings and fields in interacting quantum field theory are the renormalized ones whereas the "bare" couplings which are infinite are not physical. They are merely mathematical constructs to define the theory. We will come back to to this point in Chapter 13.

Problems

Problem 9.1 Show that if $\Psi(\mathbf{x})$ and $\Psi^\dagger(\mathbf{x})$ satisfy anticommutation relations, then for a Hamiltonian, H, of the form given in Section 9.1.2 and a given state, $\mid E_N, N\rangle$ with

$$\mathsf{H} \mid E_N, N\rangle = E_N \mid E_N, N\rangle\,,$$

the wave function $\Psi(\mathbf{x}_1, \ldots, \mathbf{x}_N) = 1/\sqrt{N!}\langle 0 \mid \Psi(\mathbf{x}_1)\ldots\Psi(\mathbf{x}_N) \mid E_N, N\rangle$ satisfies the Schrödinger equation

$$\left(\sum_{i=1}^{N}\left(-\frac{\hbar^2}{2m}\nabla_i^2\right) + \sum_{i<j} V(|\mathbf{x}_i - \mathbf{x}_j)\right)\Psi(\mathbf{x}_1, \ldots, \mathbf{x}_N) = E_N\Psi(\mathbf{x}_1, \ldots, \mathbf{x}_N)\,.$$

Problem 9.2 Show that the evolution operator in the Schrödinger picture

$$\mathsf{U}_S(t_2, t_1) \equiv \exp\left(-\frac{\mathrm{i}(t_2 - t_1)}{\hbar}\mathsf{H}\right),$$

and in the interaction picture, $\mathsf{U}(t_2, t_1)$ defined in the text are related by $\mathsf{U}_S(t_2, t_1) = \mathsf{U}_0(t_2, t_1)\mathsf{U}(t_2, t_1)$.

Problem 9.3 Show that if $:\mathsf{A}_i:(t_i)$ be bosonic normal-ordered local products of operators, then

$$\langle 0 \mid T\left(:\mathsf{A}_1:(t_1)\cdots:\mathsf{A}_n:(t_n)\right) \mid 0\rangle$$

is given by the product of all possible expectation values of time-ordered products of pairs of the fundamental operators appearing in the normal-ordered products $:\mathsf{A}_i:(t_i)$ with the restriction that no pairings of two operators within the same normal-ordered monomial $:\mathsf{A}_i:(t_i)$ appear. This then proves Corollary 9.3.

Problem 9.4 Express the time-ordered three-point function $T(\mathsf{A}_1(t_1)\mathsf{A}_2(t_2)\mathsf{A}_3(t_3))$ in terms of θ functions.

Problem 9.5 Draw all Feynman diagrams for the second-order contribution to the helium scattering problem treated in Section 9.6. At this order the evolution operator is given by

$$U^{(2)}(T,-T) = \left(\frac{\mathrm{i}}{\hbar}\right)^2 \frac{1}{2!}\int_{-T}^{T} \mathrm{d}\tau_1 \int_{-T}^{T} \mathrm{d}\tau_2 T(H_I(\tau_1)H_I(\tau_2)).$$

Problem 9.6 As an example of an interacting fermionic system, consider the two-dimensional lattice Hamiltonian (Hubbard model)

$$\mathsf{H} = -\sum_{s,\langle i,j\rangle} t_{ij}\mathsf{b}^\dagger_{is}\mathsf{b}_{js} + U\sum_i \mathsf{n}_{i\uparrow}\mathsf{n}_{i\downarrow}$$

where $s = \uparrow, \downarrow$, and $\mathsf{b}^\dagger_{is}$ creates an electron with spin s. The symbol $\mathsf{n}_{is} \equiv \mathsf{b}^\dagger_{is}\mathsf{b}_{js}$ is the occupation number operator. The symbol $\langle i, j\rangle$ indicates summation over nearest neighbors and t_{ij} is given by

$$t_{ij} = \begin{cases} t & i, j \text{ nearest neighbours} \\ 0 & \text{otherwise} \end{cases}$$

(1) Diagonalize H for weak coupling ($U = 0$).
(2) In the strong coupling limit, ($t = 0$) the electrons have no dynamics and for $U > 0$ the ground state will have no double occupancies. In particular, at "half filling" the ground state will contain exactly one electron per site. Express the interaction term in terms of the spin operators

$$\mathbf{S}_i \equiv \frac{1}{2}\sum_{ss'} \mathsf{b}^\dagger_{is}\boldsymbol{\sigma}_{ss'}\mathsf{b}^\dagger_{is'}$$

where $\boldsymbol{\sigma} = (\sigma_1, \sigma_2, \sigma_3)$ are the Pauli matrices.

Problem 9.7 Show that for $\mathsf{A}_i(\tau_i) \in \{\mathsf{a}_{\mathbf{k}_i}(\tau_i), \mathsf{a}^+_{\mathbf{k}_i}(\tau_i), \mathsf{b}_{\mathbf{k}_i}(\tau_i), \mathsf{b}^+_{\mathbf{k}_i}(\tau_i)\}$, the thermal n-point function

$$\frac{1}{Z_0}\mathrm{Tr}(\rho_0 T(\mathsf{A}_1(\tau_1)\cdots\mathsf{A}_n(\tau_n))),$$

is given by the sum over all completely contracted terms, where each contraction is replaced by a temperature-dependent Green function (Theorem 9.4).

Problem 9.8 Compute the first perturbative correction to the partition function for fermions with a two-body interaction potential $V_{ij} = V(|\mathbf{y}_i - \mathbf{y}_j|)$.

Further reading

A comprehensive text which is standard in solid state physics and accessible to advanced undergraduate students is G. D. Mahan, *Many-Particle Physics*, Plenum (1990). Another very good book with a thorough discussion of Green functions and many applications including plasma physics is A. L. Fetter and J. D. Walecka, *Quantum Theory of Many-Particle Systems*, McGraw-Hill (1971). Many applications of Feynman diagrams to condensed matter physics can be found in A. A. Abrikosov, L. P. Gorkov, and I. E. Dzyaloshinskii, *Methods of Quantum Field Theory in Statistical Physics*, Dover (1975). Further classic texts which include discussions of the renormalization group are L. E. Reichl, *A Modern Course in Statistical Physics*, Edward Arnold (1980), and at a more advanced level, M. Le Bellac, *Quantum and Statistical Field Theory*, Oxford University Press (1991); J. W. Negele and H. Orland, *Quantum Many-Particle Systems*, Perseus, (1998) and G. Parisi, *Statistical Field Theory*, Addison-Wesley (1988). A thorough discussion of relativistic quantum field theory with a detailed discussion of renormalization and applications to particle physics can be found in C. Itzykson and J. B. Zuber, *Quantum Field Theory*, McGraw-Hill (1985) and in E. Peskin and D. V. Shroeder, *An Introduction to Quantum Field Theory*, Addison-Wesley (1995).

10
Superfluidity

The goal of this chapter will be to briefly describe the remarkable properties of helium at low temperatures. After stating some of these properties we will see how they can be understood in terms of the phenomenon of Bose–Einstein condensation described in Chapter 7. We will give the main argument in two different formulations, once using the quasi-particle method of Bogoliubov, and then using a Green function approach.

We start with some experimental facts. Helium is a remarkable element. It was predicted to exist from observations of the Sun before it was found on Earth. It is the only element which remains a liquid at zero temperature and atmospheric pressure. Experimentally the phase diagram of ^{4}He is shown in Figure 10.1. Helium I is a normal fluid and has a normal gas–liquid critical point. Helium II is a mixture of a normal fluid and a superfluid. The superfluid is characterized by the vanishing of its viscosity. Helium I and helium II are separated by a line known as the λ-transition line. At $T_\lambda = 2.18$ K, $P_\lambda = 2.29$ Pa, helium I, helium II, and helium gas coexist. The specific heat of liquid helium along the vapor transition line forms a logarithmic discontinuity shown in Figure 10.2. The form of this diagram resembles the Greek letter λ and is the reason for calling the transition a λ-transition.

The lack of viscosity of helium II leads to some remarkable experimental consequences, one of which we briefly describe. Let two containers A and B be linked by a thin capillary through which only a fluid with zero (or very low) viscosity can pass freely. Originally T_{A}, T_{B} are both temperatures below T_λ. If the temperature of A is raised, thereby decreasing the amount of helium II in it, then tank B now contains a larger proportion of helium II. As a result, more liquid helium will flow from B to A than from A to B leading to a difference in the level of liquids between the two tanks.

The excitation spectrum $E(\mathbf{p})$ of helium II can be measured experimentally through elastic neutron scattering. It is found to consist of two parts, the *phonon*

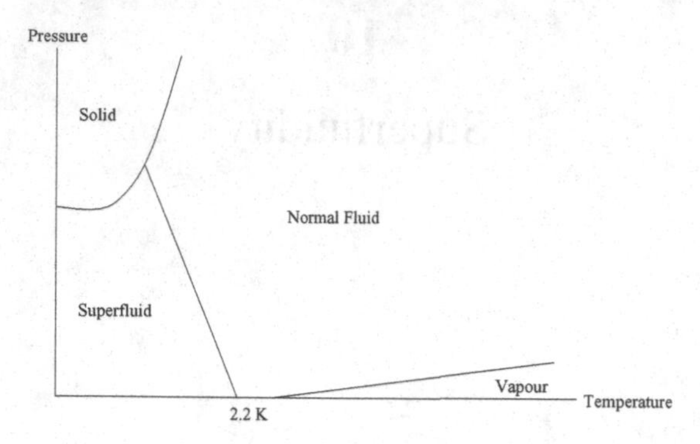

Figure 10.1 The phase structure of ^{4}He at low temperature.

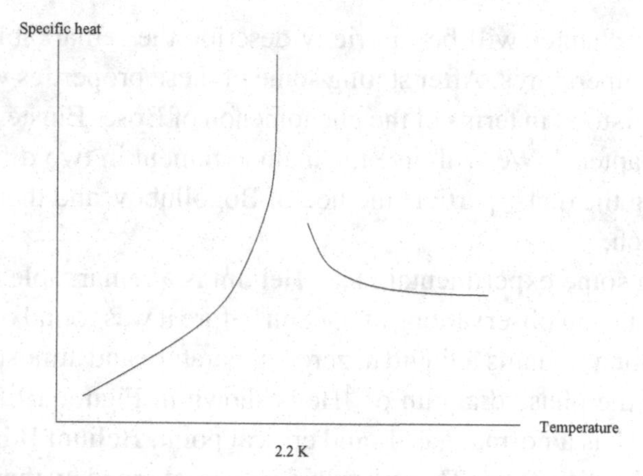

Figure 10.2 The specific heat C of helium as a function of temperature.

region

$$E(\mathbf{p}) = c|\mathbf{p}|, \quad \text{when } |\mathbf{p}| \ll |\mathbf{p}_0|,$$

and the *roton* region:

$$E(\mathbf{p}) = \Delta + \frac{1}{2\mu}|\mathbf{p} - \mathbf{p}_0|^2, \quad \text{when } |\mathbf{p}| \sim |\mathbf{p}_0|,$$

where $c = 226$ m/s is the velocity of sound, $\Delta/k_B = 9$ K are the roton parameters, and $\mu = 0.25 m_{He}$. There is another velocity parameter known as the critical velocity v_0. It is only when helium II moves with velocity greater than v_0 that viscous effects arise. At low temperature the roton excitations are damped by the Boltzmann factor $\mathrm{e}^{-\Delta/k_B T}$.

Our theoretical goal is to understand qualitatively the features of the excitation spectrum described. We will show how the linear phonon region can be theoretically understood. We will also show how such a linear phonon spectrum can explain the lack of viscosity of helium II. Our treatment will be qualitative. We will not explain the roton part of the spectrum but merely state that such excitations represent collective vortex excitations in helium II.

Before we turn to a systematic quantum field theory approach to the problem we briefly state the physical picture of superfluidity that we want to develop. We first note that the fact that helium is a liquid at low temperature can be understood to be a consequence of the smallness of the inter-atomic potential between helium atoms and the low mass of helium. Let us explain how this happens. Experimentally the minimum energy of the interaction potential for helium is -9 K. The small mass of helium means the energy uncertainty of helium atoms ΔE is large and is of order

$$\Delta E \simeq \frac{1}{2m}\left(\frac{\hbar}{\Delta x}\right)^2 \simeq 10\ \mathrm{K}$$

taking $\Delta x \simeq 0.5$ nm. Thus ΔE is comparable to the depth of the potential well and therefore localization of helium atoms to form a solid is impossible. Since localization of helium atoms is not possible the wave functions of individual atoms overlaps and the symmetry features of the wave function become important. Thus the quantum phenomenon of Bose–Einstein condensation, which we discussed in Chapter 7, is expected to play an important role. Our qualitative reason for helium remaining a liquid at low temperatures also explains why hydrogen, although lighter than helium and hence with greater energy uncertainty, does not remain a liquid at low temperatures. This is because the attractive force between hydrogen atoms is much stronger than the attractive force between helium atoms. This allows hydrogen to form a solid at low temperature with a consequent loss of quantum coherence in its ground state wave function. Once we accept that ^{4}He is a liquid at low temperatures the dispersion relations for low energies can be inferred from general considerations: for small momenta these excitations correspond to long wave oscillations, i.e. sound waves. This explains the existence of the "phonon region". The roton region cannot be explained quite so easily. We now turn to a quantum field theory formulation of the problem.

10.1 Quantum field theory formulation

Let us start by writing down the operator describing helium atoms in statistical mechanics. We write the momentum space form for this Hamiltonian and assume that the interatomic interactions are described by a two-body potential $\tilde{V}(\mathbf{q})$

$$\mathsf{H} = \sum_{\mathbf{k}} E_0(\mathbf{k})\mathsf{a}^\dagger_{\mathbf{k}}\mathsf{a}_{\mathbf{k}} + \frac{1}{2V}\sum_{\mathbf{k}_1,\mathbf{k}_2,\mathbf{q}} \tilde{V}(\mathbf{q})\mathsf{a}^\dagger_{\mathbf{k}_1+\mathbf{q}}\mathsf{a}^\dagger_{\mathbf{k}_2-\mathbf{q}}\mathsf{a}_{\mathbf{k}_1}\mathsf{a}_{\mathbf{k}_2}$$

where $E_0(\mathbf{k}) = (|\mathbf{k}|^2/2m) - \mu$, and μ is the chemical potential. To study the statistical mechanics properties of the system we need to construct the grand canonical partition function,

$$Z_\Omega = \mathrm{Tr}[\mathrm{e}^{-\beta \mathsf{K}}].$$

At low temperature, states with low energy value become dominant. These are expected to be states with low values of momentum. Let us consider the system close to $T \simeq 0\,\mathrm{K}$. We can then assume that the state of lowest energy corresponds to atoms of low momentum with a sizable fraction of molecules in the zero momenta state, leading to Bose–Einstein condensation. Thus if the system has on average N atoms then a significant number N_0 of the atoms are in the lowest energy state. More precisely we suppose that the ratio N_0/N converges to a constant in the limit $N \to \infty$. We can implement these observations as follows. Let us suppose that $|C; N, N_0>$ is a superfluid state with a total of N helium atoms, N_0 of which are in the zero momentum plane wave state (which we take to be the lowest energy state possible for a helium atom). If $\mathsf{a}_0^\dagger$ and a_0 are creation and destruction operators for a state of zero momentum, we have

$$\begin{aligned} \mathsf{a}_0|C; N, N_0> &= \sqrt{N_0}|C; N, N_0 - 1> \\ \mathsf{a}_0^\dagger|C; N, N_0> &= \sqrt{N_0 + 1}|C; N, N_0 + 1>, \end{aligned}$$

so that

$$\begin{aligned} \mathsf{a}_0^\dagger \mathsf{a}_0|C; N, N_0> &= N_0|C; N, N_0> \\ \mathsf{a}_0 \mathsf{a}_0^\dagger|C; N, N_0> &= (N_0 + 1)|C; N, N_0>. \end{aligned}$$

For large N_0 we can approximate $N_0 + 1$ by N_0 so that on the state $|C; N, N_0>$ we can replace both the operators $\mathsf{a}_0\mathsf{a}_0^\dagger$ and $\mathsf{a}_0^\dagger\mathsf{a}_0$ by a single c-number, N_0. This introduces a natural large parameter in the problem. In particular the parameter N_0 will be present in the Hamiltonian operator when it is restricted to act on the superfluid state $|C; N, N_0>$ where it can be utilized to introduce an appropriate approximation scheme. This is the quasi-particle approach of Bogoliubov.

10.2 The quasi-particle approach

We ask what does the number operator N look like in the state $|C; N, N_0>$. We have $\mathsf{N}|C; N, N_0> = N|C; N, N_0>$, where $\mathsf{N} = \sum \mathsf{a}_\mathbf{k}^\dagger \mathsf{a}_\mathbf{k}$. We also have assumed that $\mathsf{a}_0^\dagger\mathsf{a}_0|C; N, N_0> = N_0|C; N, N_0>$ where N_0 is the number of zero momentum states in $|C>$. Writing

$$\mathsf{N} = \mathsf{a}_0^\dagger \mathsf{a}_0 + \sum_{\mathbf{k}\neq 0} \mathsf{a}_{\mathbf{k}}^\dagger \mathsf{a}_{\mathbf{k}}.$$

We have for $|C; N, N_0>$

$$\mathsf{N} = N_0 + \sum_{\mathbf{k}\neq 0} \mathsf{a}_{\mathbf{k}}^\dagger \mathsf{a}_{\mathbf{k}}$$

and

$$\mathsf{N}^2 = \left(N_0 + \sum_{\mathbf{k}\neq 0} \mathsf{a}_{\mathbf{k}}^\dagger \mathsf{a}_{\mathbf{k}} \right)^2.$$

Neglecting terms of order N_0^0 we have

$$\mathsf{N}^2 \simeq N_0^2 + 2N_0 \sum_{\mathbf{k}\neq 0} \mathsf{a}_{\mathbf{k}}^\dagger \mathsf{a}_{\mathbf{k}}$$

as a constraint on the operator $\sum_{k\neq 0} \mathsf{a}_{\mathbf{k}}^\dagger \mathsf{a}_{\mathbf{k}}$ acting on the state $|C; N, N_0>$. We next examine the interaction part of H, that is

$$\mathsf{H}_I = \frac{1}{2V} \sum_{\mathbf{k}_1,\mathbf{k}_2,\mathbf{q}} \tilde{V}(\mathbf{q}) \mathsf{a}_{\mathbf{k}_1+\mathbf{q}}^\dagger \mathsf{a}_{\mathbf{k}_2-\mathbf{q}}^\dagger \mathsf{a}_{\mathbf{k}_1} \mathsf{a}_{\mathbf{k}_2}$$

when restricted to $|C>$.

We proceed to isolate terms in H_I which contain either $\mathsf{a}_0^\dagger$ or a_0 and to replace these operators by $\sqrt{N_0}$. We order terms in H_I according to the number of a_0, $\mathsf{a}_0^\dagger$ factors they contain. When all four operators in H_I have zero momentum, we have the term

$$\begin{aligned}
\mathsf{H}_I^{(0)} &= \frac{1}{2V} \tilde{V}(0) \mathsf{a}_0^\dagger \mathsf{a}_0^\dagger \mathsf{a}_0 \mathsf{a}_0 \\
&= \frac{1}{2V} \tilde{V}(0) N_0^2 \\
&\simeq \frac{1}{2V} \tilde{V}(0) \left[\mathsf{N}^2 - 2N_0 \sum_{\mathbf{k}\neq 0} \mathsf{a}_{\mathbf{k}}^\dagger \mathsf{a}_{\mathbf{k}} \right]
\end{aligned}$$

where we have used the equation relating N to N_0. The next term is of order N_0 and is the part of H_I containing two operators (carrying zero momentum). There are six ways in which this can happen. These are displayed with the momentum variables which are set to zero as shown

$$\mathbf{k}_1 + \mathbf{q} = \mathbf{k}_2 - \mathbf{q} = 0: \quad \frac{N_0}{2V} \sum_{\mathbf{q}\neq 0} \tilde{V}(\mathbf{q}) \mathsf{a}_{-\mathbf{q}} \mathsf{a}_{\mathbf{q}}$$

$$\mathbf{k}_1 + \mathbf{q} = \mathbf{k}_1 = 0: \quad \frac{N_0}{2V} \sum_{\mathbf{k}_2\neq 0} \tilde{V}(0) \mathsf{a}^\dagger_{\mathbf{k}_2} \mathsf{a}_{\mathbf{k}_2}$$

$$\mathbf{k}_1 + \mathbf{q} = \mathbf{k}_2 = 0: \quad \frac{N_0}{2V} \sum_{\mathbf{q}\neq 0} \tilde{V}(\mathbf{q}) \mathsf{a}^\dagger_{-\mathbf{q}} \mathsf{a}_{-\mathbf{q}}$$

$$\mathbf{k}_2 - \mathbf{q} = \mathbf{k}_1 = 0: \quad \frac{N_0}{2V} \sum_{\mathbf{q}\neq 0} \tilde{V}(\mathbf{q}) \mathsf{a}^\dagger_{\mathbf{q}} \mathsf{a}_{\mathbf{q}}$$

$$\mathbf{k}_2 - \mathbf{q} = \mathbf{k}_2 = 0: \quad \frac{N_0}{2V} \sum_{\mathbf{k}_1\neq 0} \tilde{V}(0) \mathsf{a}^\dagger_{\mathbf{k}_1} \mathsf{a}_{\mathbf{k}_1}$$

$$\mathbf{k}_1 = \mathbf{k}_2 = 0: \quad \frac{N_0}{2V} \sum_{\mathbf{q}\neq 0} \tilde{V}(\mathbf{q}) \mathsf{a}^\dagger_{\mathbf{q}} \mathsf{a}^\dagger_{-\mathbf{q}}.$$

Now we make an additional assumption that, for $|\mathbf{k}|$ small, $\tilde{V}(\mathbf{k}) \simeq \tilde{V}(0)$. Since at low temperature we expect only small momenta excitations to be important we replace $\tilde{V}(\mathbf{k})$ by $\tilde{V}(0)$ in H_I. Therefore, on the state $|C; N, N_0 >$, the interacting Hamiltonian, keeping terms of $O(N_0)$ is given by

$$\begin{aligned} \mathsf{H}_I \simeq & \frac{1}{2V} \tilde{V}(0) \left[N^2 - 2N_0 \sum_{k\neq 0} \mathsf{a}^\dagger_{\mathbf{k}} \mathsf{a}_{\mathbf{k}} \right] \\ & + \frac{N_0}{2V} \sum_{\mathbf{k}\neq 0} \tilde{V}(0) [\mathsf{a}^\dagger_{\mathbf{k}} \mathsf{a}^\dagger_{-\mathbf{k}} + \mathsf{a}_{\mathbf{k}} \mathsf{a}_{-\mathbf{k}}] \\ & + \frac{\tilde{V}(0)}{2V} 4N_0 \sum_{k\neq 0} \mathsf{a}^\dagger_{\mathbf{k}} \mathsf{a}_{\mathbf{k}} \equiv \mathsf{H}_I^B, \end{aligned}$$

and the total Hamiltonian, H, can be approximated by the *Bogoliubov Hamiltonian* H^B given by

$$\mathsf{H}^B = \sum_{\mathbf{k}} \left(\frac{|\mathbf{k}|^2}{2m} - \mu \right) \mathsf{a}^\dagger_{\mathbf{k}} \mathsf{a}_{\mathbf{k}} + H_I^B.$$

Observe that H originally had the property $[\mathsf{H}, \mathsf{N}] = 0$ so the system had a fixed number of particles and a well-defined energy eigenvalue. But now H has an interaction term H_I^B which contains the operator $[\mathsf{a}^\dagger_{\mathbf{k}} \mathsf{a}^\dagger_{-\mathbf{k}} + \mathsf{a}_{\mathbf{k}} \mathsf{a}_{-\mathbf{k}}]$. This operator changes the particle number. Thus $[\mathsf{H}_I^B, \mathsf{N}] \neq 0$. Since the energy states of H^B do not have well-defined particle numbers we must then set $\mu = 0$. The effective Hamiltonian we use is then, dropping non-operator parts of H_I^B,

$$\mathsf{H}^B = \sum_{\mathbf{k}\neq 0}\left(\frac{|\mathbf{k}|^2}{2m} + \frac{\tilde{V}(0)}{V}N_0\right)\mathsf{a}_{\mathbf{k}}^\dagger \mathsf{a}_{\mathbf{k}} + \frac{\tilde{V}(0)N_0}{2V}\sum_{\mathbf{k}\neq 0}[\mathsf{a}_{\mathbf{k}}^\dagger \mathsf{a}_{-\mathbf{k}}^\dagger + \mathsf{a}_{\mathbf{k}}\mathsf{a}_{-\mathbf{k}}].$$

We turn next to the problem of determining the energy eigenvalues of this Hamiltonian. This we do by using the method of the *Bogoliubov–Valatin transform*. The idea is this: H^B is a quadratic function of the operators $\mathsf{a}_{\mathbf{k}}$ and $\mathsf{a}_{\mathbf{k}}^\dagger$. By taking appropriate linear combinations of these operators we can form new operators $\hat{b}_k$ and $\hat{b}_{\mathbf{k}}^\dagger$ which "diagonalize" H_B, i.e. which lead to

$$\mathsf{H}^B = \sum_{\mathbf{k}\neq 0}\mathcal{E}(\mathbf{k})\mathsf{b}_{\mathbf{k}}^\dagger \mathsf{b}_{\mathbf{k}}.$$

The function $\mathcal{E}(\mathbf{k})$ will then determine the different excitations of the system while $\mathsf{b}_{\mathbf{k}}, \mathsf{b}_{\mathbf{k}}^\dagger$ will be destruction and creation operators for these excitations or "quasi-particles", provided they satisfy the commutation rules

$$[\mathsf{b}_{\mathbf{k}_1}, \mathsf{b}_{\mathbf{k}_2}^\dagger] = \delta_{\mathbf{k}_1,\mathbf{k}_2}.$$

We write

$$\mathsf{b}_{\mathbf{k}} = \alpha(\mathbf{k})\mathsf{a}_{\mathbf{k}} - \beta(\mathbf{k})\mathsf{a}_{-\mathbf{k}}^\dagger$$
$$\mathsf{b}_{\mathbf{k}}^\dagger = \alpha(\mathbf{k})\mathsf{a}_{\mathbf{k}}^\dagger - \beta(\mathbf{k})\mathsf{a}_{-\mathbf{k}}$$

where $\alpha(\mathbf{k})$, $\beta(\mathbf{k})$, are real valued functions of $|\mathbf{k}|$. These functions are called *Bogoliubov coefficients*. Note that $\mathsf{a}_{\mathbf{k}}$ acting on a state removes momentum $\mathbf{k}$ from the system while $\mathsf{a}_{-\mathbf{k}}^\dagger$ adds momentum $-\mathbf{k}$ to the system so that both terms influence momentum states in the same way. Requiring

$$[\mathsf{b}_{\mathbf{k}_1}, \mathsf{b}_{\mathbf{k}_2}^\dagger] = \delta_{\mathbf{k}_1,\mathbf{k}_2}$$

leads to the constraint

$$\alpha(\mathbf{k})^2 - \beta(\mathbf{k})^2 = 1, \qquad \text{for all } \mathbf{k}.$$

It is now easy to see that

$$\mathsf{a}_{\mathbf{k}} = \alpha(\mathbf{k})\mathsf{b}_{\mathbf{k}} + \beta(\mathbf{k})\mathsf{b}_{-\mathbf{k}}^\dagger$$
$$\mathsf{a}_{\mathbf{k}}^\dagger = \alpha(\mathbf{k})\mathsf{b}_{\mathbf{k}}^\dagger + \beta(\mathbf{k})\mathsf{b}_{-\mathbf{k}}.$$

Substituting these expressions we get

$$\begin{aligned}\mathsf{H}_B = &\sum_{\mathbf{k}\neq 0}[g(\mathbf{k})(\beta(\mathbf{k})^2 + \alpha(\mathbf{k})^2) + 4h(\mathbf{k})\alpha(\mathbf{k})\beta(\mathbf{k})]\mathsf{b}_{\mathbf{k}}^\dagger \mathsf{b}_{\mathbf{k}}\\ &+\sum_{k\neq 0}[g(\mathbf{k})\alpha(\mathbf{k})\beta(\mathbf{k}) + h(\mathbf{k})(\alpha(\mathbf{k})^2 + \beta(\mathbf{k})^2)][\mathsf{b}_{\mathbf{k}}\mathsf{b}_{-\mathbf{k}} + \mathsf{b}_{\mathbf{k}}^\dagger \mathsf{b}_{-\mathbf{k}}^\dagger]\\ &+ \quad \text{terms not involving operators}\end{aligned}$$

where

$$h(\mathbf{k}) = \frac{N_0 \tilde{V}(0)}{2V}$$
$$g(\mathbf{k}) = \frac{|\,\mathbf{k}\,|^2}{2m} + \frac{N_0 \tilde{V}(0)}{V}.$$

In order to get $\mathsf{H}_B = \sum_{k\neq 0} \mathcal{E}(\mathbf{k})\mathsf{b}_{\mathbf{k}}^{\dagger}\mathsf{b}_{\mathbf{k}}$ we must set the coefficient of the operator $\mathsf{b}_{\mathbf{k}}\mathsf{b}_{-\mathbf{k}} + \mathsf{b}_{\mathbf{k}}^{\dagger}\mathsf{b}_{-\mathbf{k}}$ equal to zero. Then we have

$$\mathsf{H}_B = \sum_{\mathbf{k}\neq 0} \mathcal{E}(k)\mathsf{b}_{\mathbf{k}}^{\dagger}\mathsf{b}_{\mathbf{k}} + \quad \text{non-operator terms.}$$

with

$$\mathcal{E}(\mathbf{k}) = g(\mathbf{k})(\beta(\mathbf{k})^2 + \alpha(\mathbf{k})^2) + 4h(\mathbf{k})\alpha(\mathbf{k})\beta(\mathbf{k})$$

while

$$\alpha(\mathbf{k})^2 - \beta(\mathbf{k})^2 = 1$$
$$g(\mathbf{k})\alpha(\mathbf{k})\beta(\mathbf{k}) + h(\mathbf{k})(\alpha(\mathbf{k})^2 + \beta(\mathbf{k})^2) = 0.$$

Our problem is to determine the unknown functions $\alpha(\mathbf{k})$, $\beta(\mathbf{k})$ in terms of the known functions $g(\mathbf{k})$, $h(\mathbf{k})$ by solving these two equations. The solutions obtained are

$$\beta(\mathbf{k})^2 = \frac{1}{2}\left\{\sqrt{\frac{g^2(\mathbf{k})}{g^2(\mathbf{k}) - 4h^2(\mathbf{k})}} - 1\right\}$$
$$\alpha(\mathbf{k})^2 = \frac{1}{2}\left\{\sqrt{\frac{g^2(\mathbf{k})}{g^2(\mathbf{k}) - 4h^2(\mathbf{k})}} + 1\right\}.$$

Thus

$$g(\mathbf{k})(\alpha(\mathbf{k})^2 + \beta(\mathbf{k})^2) = \frac{g(\mathbf{k})^2}{\sqrt{g^2(\mathbf{k}) - 4h^2(\mathbf{k})}}$$
$$\alpha(\mathbf{k})^2\beta(\mathbf{k})^2 = \frac{h^2(\mathbf{k})}{(g^2(\mathbf{k}) - 4h^2(\mathbf{k}))}.$$

To ensure the coefficient of $\mathsf{b}_{\mathbf{k}}\mathsf{b}_{-k} + \mathsf{b}_{\mathbf{k}}^{\dagger}\mathsf{b}_{-k}^{\dagger}$ is zero we must choose

$$\alpha(\mathbf{k})\beta(\mathbf{k}) = -\frac{h(\mathbf{k})}{\sqrt{g^2(\mathbf{k}) - 4h^2(\mathbf{k})}}.$$

We thus have

$$g(\mathbf{k})(\alpha(\mathbf{k})^2 + \beta(\mathbf{k})^2) + 4h(k)\alpha(\mathbf{k})\beta(\mathbf{k}) = \sqrt{g(\mathbf{k})^2 - 4h^2(\mathbf{k})}\,,$$

so that finally

$$\mathsf{H}^B = \sum_{\mathbf{k}\neq 0} \mathcal{E}(\mathbf{k})\mathsf{b}_{\mathbf{k}}^{\dagger}\mathsf{b}_{\mathbf{k}}$$

with

$$\mathcal{E} = \sqrt{\frac{\mathbf{k}^2}{2m}(\frac{\mathbf{k}^2}{2m} + \frac{2N_0\tilde{V}(0)}{V})}.$$

The energy $\mathcal{E}(\mathbf{k})$ is called the quasi-particle energy and the operators $\mathsf{b}_{\mathbf{k}}^{\dagger}$, and $\mathsf{b}_{\mathbf{k}}$ are quasi-particle creation and destruction operators. Our analysis has thus confirmed the picture presented at the beginning of this chapter and in which the excitations of a system of bosons where a large number, N_0, of particles are in the state with $\mathbf{k} = 0$ are quasi-particle excitation. Indeed, for small values of $\mathbf{p} = \mathbf{k}$ we have

$$\mathcal{E}(\mathbf{k}) \simeq \frac{|\mathbf{p}|}{m}\sqrt{\left(\frac{N_0}{V}\right)\tilde{V}(0)m}\,.$$

Observe that for $\mathcal{E}(\mathbf{k})$ to be real we must have $\tilde{V}(0) > 0$. Recall

$$\tilde{V}(\mathbf{q}) = \int \mathrm{d}^3x e^{\mathrm{i}\mathbf{q}\cdot\mathbf{x}} V(\mathbf{x}),$$

so that the condition

$$\tilde{V}(0) = \int \mathrm{d}^3x\, V(\mathbf{x}) \quad > \quad 0$$

implies that for a short range potential $V(\mathbf{x})$, which represents the interaction between the helium atoms, there is a repulsive region for $V(x)$ which must dominate the integral. Observe also that $|\mathbf{k}|/m = |\mathbf{v}|$ is a velocity, and $N_0 m/V = \rho$, is the density of the superfluid helium so that the quasi-particle energy can be written as

$$\mathcal{E}(\mathbf{p} = m\mathbf{v}) \simeq |\mathbf{v}|\sqrt{\rho\tilde{V}(0)}\,.$$

We now show that a system with such an energy spectrum represents a superfluid, i.e. a system with no friction. Friction in a system represents dissipation of energy. Consider a molecule of mass M_{A} moving in a medium. If this molecule can change its energy through collisions with the excitations of the medium, then the system has friction. We will find that a molecule of mass M_{A} and velocity $\mathbf{V}_{\mathrm{A}}$ moving through a system consisting of quasi-particles of energy $\hat{E}(\mathbf{k})$ cannot change its energy by scattering off quasi-particles if $|\mathbf{V}_{\mathrm{A}}| < |\mathbf{v}_0|$ where $|\mathbf{v}_0|$ is a critical velocity

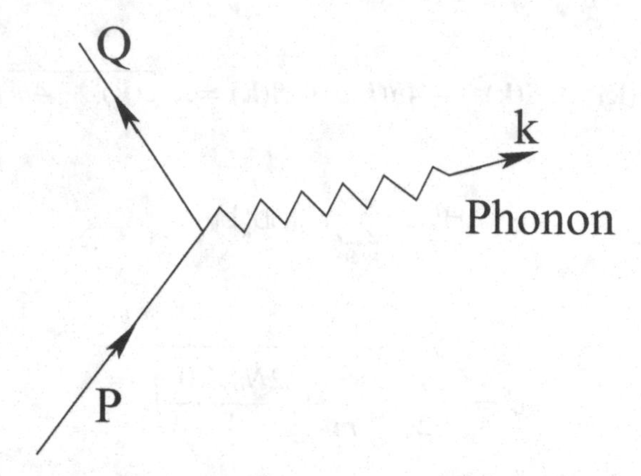

Figure 10.3 Scattering of a molecule with momentum P with a phonon.

determined by $\tilde{V}(0)$ and ρ. Thus for $|\mathbf{V}_\mathrm{A}| < |\mathbf{v}_0|$ the system of quasi-particles behaves as a frictionless superfluid.

To see this, let us consider the collision of a molecule of mass M_A and velocity $\mathbf{V}_\mathrm{A}$ with a quasi-particle at rest (see Fig. 10.3). If the final momentum of the molecule is $\mathbf{Q}_\mathrm{A}$ and that of the quasi-particle is $\mathbf{k}$, we have, from momentum conservation,

$$\mathbf{P}_\mathrm{A} = \mathbf{Q}_\mathrm{A} + \mathbf{k}\,,$$

where $\mathbf{P}_\mathrm{A} = M_\mathrm{A}\mathbf{V}_\mathrm{A}$. Conservation of energy gives on the other hand

$$\frac{|\mathbf{P}_\mathrm{A}|^2}{2M_\mathrm{A}} = \frac{|\mathbf{Q}_\mathrm{A}|^2}{2M_\mathrm{A}} + \mathcal{E}(\mathbf{k})\,.$$

Taking the square of momentum conservation law gives

$$|\mathbf{Q}_\mathrm{A}|^2 = |\mathbf{P}_\mathrm{A}|^2 + |\mathbf{k}|^2 - 2\mathbf{P}_\mathrm{A}\cdot\mathbf{k}\,.$$

If we denote by θ the angle between $\mathbf{P}$ and $\mathbf{k}$, we then have

$$|\cos\theta| = \frac{|\mathbf{P}_\mathrm{A}|^2 - |\mathbf{Q}_\mathrm{A}|^2 + |\mathbf{k}|^2}{2|\mathbf{P}_\mathrm{A}||\mathbf{k}|}\,,$$

so that, in particular,

$$\frac{|\mathbf{P}_\mathrm{A}|^2 - |\mathbf{Q}_\mathrm{A}|^2}{|\mathbf{P}_\mathrm{A}||\mathbf{k}|} \leq |\cos\theta| \leq 1.$$

Combining this with the energy conservation condition we end up with

$$\frac{2M_\mathrm{A}\mathcal{E}(\mathbf{k})}{2\mid \mathbf{P}_\mathrm{A}\mid\mid \mathbf{k}\mid} \leq 1.$$

Recalling now that

$$\mathcal{E}(\mathbf{k}) \simeq \frac{|\,\mathbf{k}\,|}{m}\sqrt{\frac{\tilde{V}(0)N_0 m}{V}}$$

we get

$$v_0 = \frac{1}{m}\sqrt{\frac{\tilde{V}(0)N_0 m}{V}} \leq \frac{|\,\mathbf{P}_{\mathrm{A}}\,|}{M_{\mathrm{A}}} = |\,\mathbf{V}_{\mathrm{A}}\,| \ .$$

Thus the process of changing energy for the molecule is not allowed if $|\mathbf{V}_{\mathrm{A}}| < v_0$ and the system of quasi-particles behaves like a superfluid.

10.3 Green function approach

We will now rederive the same result using the method of Green functions. The purpose of this section is to illustrate how the spectrum of a theory can be extracted from its Green functions.

Our starting point is the Bogoliubov Hamiltonian defined in the previous section.

$$\mathsf{H}^B = \sum_{\mathbf{k}} \hat{E}(\mathbf{k})\mathsf{a}^\dagger_{\mathbf{k}}\mathsf{a}_{\mathbf{k}} + \frac{\tilde{V}(0)N_0}{2V}\sum_{\mathbf{k}\neq 0}[\mathsf{a}^\dagger_{\mathbf{k}}\mathsf{a}^\dagger_{-\mathbf{k}} + \mathsf{a}_{\mathbf{k}}\mathsf{a}_{-\mathbf{k}}]\,,$$

with

$$\hat{E}(\mathbf{k}) = \left(\frac{\mathbf{k}^2}{2m} + \frac{\tilde{V}(0)N_0}{V}\right).$$

We can take all operators to be Heisenberg operators with time dependence governed by H^B, i.e. $\mathsf{a}_{\mathbf{k}} = \mathsf{a}_{\mathbf{k}}(t)$. The Green function for this system is defined to be

$$G_{\mathbf{k}}(t) = \langle 0|T(\mathsf{a}^\dagger_{\mathbf{k}}(t)\mathsf{a}_{\mathbf{k}}(0))|0\rangle.$$

The Green function $G_{\mathbf{k}}(t)$ contains in it, we claim, information regarding all the excitations of the system. If we determine the Green function $G_{\mathbf{k}}(t)$ then we will know what these excitations are. For this it is useful to consider its Fourier transform

$$\hat{G}_{\mathbf{k}}(w) = \int_{-\infty}^{\infty}\frac{\mathrm{d}t}{2\pi}\mathrm{e}^{-iwt}G_{\mathbf{k}}(t).$$

We will argue that the poles of $\tilde{G}_{\mathbf{k}}(w)$ in w directly represent the excitations of the system and we will show that, for operators evolving in time according to H_B, the poles of $\tilde{G}_{\mathbf{k}}(w)$ are precisely the quasi-particle excitations obtained in the last section.

Let us first proceed with our calculations of $\tilde{G}_{\mathbf{k}}(w)$. Later we will argue why the poles of $\tilde{G}_{\mathbf{k}}(w)$ represent excitations of the system. We note that

$$\frac{\partial G_{\mathbf{k}}(t)}{\partial t} = \langle 0|[\mathsf{a}_{\mathbf{k}}^{\dagger}(t), \mathsf{a}_{\mathbf{k}}(0)]|0\rangle \delta(t) + \langle 0|T(\frac{\partial \mathsf{a}_{\mathbf{k}}^{\dagger}(t)}{\partial t}\mathsf{a}_{\mathbf{k}}(0))|0\rangle .$$

Now

$$\begin{aligned} \hbar \frac{\partial \mathsf{a}_{\mathbf{k}}^{\dagger}(t)}{\partial t} &= \mathrm{i}[\mathsf{H}_B, \mathsf{a}_{\mathbf{k}}^{\dagger}(t)] \\ &= \mathrm{i}\hat{E}(\mathbf{k})\mathsf{a}_{\mathbf{k}}^{\dagger}(t) + \mathrm{i}\frac{\tilde{V}(0)}{V} N_0 \mathsf{a}_{-\mathbf{k}}(t) \end{aligned}$$

therefore

$$\hbar \frac{\partial G_{\mathbf{k}}(t)}{\partial t} = -\hbar\delta(t) + \mathrm{i}\hat{E}(\mathbf{k})G_{\mathbf{k}}(t) + \mathrm{i}\frac{\tilde{V}(0)N_0}{V} G_{\mathbf{k}}^{(-)}(t)$$

where

$$G_{\mathbf{k}}^{(-)}(t) \equiv \langle 0|T(\mathsf{a}_{-\mathbf{k}}(t)\mathsf{a}_{\mathbf{k}}(0))|0\rangle .$$

Differentiating $G_{\mathbf{k}}^{(-)}(t)$ with respect to t similarly gives

$$\hbar \frac{\partial G_{\mathbf{k}}^{(-)}(t)}{\partial t} = -\mathrm{i}\hat{E}(\mathbf{k})G_{\mathbf{k}}^{(-)}(t) - \mathrm{i}\frac{\tilde{V}(0)N_0}{V} G_{\mathbf{k}}(t) .$$

Writing

$$\begin{aligned} G_{\mathbf{k}}(t) &= \int_{-\infty}^{\infty} \frac{\mathrm{d}w}{2\pi} \mathrm{e}^{\mathrm{i}wt} \tilde{G}_{\mathbf{k}}(w) \\ \delta(t) &= \int_{-\infty}^{\infty} \frac{\mathrm{d}w}{2\pi} \mathrm{e}^{\mathrm{i}wt} \\ G_{\mathbf{k}}^{(-)}(t) &= \int_{-\infty}^{\infty} \frac{\mathrm{d}w}{2\pi} \mathrm{e}^{\mathrm{i}wt} \tilde{G}_{\mathbf{k}}^{(-)}(w) \end{aligned}$$

we get

$$\mathrm{i}\hbar w \tilde{G}_{\mathbf{k}}(w) = \hbar + \mathrm{i}\hat{E}(\mathbf{k})\tilde{G}_{\mathbf{k}}(w) + \mathrm{i}\frac{\tilde{V}(0)N_0}{V}\tilde{G}_{\mathbf{k}}^{(-)}(w)$$

and

$$\mathrm{i}\hbar w \tilde{G}_{\mathbf{k}}^{-}(w) = -\mathrm{i}\hat{E}(\mathbf{k})\tilde{G}_{\mathbf{k}}^{(-)}(w) - \mathrm{i}\frac{\tilde{V}(0)N_0}{V}\tilde{G}_{\mathbf{k}}(w).$$

Solving for $\tilde{G}_{\mathbf{k}}$ and $\tilde{G}_{\mathbf{k}}^{(-)}$ we find,

$$\tilde{G}_{\mathbf{k}}(w) = \hbar \left((\hbar w - \hat{E}(\mathbf{k})) + (\frac{\tilde{V}(0)N_0}{V})^2 \frac{1}{\hbar w + \hat{E}(\mathbf{k})} \right)^{-1} .$$

From this expression we see that $\tilde{G}_{\mathbf{k}}(w)$ has a pole located at the point

$$\hbar^2 w^2 = \hat{E}^2(\mathbf{k}) - \left(\frac{N_0 \tilde{V}(0)}{V}\right)^2 .$$

This gives

$$\mathcal{E}(k) = \sqrt{\frac{|\mathbf{k}|}{2m}\left(\frac{|\mathbf{k}|}{2m} + 2\frac{N_0 \tilde{V}(0)}{V}\right)}$$

for the excitation energy spectrum of the system. But this is exactly the quasi-particle energy spectrum we calculated previously with the Bogoliubov method and thus supports our statement regarding the interpretation of the poles of $\tilde{G}_{\mathbf{k}}(a)$.

Let us now explain why the poles of $G_{\mathbf{k}}(w)$ are related to the excitations of the system in a qualitative way. Our discussion will be within the framework of perturbation theory. For a system governed by the Hamiltonian

$$\mathsf{H}_0 = \sum_{\mathbf{k}} E(\mathbf{k}) \mathsf{a}_{\mathbf{k}}^{\dagger} \mathsf{a}_{\mathbf{k}}$$

the Green function

$$G_{\mathbf{k}}^0(t) = \langle 0 | T(\mathsf{a}_{\mathbf{k}}^{\dagger}(t) \mathsf{a}_{\mathbf{k}}(0)) | 0 \rangle$$

satisfies the equation:

$$\hbar \frac{\partial}{\partial t} G_{\mathbf{k}}^0(t) = -\hbar \delta(t) + \mathrm{i} E(\mathbf{k}) G_{\mathbf{k}}^0(t).$$

Taking Fourier transforms this equation becomes:

$$\mathrm{i}(\hbar w - E(\mathbf{k})) \tilde{G}_{\mathbf{k}}^0(w) = \hbar .$$

Here we see clearly that the pole of $\tilde{G}_{\mathbf{k}}(w)$ is at the point $\hbar w = E(\mathbf{k})$ which represents the energy of the excitations of the system.

Let us now consider an interacting theory with Hamiltonian

$$\mathsf{H} = \sum_{\mathbf{k}} E(\mathbf{k}) \mathsf{a}_{\mathbf{k}}^{\dagger} \mathsf{a}_{\mathbf{k}} + \frac{1}{2V} \sum_{\mathbf{k}_1, \mathbf{k}_2, \mathbf{q}} \tilde{V}(q) \mathsf{a}_{\mathbf{k}_1+\mathbf{q}}^{\dagger} \mathsf{a}_{\mathbf{k}_2-\mathbf{q}}^{\dagger} \mathsf{a}_{\mathbf{k}} \mathsf{a}_{\mathbf{k}_2} .$$

The structure of the Green function

$$G_{\mathbf{k}}(t) = \langle 0 | T(\mathsf{a}_{\mathbf{k}}^{\dagger}(t) \mathsf{a}_{\mathbf{k}}(0)) | 0 \rangle$$

can be studied using perturbation theory. We do so by using the graphical rules discussed earlier. To simplify the analysis we set $1/2V \tilde{V}(\mathbf{q}) = \lambda$, a constant. Terms which contribute to $G_{\mathbf{k}}(t)$ can be graphically represented. We give a few (random)

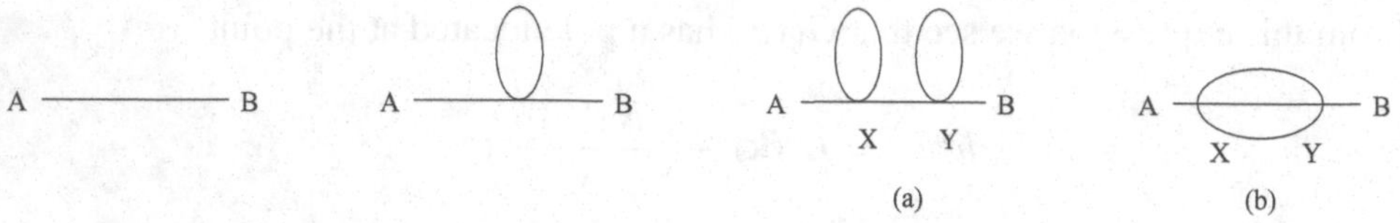

Figure 10.4 Some Feynman diagrams contributing to $G_{\mathbf{k}}(t)$.

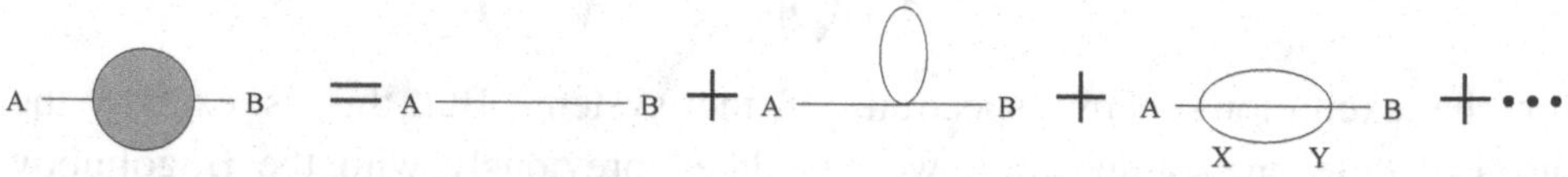

Figure 10.5 Sum of all one-particle irreducible graphs.

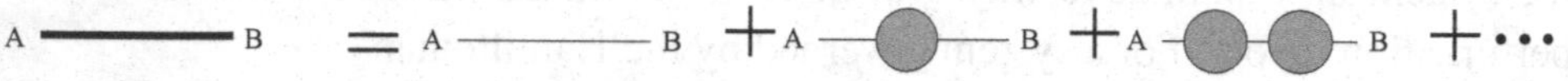

Figure 10.6 The full propagator.

examples in Figure 10.4. The graphs fall into two broad categories: those which split into disjoint pieces when one line is cut, and those that do not. For instance, Figure 10.4 (a) splits into two pieces when the line joining the two bubbles is cut, while Figure 10.4 (b) does not have this property. In our discussion the two lines joining the rest of the graph to the end points (A, B) (i.e. the lines (A, X) and (Y, B)) are to be left uncut. Graphs which remain as connected pieces when an internal line is cut are called *one-particle irreducible graphs*. The strategy followed for determining G is to first sum all possible graphs joining two points (A, B) which are one particle irreducible, and then to sum all the one particle reducible graphs. The result of the summation of all one particle irreducible graphs is represented in Figure 10.5. Recalling the discussion of Feynman diagrams in Chapter 9, we interpret the lines (A, X) and (Y, B) as the free propagators G^0 of perturbation theory. Adding up all the reducible graphs is represented in Figure 10.6.

Symbolically the sum of these graphs represents the Green function G as

$$G = G^0 + G^0 \Sigma G^0 + G^0 \Sigma G^0 \Sigma G^0 + \cdots$$

This is a geometric series involving the matrices G and Σ which we can formally sum to get

$$G = G^0 \frac{1}{1 - \Sigma G^0}.$$

When Fourier transformed, G^0 and Σ both become functions of w, i.e. they become proportional to the identity matrix, and we can write

$$\tilde{G}_{\mathbf{k}}(w) = \left(\frac{1}{\tilde{G}^0(w)} - \tilde{\Sigma}(w) \right)^{-1} .$$

For a free theory we saw $\tilde{G}_0(w) = \hbar(\hbar w - E_0(\mathbf{k}))^{-1}$, and so we find

$$\tilde{G}_{\mathbf{k}}(w) = \hbar(\hbar w - (E_0(\mathbf{k}) + \tilde{\Sigma}(w)))^{-1} .$$

The effect of the interactions, in this perturbation theory approach, is thus to shift the pole of $\tilde{G}_{\mathbf{k}}(w)$ from its free value $E_0(\mathbf{k})$ to $E_0(\mathbf{k}) + \hat{\Sigma}_{\mathbf{k}}(w)$. This shifted energy value represents the new spectrum of excitations present in the system which takes the interaction of the system into account. It can happen that the sum of the one-particle irreducible graph $\hat{\Sigma}_{\mathbf{k}}(w)$ is a complex quantity. Then the real part, $\mathrm{Re}(\hat{\Sigma}_{\mathbf{k}}(w))$, added to $E_0(\mathbf{k})$ gives the energy of the excitations of the system (the quasi-particle spectrum) while the imaginary part, $\mathrm{Im}(\hat{\Sigma}_{\mathbf{k}}(w))$, is interpreted as the inverse of the quasi-particle lifetime. Such an interpretation is natural if we remember that $\mathrm{e}^{-\mathrm{i}Et}$ is the characteristic time of oscillations associated with an energy state E. If E is complex we write $E = E_0 - \mathrm{i}\Gamma$ then $\mathrm{e}^{-\mathrm{i}(E_0 - \mathrm{i}\Gamma)t} = \mathrm{e}^{-\mathrm{i}E_0 t} e^{-t\Gamma}$ represents a state of energy E_0 and an amplitude which decays in time with a lifetime $\tau = 1/\Gamma$.

10.4 Summary

In this chapter we have made use of non-perturbative properties of non-relativistic quantum field theory to derive an important emergent phenomenon, superfluidity. This example nicely illustrates the power of quantum field theory. Our aim was to present two useful approaches to the problem of determining the low-energy excitations of interacting many-body systems. Both approaches, as we presented them, involved reducing the problem to one involving quadratic functions of operators and then studying this quadratic operator problem either directly, using a diagonalization method, or indirectly by looking at the poles of an associated Green function. We restricted our discussion in this chapter to ^{4}He which is a bosonic system. Superfluidity has also been observed at very low temperature in ^{3}He, which is a fermionic system. This phenomenon can also be understood within the framework of Bose–Einstein condensation in which a "pairing" of fermions first takes place due, in this case, to a weak attractive force when two helium atoms both have parallel spins. These correlated pairs (which are bosons) can now undergo Bose–Einstein condensation. The reason very low temperatures are needed before such superfluid states can be created is because the "binding energy", B, of the pairs is small and the pairs do not remain correlated unless $kT < B$. Although the

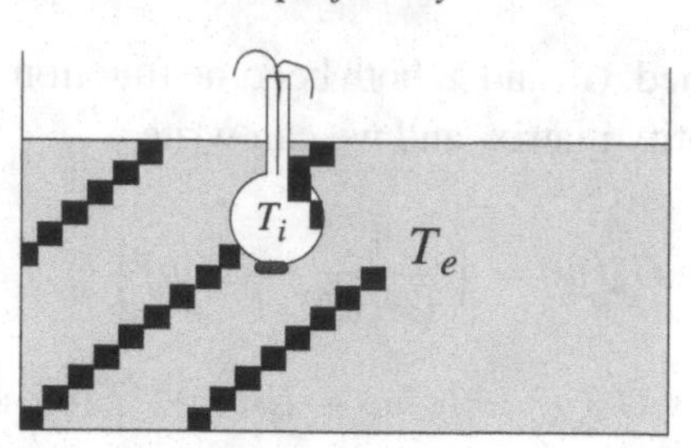

Figure 10.7 Experimental set-up for the fountain effect.

theory of Bogoliubov is in qualitative agreement with experimental facts, in detail it makes predictions which do not agree with experimental results. In particular, the limiting velocity for frictionless flow is off by a factor of ten! The problem lies in the restriction to two-body interactions made at the beginning. This amounts to the assumption that the gas is dilute or weakly non-ideal. The present day conclusion is that the Bogoliubov theory is not suitable for a quantitative description of the excitation properties of superfluid ^{4}He. The idea of Bose–Einstein condensation is valid but implementing this at the level of the Hamiltonian operator labeled by momentum states does not seem to work. The current theory of superfluidity in ^{4}He treats the liquid as a highly correlated and strongly interacting system. In the modern approach the ground state wave function has the form

$$\Psi_0(\mathbf{x}_1, \ldots, \mathbf{x}_N) = \text{const} \times \exp(-\sum_{i<j} u(\mathbf{x}_i - \mathbf{x}_j)).$$

This approach is not based on plane waves, and is quite successful. We will not pursue this theory any further at present but instead refer the interested reader to the literature, in particular the book by Feynman.

Problems

Problem 10.1 Fountain effect: liquid helium shows some unusual properties at low temperatures. Consider the experimental set-up in Figure 10.7, where a container with two openings is immersed in liquid helium at $T < 2$ K. At the bottom of the container there is a small hole covered by a powder preventing viscous fluid from passing from the outside into the container. In order to analyze this system we will assume that for $0 \text{ K} < T < 2$ K, He II consists in parts of a normal liquid and a superfluid with vanishing entropy and viscosity. If the temperature T_i inside the container is increased then the portion of superfluid helium in the container will decrease. Then, to restore the equilibrium superfluid helium will pass through the powder leading to an increase in pressure inside the container which, in turn, will produce the fountain effect.

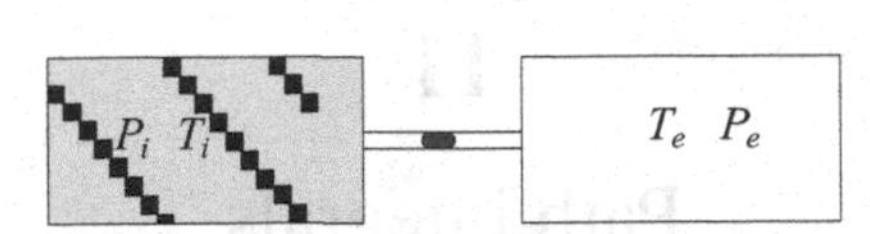

Figure 10.8 Model for the fountain effect.

To analyze this system quantitatively we model it as in Figure 10.8.

(1) Show that, allowing for exchange of superfluid helium, the Gibbs potentials inside and outside the container have to be equal at equilibrium.
(2) Derive a relation between the pressure and temperature differences.
(3) In a crude approximation we can take the entropy per unit mass to be $s_n = 1600$ J/kg K for the normal fluid. The density of the normal fluid is taken to be $\rho_n = \rho(T/T_\lambda)^4$, where $\rho = 144$ kg/m^3 and $T_\lambda = 2.2$ K is the critical temperature. Calculate the pressure difference for $T_i = 1.2$ K and $T_e = 1.1$ K.

Problem 10.2 Pomeranchuk effect: if ^{3}He in the solid phase is cooled at a pressure $P \simeq 3$ kPa below $T \simeq 1$ K, then one observes a phase transition, at some $T_0 < 1$ K, from the solid to the liquid phase! In order to get a qualitative understanding of this unusual behavior, compare the entropy of a crystal of spin 1/2 particles (^{3}He nuclei) with that of a Fermi liquid (liquid ^{3}He) at low temperatures.

Problem 10.3 As we have explained in the beginning of this chapter, the excitation spectrum of He II (super liquid ^{4}He) consists of a phonon part and a roton part. Compute the contribution of the rotons to the energy, entropy, and specific heat of He II by treating them as an ideal gas of spin 0 bosons whose particle number is not conserved and whose dispersion relation is given by $E(\mathbf{p}) = \Delta + 1/2\mu|\mathbf{p} - \mathbf{p}_0|^2$.

Further reading

A Green function approach to the excitation spectrum of ^{4}He can be found in A. A. Abrikosov, L. P. Gorkov, and I. E. Dzyaloshinski, *Methods of Quantum Field Theory in Statistical Physics*, Dover (1975), which also contains a qualitative derivation of the roton part of the dispersion relation. R. P. Feynman's book on *Statistical Mechanics*, Perseus (1998) describes the modern theory of superfluidity indicated in the text.

11

Path integrals

In the chapter on quantum statistical mechanics we showed that the partition function can be written as a trace over the Hilbert space of the density matrix. In the canonical ensemble this reads

$$Z = \mathrm{Tr}\, \mathrm{e}^{-\beta \mathsf{H}},$$

where H is the quantum mechanical Hamiltonian for our system and the trace is over the N-particle Hilbert space. This formula is completely general as we have seen, and can be applied to any quantum system. We have also seen that, in all but the simplest cases, actually calculating this trace is very difficult. The problem is that we must know the full spectrum of the operator H, and we must be able to take the trace of the exponential of this operator over the full Hilbert space of the system.

In this chapter, we will develop an alternative approach which allows us to work with commuting numbers rather than operators. As we shall see, quantum partition function calculations for Bose–Einstein systems will take a form which is reminiscent of the classical partition function we have treated in Chapter 2. We will then make use of this formalism extensively when discussing Landau theory and renormalization group methods in Chapter 13.

11.1 Quantum mechanics

To derive the basic path integral formula for the partition function we first consider the case of a quantum mechanical system confined to a one-dimensional box of size L. The Hamiltonian for the system is taken to be

$$\mathsf{H} = T(\mathsf{p}) + V(\mathsf{q}) = \frac{1}{2m}\mathsf{p}^2 + V(\mathsf{q}).$$

H, p and q are quantum mechanical operators, with p and q satisfying the usual commutation relation

$$[\mathsf{p}, \mathsf{q}] = -i\hbar.$$

In what follows we will work in the Heisenberg picture, where states are independent of time. Since the partition function we seek to evaluate involves a trace over all states of the system, we are at liberty to choose a basis for the operators and states which will be convenient for our purpose. Finally, we impose periodic boundary conditions on wave functions in our box. The basic idea underlying the path integral approach to quantum mechanics is to replace a calculation involving operators and states by an alternative but equivalent calculation involving just commuting numbers. The fundamental operators in our current problem are the momentum operator p and the coordinate operator q. We can always replace one of these operators with a commuting number if we arrange that the operator acts on one of its eigenvectors. For a particle confined to a one dimensional box of length L with periodic boundary conditions, the possible eigenvalues of the coordinate operator, q, are the continuous numbers $q \in [-L/2, L/2]$, while the possible eigenvalues of the momentum operator, p, are discrete, $p = 2\pi n\hbar/L$, with n an integer, as we have seen in Chapter 7. We have

$$\mathsf{q} \mid q\rangle = q \mid q\rangle \qquad \mathsf{p} \mid p\rangle = p \mid p\rangle$$

with the following orthogonality and completeness relations,

$$\int \mathrm{d}q \mid q\rangle\langle q \mid = 1 \qquad\qquad \langle q \mid q'\rangle = \delta(q - q')$$

$$\frac{2\pi\hbar}{L}\sum_p \mid p\rangle\langle p \mid = 1 \qquad\qquad \langle p \mid p'\rangle = \frac{L}{2\pi\hbar}\delta_{p,p'}.$$

The normalizations are chosen so that the discrete sum over p is correctly weighted to become an integral in the limit $L \to \infty$. In this same limit, the discrete δ function defining $\langle p \mid p'\rangle$ is also correctly weighted to become $\delta(p - p')$. The transition between complete basis vectors $\mid q\rangle$ and $\mid p\rangle$ is given by the scalar product

$$\langle q \mid p\rangle = \frac{1}{\sqrt{2\pi\hbar}}\mathrm{e}^{ipq/\hbar}.$$

However, p and q do not commute, and it is thus not possible to have simultaneous eigenvectors and eigenvalues of p and q.

Consider now applying H to a state $\mid p\rangle$ or a state $\mid q\rangle$. H is a sum of two terms. The kinetic term depends on the operator p only, so when applied to $\mid p\rangle$ this term becomes a commuting number term

$$\frac{1}{2m}\mathsf{p}^2 \mid p\rangle = \frac{1}{2m}p^2 \mid p\rangle.$$

Similarly the potential term depends on the operator q only, so when applied to $\mid q\rangle$ this term also becomes a commuting number term

$$V(\mathsf{q}) \mid q\rangle = V(q) \mid q\rangle.$$

The commutation relation for q and p however implies that it is not possible to find a state which is simultaneously an eigenvector of both p and q. Thus although we can arrange that either the kinetic or potential term can be made into a commuting number term by acting either on $\mid p\rangle$ or $\mid q\rangle$ it is not possible that both simultaneously be made into commuting number terms.

11.1.1 Phase space path integral

To address the problem just described we adopt an approach in which we act first on states $\mid p\rangle$ then on states $\mid q\rangle$. We begin by considering the operator $\exp(-\epsilon(\mathsf{T} + \mathsf{V}))$ which is the fundamental object of interest. We anticipate a little here by replacing β by ϵ which, as we shall see later, needs to be taken small. We would like now to act with this operator on either a state $\mid p\rangle$ or $\mid q\rangle$ and replace respectively either the operators p or q with their corresponding eigenvalues. This is not immediately possible, however, since the exponential function is not a simple function as was the Hamiltonian. We can replace an operator with its eigenvalue only so long as it is the rightmost operator acting on the corresponding eigenvector. The exponential function (when expanded in power series) will give us many different orderings of operators, and only a few of the many terms which occur in this series will be in the correct rightmost positions. The simplest solution to this problem is to use the Baker–Campbell–Hausdorff formula which we generated in Chapter 6 to reorder operators in the exponential. We express the result we require as a theorem

Theorem 11.1 The term $\mathsf{e}^{-\epsilon(\mathsf{T}+\mathsf{V})}$ is given by

$$\mathsf{e}^{-\epsilon(\mathsf{T}+\mathsf{V})} = \mathsf{e}^{-\frac{\epsilon}{2}\mathsf{V}}\mathsf{e}^{-\epsilon\mathsf{T}}\mathsf{e}^{-\frac{\epsilon}{2}\mathsf{V}} + O(\epsilon^3).$$

Proof. The proof that follows is along the same lines as the theorem proved in Chapter 6. We will thus not repeat it here. □

This theorem gives us a particular splitting of the exponential of the Hamiltonian where terms depending only on p are cleanly separated from terms depending only on q. The error we make in so separating terms is of $O(\epsilon^3)$. In order to make this error small we must make ϵ small. Our basic goal is to evaluate $Z = \mathrm{Tr}\,\mathsf{e}^{-\beta\mathsf{H}}$. A direct splitting following the theorem is of no use since β is not necessarily small, and the error we make will also not necessarily be small. However, we can proceed in steps. Define ϵ as

$$\epsilon = \frac{\beta}{n}$$

where n is an integer. Then

$$\begin{aligned} \mathrm{Tr}\, \mathrm{e}^{-\beta \mathsf{H}} &= \mathrm{Tr}\, (\mathrm{e}^{-\epsilon \mathsf{H}})^n \\ &= \mathrm{Tr}\, \left(\mathrm{e}^{-\frac{\epsilon}{2}\mathsf{V}} \mathrm{e}^{-\epsilon \mathsf{T}} \mathrm{e}^{-\frac{\epsilon}{2}\mathsf{V}}\right)^n + n O(\epsilon^3) \\ &= \mathrm{Tr}\, \left(\mathrm{e}^{-\epsilon \mathsf{T}} \mathrm{e}^{-\epsilon \mathsf{V}}\right)^n + n O(\epsilon^3). \end{aligned}$$

These manipulations result in splitting the exponential $\mathrm{e}^{-\beta\mathsf{H}}$ into an interleaved product of n factors $\exp(-\epsilon\mathsf{T})$ and n factors $\exp(-\epsilon\mathsf{V})$. The error we make in this process is $nO(\epsilon^3) = \beta^3 O(n^{-2})$. Since n is a free parameter in this procedure, we are free to take it as large as we like. In the limit $n \to \infty$ the error term will go to zero, and we have achieved a splitting of the original exponential operator

$$\mathrm{Tr}\, \mathrm{e}^{-\beta \mathsf{H}} = \lim_{n \to \infty} \mathrm{Tr}\, \left(\mathrm{e}^{-\frac{\beta}{n}\mathsf{T}} \mathrm{e}^{-\frac{\beta}{n}\mathsf{V}}\right)^n .$$

At this point we are still working with operators, but we can now insert a complete set of states between each term in the product, and convert the problem to one with just commuting numbers. Immediately to the right of each factor $\exp(-\epsilon\mathsf{T})$ we insert the complete set of states

$$\frac{2\pi\hbar}{L} \sum_p \mid p\rangle\langle p \mid,$$

while immediately to the right of each factor $\exp(-\epsilon\mathsf{V})$ we insert the complete set of states,

$$\int_{-L/2}^{L/2} \mid q\rangle\langle q \mid .$$

Since we actually have n factors of each kind, we have to be careful to label the different insertions to the right of each different term,

$$\begin{aligned} \mathrm{Tr}\, \mathrm{e}^{-\beta \mathsf{H}} = \left(\frac{2\pi\hbar}{L}\right)^n \sum_{p_1, \cdots, p_{n-1}} \int_{-L/2}^{L/2} \mathrm{d}q_1 \cdots \mathrm{d}q_{n-1}\, \mathrm{Tr}\, \big(\mathrm{e}^{-\epsilon \mathsf{H}} \mid p_{n-1}\rangle\langle p_{n-1} \mid \mathrm{e}^{-\epsilon \mathsf{V}} \mid q_{n-1}\rangle \\ \cdot \langle q_{n-1} \mid \ldots \mathrm{e}^{-\epsilon \mathsf{H}} \mid p_0\rangle\langle p_0 \mid \mathrm{e}^{-\epsilon \mathsf{V}} \mid q_0\rangle\langle q_0 \mid \big). \end{aligned}$$

To simplify the notation at this point, we introduce the quantity $\int [\mathrm{d}p\mathrm{d}q]$ as a shorthand way of indicating the integrations over complete sets of states which we have to perform,

$$\int \left[\frac{\mathrm{d}p\mathrm{d}q}{2\pi\hbar}\right] \equiv \prod_{k=0}^{n-1} \left(\frac{1}{L}\sum_{p_k}\int_{-L/2}^{L/2} \mathrm{d}q_k\right).$$

The extra factor $1/2\pi\hbar$ included here for each pair p_k, q_k produces a dimensionless integration measure for that pair.

Consider now the integrand. This still involves operators, but now each operator acts on an eigenstate immediately to its right. We can therefore replace each operator with its corresponding eigenvalue.

$$\begin{aligned}\mathrm{Tr}\, \mathrm{e}^{-\beta \mathsf{H}} = (2\pi\hbar)^n \int \left[\frac{\mathrm{d}p\mathrm{d}q}{2\pi\hbar}\right] \\ \times \mathrm{Tr}\left(\mathrm{e}^{-\epsilon T(p_{n-1})} \mid p_{n-1}\rangle\langle p_{n-1} \mid \mathrm{e}^{-\epsilon V(q_{n-1})} \mid q_{n-1}\rangle\langle q_{n-1} \mid \right. \\ \left. \ldots \mathrm{e}^{-\epsilon T(p_0)} \mid p_0\rangle\langle p_0 \mid \mathrm{e}^{-\epsilon V(q_0)} \mid q_0\rangle\langle q_0 \mid \right).\end{aligned}$$

Since the exponentials in the integrand are now just numbers we can collect them into a single exponential $\exp[\sum_{k=0}^{n-1} T(p_k) + V(q_k)]$. In addition we substitute the scalar product for the brakets $\langle p_i \mid q_j\rangle$. This then leads to the simple expression

$$\mathrm{Tr}\, \mathrm{e}^{-\beta \mathsf{H}} = \int \left[\frac{\mathrm{d}p\mathrm{d}q}{2\pi\hbar}\right] \mathrm{e}^{-\epsilon \sum\limits_{k=0}^{n-1}(T(p_k)+V(q_k))} \mathrm{e}^{\frac{\mathrm{i}}{\hbar} p_{n-1}(q_0-q_{n-1})} \ldots \mathrm{e}^{\frac{\mathrm{i}}{\hbar} p_0(q_1-q_0)}.$$

The factor $(2\pi\hbar)^n$ has cancelled with the denominator in the scalar product $\langle p_i \mid q_j\rangle$. It is convenient to introduce a new redundant variable $q_n \equiv q_0$, so that we can rewrite the above expression in the compact form

$$\mathrm{Tr}\, \mathrm{e}^{-\beta \mathsf{H}} = \int \left[\frac{\mathrm{d}p\mathrm{d}q}{2\pi\hbar}\right] \mathrm{e}^{\sum\limits_{k=0}^{n-1}\left(\frac{\mathrm{i}}{\hbar} p_k(q_{k+1}-q_k) - \epsilon H(p_k,q_k)\right)}.$$

In generating this form for the partition function we have fulfilled our basic aim to replace a problem involving operators and states with a problem involving commuting numbers. Our calculation of the partition function is now seen to reduce to that of evaluating a large multidimensional integral, and we find that the quantum problem we began with is now reduced to a problem in the same basic form as that of evaluating a canonical partition function for a classical system.

11.1.2 Feynman–Kac formula

If the volume of our quantum mechanical system tends to infinity, so that the sum over the momenta p_k can be replaced by integrals, we can further simplify the above expression for the partition function. Indeed since $T(p) = 1/2m\,p^2$, the p_k integral is just a Gaussian integral of the form

$$\int \frac{dx}{2\pi\hbar} e^{-\frac{a}{2}x^2 - ixb} = \frac{1}{\sqrt{2\pi\hbar^2 a}} e^{-\frac{b^2}{2a}},$$

for each p_k, with a and b real numbers and $a > 0$. This formula follows directly from the Gauss integral formula by a shift $x \to x + \frac{i}{a}b$. Thus we can perform the p_k integral explicitly for each k leading to

$$\mathrm{Tr}\, e^{-\beta H} = \left(\frac{m}{2\pi\epsilon\hbar^2}\right)^{\frac{n}{2}} \int [dq]\, e^{-\sum\limits_{k=0}^{n-1}\left(\frac{m}{2\epsilon\hbar^2}\left(q_{k+1}-q_k\right)^2 + \epsilon V(q_k)\right)}.$$

This is the celebrated *Feynman–Kac formula* for the path integral representation of the partition sum.

To illustrate how the path integral works let us apply it to the by now familiar one-dimensional harmonic oscillator with Hamiltonian $H = (1/2m)p^2 + (\kappa^2/2)q^2$. Substituting this Hamiltonian into the phase space path integral expression for the partition sum and performing the Gaussian integral over the momenta p_k we end up with

$$\mathrm{Tr}\, e^{-\beta H} = \left(\frac{m}{2\pi\epsilon\hbar^2}\right)^{\frac{n}{2}} \int [dq]\, e^{-\frac{m}{2}\sum\limits_{k=0}^{n-1}\left(\frac{1}{\epsilon\hbar^2}(q_{k+1}-q_k)^2 + \epsilon\omega^2 q_k^2\right)},$$

where $\omega^2 = \kappa^2/m$. Since this action is quadratic in q_k we can write it as a bilinear form in a $\mathbf{R^n}$. For this we introduce the vector $q = (q_0, \cdots, q_{n-1})$ and a $n \times n$ matrix

$$\mathsf{M} = \frac{m}{\epsilon\hbar^2}\begin{pmatrix} a & -1 & 0 & \cdots & \cdots & \cdots & 0 & -1 \\ -1 & a & -1 & 0 & \cdots & \cdots & 0 & 0 \\ 0 & -1 & a & -1 & 0 & \cdots & 0 & 0 \\ \cdot & & & & & & & \\ \cdot & & & & & & & \\ \cdot & & & & & & & \\ 0 & 0 & \cdots & \cdots & 0 & -1 & a & -1 \\ -1 & 0 & \cdots & \cdots & \cdots & 0 & -1 & a \end{pmatrix}$$

with

$$a = 2 + \epsilon^2\hbar^2\omega^2.$$

The path integral representation of the partition sum of the one-dimensional harmonic oscillator is thus just the multidimensional Gauss integral

$$\mathrm{Tr}\, e^{-\beta H} = \left(\frac{m}{2\pi\epsilon\hbar^2}\right)^{\frac{n}{2}} \int d^n q\, e^{-\frac{m}{2\epsilon\hbar^2}(\mathbf{q}, \mathsf{M}\mathbf{q})}.$$

This integral can be evaluated in the same way as the usual Gauss integral. Here we prove a slightly more general formula which will be used in the sequel.

Result 11.2

$$I(\mathsf{M},\mathbf{B}) = \int_{-\infty}^{\infty} \frac{\mathrm{d}^n x}{\sqrt{(2\pi)^n}}\, \mathrm{e}^{-\frac{1}{2}(\mathbf{x},\mathsf{M}\mathbf{x})+(\mathbf{B},\mathbf{x})}$$

$$= (\det \mathsf{M})^{-\frac{1}{2}} \mathrm{e}^{\frac{1}{2}(\mathbf{B},\mathsf{M}\mathbf{B})}$$

where $M_{ij} = M_{ji}$ and M_{ij} is assumed to be real.

Proof. The integral is invariant under the substitution $\mathbf{x} \to \mathbf{x}' = \mathsf{R}\mathbf{x}$, where the matrix R satisfies the condition $\mathsf{R}^T\mathsf{R} = \mathsf{I}$ where R^T is the transpose of R. Since the matrix M in the problem is symmetric there is a matrix R such that $\mathsf{R}^T\mathsf{M}\mathsf{R} = \mathsf{D}$ with D diagonal. We suppose all the diagonal entries of D are non-zero. Then,

$$I(\mathsf{M},\mathbf{B}) = \prod_{i=1}^{n} \left(\int \frac{\mathrm{d}x_i'}{\sqrt{2\pi}} \mathrm{e}^{-\frac{\lambda_i}{2}(x_i')^2 + B_i' x_i'} \right)$$

where $B_i' = \sum_j R_{ij} B_j$, $x_i' = \sum_j R_{ij} x_j$, and $\lambda_i, = 1, \ldots, n$ are the eigenvalues of M. But we have already seen that

$$\int_{-\infty}^{\infty} \mathrm{e}^{-\frac{1}{2}x^2\lambda + Bx} \frac{\mathrm{d}x}{\sqrt{2\pi}} = \frac{1}{\sqrt{\lambda}} \mathrm{e}^{\frac{1}{2}B^2\frac{1}{\lambda}}.$$

Thus the result holds with $\sqrt{\det \mathsf{M}} = \sqrt{\lambda_1 \ldots \lambda_n}$ and $(\mathbf{B}\mathsf{M}^{-1}\mathbf{B}) = \sum_{i=1}^{N} (B_i)^2 \lambda_i^{-1}$.

□

To complete our calculation we need to evaluate the determinant of M. For this we expand the determinant in the first column

$$\det \mathsf{M} = a \det \mathsf{I}_{n-1} - 2 \det \mathsf{I}_{n-2} - 2\,,$$

where I_{n-p} is obtained from M by deleting the first p rows and columns. Let us begin by calculating $\det \mathsf{I}_k$, $k \le n-1$. For this we expand $\det \mathsf{I}_k$ in the first row which leads to the recursion relation $\det \mathsf{I}_k = a \det \mathsf{I}_{k-1} - \det \mathsf{I}_{k-2}$ with initial conditions $\det \mathsf{I}_0 = 1$ and $\det \mathsf{I}_1 = a$. This relation is solved by

$$\det \mathsf{I}_{n-1} = \frac{\sinh(n\mu/2)}{\sinh(\mu/2)}\,, \quad \text{where} \quad \cosh\frac{\mu}{2} = \frac{a}{2}\,.$$

Substituting this result into the expression for $\det \mathsf{M}$ and using the identity $2\cosh a \sinh b = \cosh(a+b) + \sinh(b-a)$, we then end up with the simple result

$$\det \mathsf{M} = 4\sinh^2\left(\frac{n\mu}{4}\right).$$

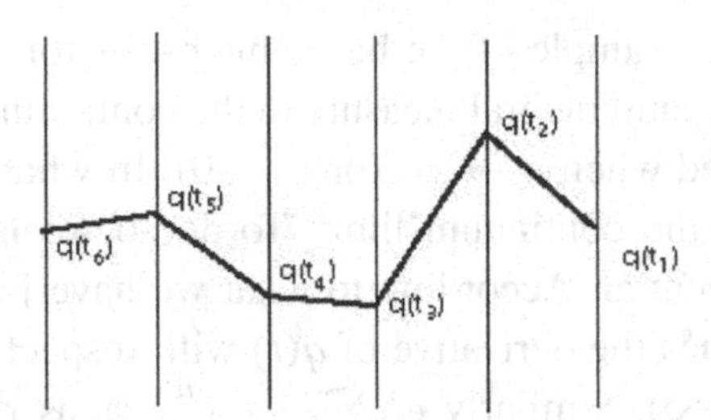

Figure 11.1 A broken path for $n = 6$.

We are now ready to evaluate the path integral completely. Performing the Gaussian integral it is not hard to see the factors of ϵ, $\hbar$, m, and 2π all cancel out so that the final result reads simply

$$\mathrm{Tr}\, \mathrm{e}^{-\beta \mathsf{H}} = \frac{1}{2\sinh(n\mu/4)}\,.$$

However, we can bring this result into a more familiar form by expressing μ in terms of a. We have

$$\cosh\frac{\mu}{2} \simeq 1 + \frac{\mu^2}{8} = 1 + \frac{\epsilon^2\hbar^2\omega^2}{2}$$

and thus

$$\begin{aligned}\mathrm{Tr}\, \mathrm{e}^{-\beta \mathsf{H}} &= \frac{1}{2\sinh(\hbar\omega\beta/2)}\\ &= \frac{\mathrm{e}^{-\frac{\hbar\omega\beta}{2}}}{1-\mathrm{e}^{-\hbar\omega\beta}}\\ &= \sum_{n=0}^{\infty} \mathrm{e}^{-\hbar\omega\beta\left(n+\frac{1}{2}\right)}\,,\end{aligned}$$

which is just the result we obtained for the harmonic oscillator in Chapter 7 by explict evaluation of the trace.

In order to develop a geometric interpretation of the Feynman path integral we recall that $\epsilon = \beta/n$ where n is the number of steps introduced to discretize the 1-parameter family of operator $\mathrm{e}^{-\beta \mathsf{H}}$. We have seen in Section 9.8 that β can be interpreted as "imaginary time" through the identification $T = \mathrm{i}\beta\hbar$. If we carry this interpretation over to our path integral expression we would identify $\epsilon\hbar$ with an infinitesimal imaginary time while q_k is interpreted as the value of $q(t)$ at $t_k = -\mathrm{i}k\epsilon\hbar$. The above representation of the partition sum can thus be interpreted as integrating over all possible discretized periodic paths, $q(t)$, in imaginary time (see Fig. 11.1). This then explains the usage of the word *path integral*.

We have seen in the example of the harmonic oscillator that although it is not clear how to define the path integral measure in the continuum limit the final result is perfectly well defined when $n \to \infty$ (or $\epsilon \to 0$). In what follows we will give a formal definition of the continuum limit. To do so we first focus on the first argument of the exponential. According to what we have just said the difference $(q_{k+1} - q_k)/\mathrm{i}\epsilon\hbar$ becomes the derivative of $q(t)$ with respect to imaginary time as ϵ tends to zero ($n \to \infty$). Similarly $\epsilon\hbar \sum_k \to \int_0^{\hbar\beta} \mathrm{d}\tau$ as $\epsilon \to 0$. Thus the path integral expression for $\mathrm{Tr}\,\mathrm{e}^{-\beta \mathrm{H}}$ takes the simple form

$$\mathrm{Tr}\,\mathrm{e}^{-\beta\mathrm{H}} = C \int [\mathrm{d}q]\ \mathrm{e}^{-\frac{1}{\hbar}S_E[q(\tau)]}$$

where

$$S_E[q(\tau)] = \int_0^{\hbar\beta} \left(\frac{m}{2}\dot{q}(\tau)^2 + V(q(\tau))\right) \mathrm{d}\tau$$

is the action of the point particle in imaginary time $t = -\mathrm{i}\tau$. As an illustration we can again refer to the harmonic oscillator treated above. The Lagrange function for this system is found as usual via the Legendre transform

$$\begin{aligned}\mathcal{L}(q,\dot{q}) &= H(p,q) - p\dot{q}\\ &= \frac{m}{2}(-\dot{q}^2 + \omega^2 q^2).\end{aligned}$$

Note that we have defined the action with a global minus sign compared to the usual convention. This is just for later convenience and does not affect the dynamics of the system. Next we analytically continue the time variable to the negative imaginary axis, $t = -\mathrm{i}\tau$. Using the identity $\mathrm{d}/\mathrm{d}\tau = -\mathrm{i}\,\mathrm{d}/\mathrm{d}t$, the imaginary time Lagrange function is then given by

$$\mathcal{L}(q,\dot{q}) = \frac{m}{2}(\dot{q}^2 + \omega^2 q^2),$$

where the dot now symbolizes the derivative with respect to τ which, in turn, ranges for 0 to $\hbar\beta$. In order to recover the discrete version of the path integral we discretize the parameter τ into n steps with step size $\epsilon = \hbar\beta/n$. The action functional $S_E[q] = \int \mathrm{d}\tau\,\mathcal{L}(q,\dot{q})$ then takes the form of a discrete sum

$$S[q] = \frac{m}{2}\sum_{k=0}^{n-1}\left(\frac{1}{\epsilon\hbar^2}(q_{k+1} - q_k)^2 + \epsilon\omega^2 q_k^2\right).$$

The constant C in the continuum version of the path integral is formally infinite but can be absorbed in the measure $[\mathrm{d}q]$ as we have seen in the example of the harmonic oscillator.

11.2 Quantum field theory

Let us turn now to the path integral formulation for a system of non-interacting Bose–Einstein particles described by the field theory Hamiltonian,

$$\mathsf{H} = \int d^3y\, \Psi^\dagger(\mathbf{x}) \left(-\frac{\hbar^2}{2m} \nabla^2 \right) \Psi(\mathbf{y}).$$

The fields $\Psi(\mathbf{x})$ and $\Psi^\dagger(\mathbf{x})$ are the quantum fields introduced in Chapter 9. They satisfy the commutation relations

$$[\Psi(\mathbf{x}), \Psi(\mathbf{y})] = 0\,, \quad [\Psi(\mathbf{x}), \Psi^\dagger(\mathbf{y})] = \delta^3(\mathbf{x}-\mathbf{y})\,, \quad [\Psi^\dagger(\mathbf{x}), \Psi^\dagger(\mathbf{y})] = 0.$$

The derivation of this Hamiltonian in Chapter 9 was based entirely on quantum considerations. We never developed a classical view of this system. However, our prescription to develop the path integral just needs us to identify the coordinates and momenta in this system and we do this by finding operators which have the appropriate set of commutation relations.

Instead of presenting the lengthy derivation of the path integral expression for the quantum field theory partition sum we will try to guess the correct result. The commutation relations for the fields $\Psi(\mathbf{x})$ and $\Psi^\dagger(\mathbf{x})$ give us the clue to proceed. For a system with a finite number of discrete Cartesian coordinates q_k and corresponding discrete Cartesian momenta p_k where $k = 1, \ldots, n$, the fundamental quantum commutation relations are

$$[q_k, p_l] = i\hbar\delta_{k,l}.$$

Systems described by fields rather than coordinates can be considered as the limiting case which occurs when the coordinate indices k become continuous. In this limit, the quantum commutation relations must be modified suitably. For example if $k \to \mathbf{x}$, then $q_k \to q(\mathbf{x})$, and $p_k \to p(\mathbf{x})$, and we would correspondingly expect that

$$[q(\mathbf{x}), p(\mathbf{y})] = i\hbar\delta^3(\mathbf{x}-\mathbf{y}).$$

Comparing to the commutation relations for $\psi(\mathbf{x})$ and $\psi^\dagger(\mathbf{x})$, this suggests that we identify $\psi(\mathbf{x})$ as the coordinate field in our Hamiltonian and $i\hbar\psi^\dagger(\mathbf{x})$ as the momentum field. Our naive suggestion for the generalization of the quantum mechanical path integral to quantum field theory is then simply to replace in the quantum mechanical phase space integral

$$\mathrm{Tr}\, e^{-\beta\mathsf{H}} = \int \left[\frac{dp\,dq}{2\pi\hbar} \right] e^{\sum_{k=0}^{n-1} \left(\frac{i}{\hbar} p_k (q_{k+1} - q_k) - \epsilon H(p_k, q_k) \right)}$$

the momentum p_k by $\mathrm{i}\hbar\psi^\dagger(\mathbf{x})$ and to replace the sum over the "continuous index" $\mathbf{x}$ by an integration, that is

$$\mathrm{Tr}\,\mathrm{e}^{-\beta \mathsf{H}} = \int [\mathrm{d}\psi \mathrm{d}\psi^\dagger]\, \mathrm{e}^{-\sum_{k=0}^{n-1} \int [\psi^\dagger(\mathbf{x},\tau_k)(\psi(\mathbf{x},\tau_{k+1}) - \psi(\mathbf{x},\tau_k)) + \epsilon H(\psi^\dagger(\mathbf{x},\tau_k),\psi(\mathbf{x},\tau_k))] \mathrm{d}^3x} .$$

In arriving at this form for the partition function, we have not at any time explicitly defined how to go from a discrete formulation in which the path integrals are well-defined to the continuous formulation we have just written down. Mathematically, the continuous form is not well-defined. To be explicit, we must replace all continuous variables by approximate discrete indices. The extra complication that arises is that now that the coordinates and momenta describing our system are labeled by two different continuous variables, $\mathbf{x}$ and τ. A simple discretization which makes the integral well-defined is to confine $\mathbf{x}$ and τ to the sites of a discrete four-dimensional cubic lattice. Let the spacing between adjacent points on this lattice be a, then the discretization in the τ direction is to allow τ to take only the values $\tau = k_\tau a$ for $k_\tau = 0, \ldots, N_\tau$ with $N_\tau a = \hbar\beta$, and to allow $\mathbf{x}$ to take only the values $\mathbf{x} = a(k_x\hat{\mathbf{i}} + k_y\hat{\mathbf{j}} + k_z\hat{\mathbf{k}})$ for k_x, k_y, k_z taking values $1, \ldots, N$ with $(Na)^3 = V$, the volume in which the system is confined. With this discretization, the formal symbol $[\mathrm{d}\psi \mathrm{d}\psi^\dagger]$ can be defined as

$$[\mathrm{d}\psi \mathrm{d}\psi^\dagger] = \prod_{\mathbf{x},\tau} \mathrm{d}\psi^\dagger(\mathbf{x},\tau)\mathrm{d}\psi(\mathbf{x},\tau).$$

To continue we then proceed as for the quantum mechanical particle and notice that the difference $(\psi(\mathbf{x},\tau_{k+1}) - \psi(\mathbf{x},\tau_k))/\epsilon$ approaches the derivative $\hbar\, \mathrm{d}/\mathrm{d}\tau\, \psi(\mathbf{x},\tau)$ as $\epsilon \to 0$. In doing so the exponential in the previous expression for the path integral will be multiplied with an overall factor ϵ allowing us to replace the sum over k by an integral over τ, that is

$$\mathrm{Tr}\,\mathrm{e}^{-\beta \mathsf{H}} = \int [\mathrm{d}\psi \mathrm{d}\psi^\dagger]\, \mathrm{e}^{-\frac{1}{\hbar} S_E[\psi,\psi^\dagger]}$$

where

$$S_E[\psi,\psi^\dagger] = \int_0^{\hbar\beta} \mathrm{d}\tau \int_V \mathrm{d}^3x\, \psi^\dagger(\mathbf{x},\tau)\left(\hbar\frac{\mathrm{d}}{\mathrm{d}\tau} - \frac{\hbar^2}{2m}\nabla^2\right)\psi(\mathbf{x},\tau)\, \mathrm{d}^3x\mathrm{d}\tau,$$

where the x-integral is over the volume of the box. In order to reproduce the trace over all states we must supplement this *functional integral* with periodic boundary conditions on $\psi(\mathbf{x},\tau)$,

$$\psi(\mathbf{x},\tau + \hbar\beta) = \psi(\mathbf{x},\tau) .$$

Having arrived at a path integral representation of $\mathrm{Tr}\,\mathrm{e}^{-\beta \mathsf{H}}$ we will now complete the path integral representation of the partition sum by including the chemical potential. This is now an easy matter. Recalling from Chapter 9 that the particle number operator is given by $\mathsf{N} = \int \mathrm{d}^3x \Psi^\dagger(\mathbf{x})\Psi(\mathbf{x})$, we can include the chemical potential simply by redefining the action as

$$S_E[\psi, \psi^\dagger] = \int \psi^\dagger(\mathbf{x}, \tau) \left(\hbar \frac{\mathrm{d}}{\mathrm{d}\tau} - \frac{\hbar^2}{2m} \nabla^2 - \mu \right) \psi(\mathbf{x}, \tau)\, \mathrm{d}^3x \mathrm{d}\tau .$$

Although our derivation of the path integral formula for the partition function in quantum field theory was heuristic, this is the correct generalization of the quantum mechanical expression as can be shown using an approach with coherent states. We will not give a proof of this claim at present (see, however, the literature at the end of this chapter) but instead will consider a concrete example, both as a test of the proposed path integral formula and also in order to gain some familiarity with the formalism. For this we consider the free non-relativistic particle with vanishing chemical potential in a cube of length L, treated in Chapter 9.

We first consider the measure, $[\mathrm{d}\psi \mathrm{d}\psi^\dagger]$. For this we expand ψ and $\psi^\dagger$ in terms of eigenfunctions of the momentum operator, that is

$$\psi(\mathbf{x}, \tau) = \frac{1}{\sqrt{V}} \sum_{n,\mathbf{k}} c_{n,\mathbf{k}}\, \mathrm{e}^{\mathrm{i}\frac{2\pi}{\hbar\beta} n\tau}\, \mathrm{e}^{\mathrm{i}\mathbf{k}\cdot\mathbf{x}} ,$$

and similarly for $\psi^\dagger(\mathbf{x}, \tau)$, where $\mathbf{k} = 2\pi \mathbf{n}/L$ as explained in Chapter 9, and $c_{n,\mathbf{k}}$ are complex numbers. In terms of the $c_{n,\mathbf{k}}$ we then define the measure simply as an infinite-dimensional Lebesgue measure, i.e.

$$[\mathrm{d}\psi \mathrm{d}\psi^\dagger] \equiv \prod_{n,\mathbf{k}} \mathrm{d}c_{n,\mathbf{k}} \mathrm{d}c^\dagger_{n,\mathbf{k}} .$$

On the other hand, substituting this expansion in the action the path integral expression for the partition function then becomes

$$\mathrm{Tr}\,\mathrm{e}^{-\beta \mathsf{H}} = \int \prod_{n,\mathbf{k}} \mathrm{d}c_{n,\mathbf{k}} \mathrm{d}c^\dagger_{n,\mathbf{k}}\, \mathrm{e}^{-\sum\limits_{n,\mathbf{k}} c^\dagger_{n,\mathbf{k}} \left(2\pi \mathrm{i} n + \beta \frac{\hbar^2 \mathbf{k}^2}{2m} \right) c_{n,\mathbf{k}}} .$$

Performing these Gaussian integrals, remembering that $\mathrm{d}z\mathrm{d}\bar{z} = 2\mathrm{dRe}(z)\mathrm{dIm}(z)$, we end up with

$$\begin{aligned}
\mathrm{Tr}\,\mathrm{e}^{-\beta \mathsf{H}} &= \prod_{n,\mathbf{k}} \left(\mathrm{i}n + \beta \frac{\epsilon(\mathbf{k})}{2\pi} \right)^{-\frac{1}{2}} \\
&= \beta \frac{\epsilon(\mathbf{k})}{2\pi} \prod_{n \geq 0,\mathbf{k}} \left(n - \mathrm{i}\beta \frac{\epsilon(\mathbf{k})}{2\pi} \right)^{-\frac{1}{2}} \prod_{n \geq 0,\mathbf{k}} \left(n + \mathrm{i}\beta \frac{\epsilon(\mathbf{k})}{2\pi} \right)^{-\frac{1}{2}} .
\end{aligned}$$

Note that this result can only be formal, since the right-hand side is divergent. However, we can define the right-hand side using analytic continuation. For this we use the *ζ-function definition* of the determinant of an operator A. That is, we use that if λ_n, $n = 1, \ldots$ are the eigenvalues of A, then the determinant of A is given by $\det \mathsf{A} = \mathrm{e}^{-\zeta'_{\mathsf{A}}(0)}$. The proof of this result is short but instructive. We leave it as an exercise to the reader. In the present situation we will first keep $\mathbf{k}$ fixed and consider

$$\begin{aligned}\zeta_{\mathsf{A}}(s) &= \sum_{n=0}^{\infty}(n+x)^{-s} \\ &\equiv \zeta_H(s;x)\,.\end{aligned}$$

This series is known as the *Hurwitz zeta function.* It converges absolutely for $\mathrm{Re}(s) > 1$, and the function so defined may be analytically continued to the entire complex plane. There is a single pole at $s = 1$. Furthermore, the derivative of the Hurwitz zeta function at s is known to be

$$\zeta'_H(0;x) = \ln(\Gamma(x)/\sqrt{2\pi})\,.$$

Substituting this result in the ζ-function definition for determinants we have thus succeeded in extracting a finite result from the formally infinite product over n. This process is called *zeta function regularization.* It is commonly used to deal with divergences in quantum field theory. Let us now substitute this result for the right hand of the partition function. We then have

$$\begin{aligned}\mathrm{Tr}\,\mathrm{e}^{-\beta \mathsf{H}} &= \prod_{\mathbf{k}} \beta\frac{\epsilon(\mathbf{k})}{2\pi}\,\mathrm{e}^{\zeta'_H\left(0;-\mathrm{i}\beta\frac{\epsilon(\mathbf{k})}{2\pi}\right)+\zeta'_H\left(0;+\mathrm{i}\beta\frac{\epsilon(\mathbf{k})}{2\pi}\right)} \\ &= \prod_{\mathbf{k}} \beta\frac{\epsilon(\mathbf{k})}{2\pi}\,\frac{\Gamma\left(-\mathrm{i}\beta\frac{\epsilon(\mathbf{k})}{2\pi}\right)\Gamma\left(\mathrm{i}\beta\frac{\epsilon(\mathbf{k})}{2\pi}\right)}{2\pi}\,.\end{aligned}$$

On the other hand we have the identity $\Gamma(\mathrm{i}y)\Gamma(-\mathrm{i}y) = \pi(y\sinh(\pi y))^{-1}$, so that

$$\begin{aligned}\mathrm{Tr}\,\mathrm{e}^{-\beta \mathsf{H}} &= \prod_{\mathbf{k}} \frac{1}{2\sinh\left(\beta\frac{\epsilon(\mathbf{k})}{2}\right)} \\ &= \prod_{\mathbf{k}} \frac{\mathrm{e}^{\beta\frac{\epsilon(\mathbf{k})}{2}}}{1-\mathrm{e}^{-\beta\epsilon(\mathbf{k})}}\,.\end{aligned}$$

The latter form is convenient for comparison with our previous result for the partition function for a non-interacting boson in Chapter 7. Indeed recalling that the grand canonical potential, Ω is given by $-\frac{1}{\beta}\ln Z$, where $Z = \mathrm{Tr}\,\mathrm{e}^{-\beta H}$, we have

$$\Omega = \frac{1}{\beta}\sum_{\mathbf{k}} \ln\left(1 - \mathrm{e}^{-\beta\epsilon(\mathbf{k})}\right) - \frac{1}{2}\beta\epsilon(\mathbf{k})\,.$$

Thus we reproduce the grand canonical potential found in Chapter 7 up to the extra term $\sum_{\mathbf{k}} \frac{1}{2}\epsilon(\mathbf{k})$. This extra term is related to the ambiguity in the ordering of operators in the Hamiltonian discussed in Chapter 9. Indeed, comparing our last two formulas with the corresponding formulas for the Bose–Einstein system in Chapter 7, it is easily seen that the two results are related by a shift $n \to n + \frac{1}{2}$. Finally we recall that we have set the chemical potential to zero the present calculation. However, it is not hard to see how the different steps are modified for non-vanishing chemical potential. We leave this calculation as an exercise to the reader.

Having successfully guessed the correct generalization of the quantum mechanical path integral to quantum field theory we may now try to find the correct expression for fermions. Recalling our treatment of fermions in the quantum field theory formulation we note that the basic modification when going from bosons to fermions was to replace commutators by anticommutators. Explicitly,

$$\{\Psi(\mathbf{x}), \Psi(\mathbf{y})\} = 0 \quad \{\Psi(\mathbf{x}), \Psi^\dagger(\mathbf{y})\} = \delta^3(\mathbf{x}-\mathbf{y}) \quad \{\Psi^\dagger(\mathbf{x}), \Psi^\dagger(\mathbf{y})\} = 0.$$

The rules developed in this section tell us that, to generate the path integral, we should switch from quantum operators to classical coordinates, momenta or fields. The simplest suggestion to obtain the functional integral expression for the partition sum for fermions is replace the commuting integration variables ψ and $\psi^\dagger$ in the bosonic path integral by anticommuting variables $\psi(\mathbf{x},\tau)$ and $\psi^\dagger(\mathbf{x},\tau)$, satisfying

$$\{\Psi(\mathbf{x},\tau), \Psi(\mathbf{y},\tau)\} = \{\Psi(\mathbf{x},\tau), \Psi^\dagger(\mathbf{y},\tau)\} = \{\Psi^\dagger(\mathbf{x},\tau), \Psi^\dagger(\mathbf{y},\tau)\} = 0.$$

Again this turns out to be the correct generalization as can be shown by using a fermionic version of coherent states. Note, however, that for fermions, the periodic boundary conditions in τ have to be replaced by anti-periodic ones, i.e.

$$\psi(\mathbf{x}, \tau + \hbar\beta) = -\psi(\mathbf{x}, \tau).$$

Of course, the path integral obtained in this way is not defined until we give a proper definition of the measure $[\mathrm{d}\psi\mathrm{d}\psi^\dagger]$ for anticommuting fields. Again we will first assume that $\mathbf{x}$ and τ are discrete so that measure can be defined point wise. To simplify the discussion we first consider the case of two anticommuting variables, α_1 and α_2, with $\{\alpha_1, \alpha_2\} = 0$. Such variables are called *Grassman variables*. The algebra generated by linear combinations of products of the generating set of

Grassman variables is the *Grassman algebra*. A general element $f(\alpha_1, \alpha_2)$ in the algebra generated by the set $\{\alpha_1, \alpha_2\}$, for example, takes the form

$$f(\alpha_1, \alpha_2) = f^{(0)} + f_1^{(1)}\alpha_1 + f_2^{(1)}\alpha_2 + f^{(2)}\alpha_1\alpha_2$$

where the coefficients $f^{(k)}$ are complex numbers. Products can contain at most one power of each different generator α_i since the anticommutation relation $\{\alpha_k, \alpha_{k'}\} = 0$ implies, in particular, $\alpha_k\alpha_k = 0$. Furthermore, products which differ only in their ordering are linearly dependent since we can reorder these products using the anticommutation relations at the cost of multiplying by a commuting number ± 1. For example we have $\alpha_1\alpha_2 = -\alpha_2\alpha_1$. We find therefore that the algebra generated by α_1, α_2 is four-dimensional since there are exactly four linearly independent products of the generators.

A product containing an even number of Grassman variables is said to be even while a product containing an odd number of Grassman variables is said to be odd. Even products commute with even and odd products. Odd products anticommute with odd products but commute with even products.

To define the path integral for Fermi–Dirac particles we first need to introduce the concepts of differentiation and of integration on Grassman algebras. Differentiation with respect to a Grassman variable α_i is defined by the basic relation

$$\frac{\partial}{\partial\alpha_i}\alpha_j = \delta_{i,j}$$

while $\partial/\partial\alpha_i\, z = 0$, if z is a complex number. When differentiation acts on products, then the rule is to allow it to act in turn on each factor in a product by first anticommuting that factor to the leftmost position in the product. Thus, for example,

$$\frac{\partial}{\partial\alpha_i}(\alpha_j f(\alpha)) = \left(\frac{\partial}{\partial\alpha_i}\alpha_j\right) f(\alpha) - \alpha_j \frac{\partial}{\partial\alpha_i} f(\alpha).$$

Integration for classical functions is defined to be the inverse of differentiation. For Grassman variables this is reversed as are so many other properties of these variables. For Grassman variables therefore we formally define integration as the linear functional

$$\int \mathrm{d}\alpha_i\, f(\alpha) = \frac{\partial}{\partial\alpha_i} f(\alpha).$$

The fundamental reason for doing it this way is simply, as we shall see presently, that it allows us to define a useful path integral.

We shall be interested in defining the equivalent of multidimensional integrals of elements of the Grassman algebra generated by the set of Grassman variables $\{\alpha_i, {\alpha^\dagger}_i | i = 1, \ldots, n\}$. To simplify the notation, we will adopt some conventions which we now define. A function (algebra element) $f(\alpha^\dagger)$ depending on the $\alpha^\dagger$ variables only has the standard expansion:

$$f(\alpha^\dagger) = f^{(0)} + \sum_i f_i^{(1)} \alpha^\dagger{}_i + \sum_{i<j} f_{ij}^{(2)} \alpha^\dagger{}_i \alpha^\dagger{}_j + \cdots + f^{(n)} \alpha^\dagger{}_1 \alpha^\dagger{}_2 \ldots \alpha^\dagger{}_n.$$

A function $g(\alpha)$ depending on the α variables only has the standard expansion:

$$f(\alpha^\dagger) = g^{(0)} + \sum_i g_i^{(1)} \alpha_i + \sum_{i<j} g_{ij}^{(2)} \alpha_j \alpha_i + \cdots + g^n \alpha_1 \alpha_2 \ldots \alpha_n.$$

Note that the ordering of the Grassman variables in the individual terms in $g(\alpha)$ is exactly opposite to that for $f(\alpha^\dagger)$. This convention makes Hermitian conjugation straightforward. We have for example, that

$$(z\alpha^\dagger{}_i \alpha^\dagger{}_j)^\dagger = z^* \alpha_j \alpha_i$$

so that if $g(\alpha) = (f(\alpha^\dagger))^\dagger$, then $g^{(0)} = (f^{(0)})^*$, $g_i^{(1)} = (f_i^{(1)})^*$, etc. Here z^* denotes the complex conjugate of the complex number z.

A multidimensional Grassman integral I of a function $g(\alpha)$ and $f(\alpha^\dagger)$ is written in the shorthand form as

$$I = \int [\mathrm{d}\alpha^\dagger \mathrm{d}\alpha] g(\alpha) f(\alpha^\dagger)$$

and defined explicitly as

$$\prod_{i=1}^{n} \left(\int \mathrm{d}\alpha^\dagger{}_i \int \mathrm{d}\alpha_i \right) g(\alpha) f(\alpha^\dagger).$$

Note that we do not explicitly need to worry about the ordering of the pairs $\mathrm{d}\alpha^\dagger{}_i \mathrm{d}\alpha_i$ in the product since the anticommutation relations imply

$$\left(\int \mathrm{d}\alpha^\dagger{}_i \int \mathrm{d}\alpha_i \right) \left(\int \mathrm{d}\alpha^\dagger{}_j \int \mathrm{d}\alpha_j \right) = \left(\int \mathrm{d}\alpha^\dagger{}_i \int \mathrm{d}\alpha_i \right) \left(\int \mathrm{d}\alpha^\dagger{}_j \int \mathrm{d}\alpha_j \right).$$

We must however be careful to keep track of the ordering of $\mathrm{d}\alpha^\dagger{}_i$ and $\mathrm{d}\alpha_i$ in an individual term.

With these conventions, the fermionic path integral is now defined point wise for a discrete set of points $\mathbf{x}$ and τ. For continuous variables the measure is then defined formally in complete analogy with the bosonic case.

11.3 Real time path integral

So far we have presented the path integral representation of the finite temperature partition sum $\mathrm{Tr}\, \mathrm{e}^{-\beta \mathrm{H}}$. On the other hand, in our presentation of quantum field theory we saw that the partition sum can be written as the trace of the evolution operator $\mathsf{U}(T,0)$ evaluated for imaginary time $T = -\mathrm{i}\hbar\beta$. The question which we will now address is whether there is a path integral expression for this evolution operator for real T. On a formal level this is easily seen to be the case. Indeed starting from the phase space path integral expression we can implement evolution in real time simply by "rotating" ϵ by a phase $\mathrm{e}^{\mathrm{i}(\frac{\pi}{2}-\delta)}$, where $\delta > 0$ and small ensures the convergence of the Gaussian integrals. It is clear that the transformed operator $\mathrm{e}^{-\mathrm{i}\epsilon \mathsf{H}}$ describes an infinitesimal evolution in real time. On the other hand we can substitute $\epsilon \to \mathrm{i}\epsilon$ in the path integral formulas. This has the effect of multiplying the exponential by an overall factor i and furthermore changes the sign of the kinetic term since this term involves two time derivatives. Thus we end up with the expression

$$C \int [\mathrm{d}q]\, \mathrm{e}^{\frac{\mathrm{i}}{\hbar} S[q(\tau)]}\,,$$

where $S[q(\tau)] = \int \mathrm{d}t (\frac{m}{2}\dot{q}^2 - V(q))$ is the familiar classical action for a point particle. A new issue that arises in the real time path integral concerns the boundary conditions for $q(t)$. Indeed for a real time evolution periodic boundary conditions are not a natural choice. Rather one would fix the initial and final position of the particle, $q(0)$ and $q(T)$. The interpretation of the path integral with these boundary conditions is now clear: it corresponds to matrix elements of the evolution operator $\mathsf{U}(T,0)$, that is

$$\langle q_f|\mathsf{U}(T,0)|q_i\rangle = C \int_{q(0)=q_i}^{q(T)=q_f} [\mathrm{d}q]\, \mathrm{e}^{\frac{\mathrm{i}}{\hbar}\int_0^T \mathrm{d}t\, \mathcal{L}(q,\dot{q})}\,.$$

This analytic continuation in the time variable is then readily generalized to the quantum field theory description. Indeed letting $\tau = it$ in our path integral formula for the partition sum we get

$$\int [\mathrm{d}\psi \mathrm{d}\psi^\dagger]\, \mathrm{e}^{\frac{\mathrm{i}}{\hbar} S[\psi,\psi^\dagger]}$$

with

$$S[\psi,\psi^\dagger] = \int \psi^\dagger(\mathbf{x},t) \left(\frac{\hbar}{i}\frac{\mathrm{d}}{\mathrm{d}t} - \frac{\hbar^2}{2m}\nabla^2\right) \psi(\mathbf{x},t)\, \mathrm{d}^3x \mathrm{d}t\,.$$

The boundary conditions for ψ and $\psi^\dagger$ at $t = 0$ and $t = T$ are dictated by the physical set up of the system. For instance for a scattering experiment one would impose that ψ and $\psi^\dagger$ approach plane waves with wave vectors corresponding to the momenta of the scattered particles.

11.4 Relativistic scalar field

We close this short introduction to path integrals with a few comments on relativistic quantum field theory. To continue with our habit of using suggestive arguments in this chapter we proceed heuristically referring the interested reader to some of the many specialized books on this subject listed at the end of this chapter. In order to guess the modifications arising when dealing with a relativistic quantum field theory we recall from our discussion in Section 9.9 that there are two key changes in a relativistic theory. First, the relativistic dispersion relation becomes $\epsilon^2(\mathbf{p}) = c^2(\mathbf{p}^2 + m^2c^2)$ and second the relativistic field operator is self-adjoint, i.e. $\Psi^\dagger(\mathbf{x}, t) = \Psi(\mathbf{x}, t)$. Both of these modifications are easily implemented in the path integral. This amounts simply to replacing the path integral by

$$\int [\mathrm{d}\psi]\, \mathrm{e}^{\frac{\mathrm{i}}{\hbar}S[\psi]}$$

with real time action

$$S[\psi] = \frac{1}{2}\int \psi(\mathbf{x}, t)\left(\frac{\hbar^2 \mathrm{d}^2}{c^2\mathrm{d}t^2} - \hbar^2\nabla^2 + m^2\right)\psi(\mathbf{x}, t)\,\mathrm{d}^3x\mathrm{d}t.$$

Indeed, comparing this action with the last formula in the previous section and recalling the expansion of the field operator in Section 9.9 confirms that this is the correct modification.

In order to obtain the relativistic generalization of the partition sum we then have to return to the imaginary time formalism, that is we let $t = -\mathrm{i}\tau$ so that

$$\mathrm{Tr}\,\mathrm{e}^{-\beta\mathsf{H}} = \int [\mathrm{d}\psi]\, \mathrm{e}^{-\frac{1}{\hbar}S_E[\psi]}$$

with imaginary time action

$$S_E[\psi] = \frac{1}{2}\int \psi(\mathbf{x}, \tau)\left(-\frac{\hbar^2 \mathrm{d}^2}{c^2\mathrm{d}\tau^2} - \hbar^2\nabla^2 + m^2\right)\psi(\mathbf{x}, \tau)\,\mathrm{d}^3x\mathrm{d}\tau,$$

where the τ-integral ranges from 0 to $\hbar\beta$. This is then the correct partition sum for a relativistic boson without conserved number of particles, or, equivalently, with vanishing chemical potential.

Problems

Problem 11.1 If $\mathsf{A}u_n = \lambda_n u_n$, $\lambda_n \neq 0$, $n = 1, \ldots, \infty$, show that formally $\det \mathsf{A} = \mathrm{e}^{-\zeta'_{\mathsf{A}}(0)}$. Here $\zeta_{\mathsf{A}}(s) = \sum 1/\lambda_n^s$ is the zeta function of the operator A and $\zeta'_{\mathsf{A}}(0) = \mathrm{d}\zeta_{\mathsf{A}}(s)/\mathrm{d}s|_{s=0}$.

Problem 11.2 Find the path integral representation of the partition function for a bosonic non-relativistic scalar field with non-vanishing chemical potential.

Problem 11.3 Evaluate the Grassman integral

$$\prod_{i=1}^{n}\left(\int \mathrm{d}\alpha^{\dagger}{}_i \int \mathrm{d}\alpha_i\right) \mathrm{e}^{\sum\limits_{i,j=1}^{n} \alpha^{\dagger}{}_i A_{ij} \alpha_j}$$

for a Hermitian matrix A_{ij} and compare the result with the corresponding Gauss integral for commuting variables.

Problem 11.4 Explain the appearance of the anti-periodic boundary condition

$$\psi(\mathbf{x}, \tau + \hbar\beta) = -\psi(\mathbf{x}, \tau) .$$

for the fermionic path integral and compute the partition function for free, non-relativistic fermions in the path integral formalism. Compare the result with the corresponding result in Chapter 7.

Problem 11.5 Find the path integral representation of the matrix element $K(t, q, q') = \langle q|\mathsf{U}(T, 0)|q'\rangle$ for a free, quantum mechanical particle in the path integral representation and show that it satisfies the Schrödinger equation $\mathrm{i}\hbar\partial_t K(t, q, q') = \mathsf{H}K(t, q, q')$.

Historical notes

The idea of introducing path integrals via the superposition principle in quantum mechanics was suggested by Dirac in 1933 and later developed by Stueckelberg and Feynman in the 1940s. At the same time Schwinger developed an equivalent approach based on functional differentiation. The adaption of the formalism to statistical mechanics as presented in this chapter is due to Kac. Initially the functional approach to quantum mechanics was regarded with some skepticism due to the problem of defining a proper measure on the space of paths. However, nowadays the use of path integrals is the most common approach in quantum field theory with the additional advantage that it provides a unified and intuitive view of quantum mechanics, field theory, and statistical mechanics.

Further reading

An expanded version of Feynman's original derivation of the path integral for quantum mechanics can be found in R. P. Feynman and A. R. Hibbs, *Path Integrals and Quantum Mechanics*, McGraw-Hill (1965). The extension to Fermi–Dirac particles is given in F. A. Berezin, *The Method of Second Quantization*,

Academic Press (1966). An advanced text with applications of path integrals in many-particle physics is J. W. Negele and H. Orland, *Quantum Many-Particle Systems*, Addison-Wesley (1987). Further applications of path integrals can be found in H. Kleinert, *Gauge Fields in Condensed Matter Vol I*, World Scientific (1989). A thorough presentation of path integrals in quantum mechanics and relativistic quantum field theory can be found in C. Itzykson and J.-B. Zuber, *Quantum Field Theory*, McGraw-Hill (1988). For a rigorous definition of the measure in functional integrals see J. Glimm and A. Jaffe, *Quantum Physics, a Functional Integral Point of View*, Springer (1981).

12
A second look

It is time now to review the progress we have made so far. Our starting point was the fundamental atomic nature of matter. We also assumed that interactions between individual atoms and molecules are governed by the laws of mechanics, either classical or quantum depending on the particular circumstances. In the first chapter we developed a simple qualitative picture of the way molecules interact in a complex system. This qualitative picture allowed us to describe classical thermodynamics. In particular we were able to introduce the key concepts of equilibrium, temperature, entropy, and we were able to point out that complex systems in equilibrium can be well described with only a very small number of state variables. Given that matter is made of very large numbers of independent atoms or molecules this is an extraordinary result.

In the second chapter we began the process of formalizing the qualitative link from mechanics to thermodynamics. The formal development starts of course with mechanics. Mechanics on its own, however, is not enough, as it does not contain the concept of thermal equilibrium. The solution we presented was to define thermal equilibrium probabilistically, and the theory which results is statistical mechanics. This solution requires a fundamentally new idea which is not present in mechanics. Once this idea is accepted, the further development of the subject is straightforward if perhaps technically challenging.

Statistical mechanics is a very successful physical theory. In this book, we have applied it to a variety of systems including non-interacting and interacting gases, paramagnetic and spin systems, quantum systems with both Bose and Fermi statistics, astrophysics, helium superfluids and solids. These applications represent only a very small fraction of all the systems which have been successfully analyzed within statistical mechanics.

The success of statistical mechanics as a theory of nature, therefore raises the question of whether the probabilistic postulate upon which it is based is fundamental, or can be understood as a consequence of the laws of mechanics. In this chapter we address this issue briefly. The particular puzzles we address will include time reversibility, proofs of whether systems ever reach equilibrium, the meaning of temperature and its molecular basis, and ergodicity and mixing which represent attempts to understand statistical mechanics on a molecular basis.

Since we are reviewing progress we will also address some important technical issues which can arise. These include the possibility of negative temperature, the subtleties in replacing sums by integrals in partition function calculations, the intriguing question of zero-point energy in quantum systems, and the complications which can arise when internal degrees of freedom are included in partition function calculations.

12.1 The connection – reversibility

There is a profound difference between the laws of mechanics and the laws of statistical mechanics. The underlying laws of mechanics are invariant under time reversal. Given two states of a mechanical system there is no intrinsic way of determining which is the earlier and which is the later configuration. This is not the case in statistical mechanics. For an isolated system, the earlier state will have smaller entropy compared to the later state. Statistical mechanics thus provides an arrow for the direction of time. In order to understand the postulates of statistical mechanics on a microscopic basis the emergence of this arrow of time has to be understood.

The chaotic nature of complex systems is one possible explanation of irreversibility. In a chaotic system, very small changes in initial conditions cause very large changes in the trajectories followed. Thus to arrange that a mechanical system follows a prescribed time-reversed path, it is necessary to describe the initial conditions defining that reversed path to infinite precision. Any slight deviation will cause the system to follow a different path, and once generated, this difference will diverge rapidly from the trajectory we would like it to follow. From this standpoint the lack of reversibility present in complex systems is due to the impossibility of specifying configurations with infinite precision. There is an "infinite information" barrier, representing the infinite precision involved, which prevents complex systems from being time-reversal invariant. Deciding whether a given complex system is governed by these ideas of "chaotic" dynamics is not easy. Indeed no working mathematical formulation of these ideas as applied to statistical mechanics has even been suggested, let alone been implemented.

However, the properties of complex systems in equilibrium are very well described by statistical mechanics. It is therefore reasonable to think of this breakdown of time-reversal symmetry in complex systems as a law of nature as fundamental as the laws of mechanics themselves.

12.2 Poincaré recurrence

An important obstacle to combining the laws of mechanics with statistical mechanics consistently is encoded in the Poincaré recurrence, which is a property of any generic classical and, in fact, quantum mechanical system. We will describe Poincaré recurrence in this section.

12.2.1 Classical recurrence

Consider a system of N – possibly interacting – molecules in a box of finite volume V. Assume that this system is in a given configuration at time $t = 0$, not necessarily in thermal equilibrium. This configuration is then described by a point $x(0)$ in the phase space $\mathcal{P}$ of this system. The dynamics of the system, which is assumed to be governed by some Hamiltonian is then described by the corresponding Hamiltonian flow

$$g^t : \mathcal{P} \rightarrow \mathcal{P}$$
$$x(0) \mapsto x(t),$$

introduced in Chapter 8. Now, if there is a direction of time, we would not expect the system to return to this initial configuration at any later time. However, this expectation is contradicted by the following theorem of Poincaré.

Theorem 12.1 Let $g(x)$ be a continuous, volume-preserving one-to-one mapping which maps a bounded region $D \in \mathbf{R}^n$ to itself, $gD = D$. Then in any neighborhood U of any point $x \in D$ there is a point $x' \in U$ which returns to U, i.e. $g^n x \in U$ for some $n > 0$.

This theorem applies to the present situation since, as we know from Liouville's theorem in Chapter 8, the Hamiltonian flow is volume-preserving. It is also invertible since we can simple replace t by $-t$. Finally the phase space $\mathcal{P}$ for N particles in a box is bounded since for any finite energy configuration the momenta of the particles are bounded. Thus Poincaré's theorem implies in particular that any non-equilibrium configuration will eventually return to a configuration which is arbitrarily close to this configuration. Note, however, that the time necessary for this recurrence to occur can be rather long. As an estimate $t_{rec} \simeq \mathrm{e}^N$. So if the number of particles in the system is large enough this recurrence time can easily be made comparable with the age of the Universe. The problem posed by Poincaré recurrences is thus of a conceptual nature.

Proof. Consider the sequence $U, gU, \ldots, g^m U, \ldots$ All elements in this sequence have the same volume. If all elements of this sequence were non-intersecting, then necessarily the volume of D would have to be infinite. Thus

$$g^k U \cap g^l U \neq \emptyset \quad \text{some } k, l$$

or, equivalently, $g^n U \cap U \neq \emptyset, n = k - l$. If y is in this intersection, then $y = g^n x$ for some $x \in U$. Thus x and $g^n x$ are both in U. □

12.2.2 Quantum recurrence

Let us now turn to Poincaré recurrence in quantum mechanics. Instead of the boundedness of the phase space in the classical system we require finiteness of the entropy. In Section 3.4 we argued that the entropy, S, is related to the density of states $w(E)$ through $S = k \ln w(E)$. On the other hand, $w(E)$ is inversely proportional to the difference between energy levels, ΔE. Thus $\Delta E \propto e^{-\frac{1}{k}S}$.

Let us now consider the Schrödinger wave function of a quantum mechanical system which, for simplicity, we take to be identical bosons with no extra quantum numbers. Expanding the wave function in terms of eigenfunctions of the Hamiltonian we have

$$\psi(t, \mathbf{x}_1, \ldots, \mathbf{x}_N) = \sum_{i=0}^{\infty} \psi_i(\mathbf{x}_1, \ldots, \mathbf{x}_N)\, e^{-\frac{i}{\hbar}\epsilon_i t},$$

where it is assumed that the wave function is normalized, i.e. $\sum_i ||\psi_i||^2 = 1$. The quantum recurrence is then a consequence of the following

Theorem 12.2 There exists a time t^* such that $\psi(t^*)$ is arbitrarily close to $\psi(t)$, that is

$$||\psi(t^*) - \psi(0)|| < \eta$$

for arbitrarily small $\eta > 0$.

Proof. To show this we expand the norm above in terms of the eigenmodes,

$$||\psi(t^*) - \psi(0)|| = \sum_{i=0}^{\infty} 2||\psi_i||^2 \left(1 - \cos\left(\frac{\epsilon_i}{\hbar} t^*\right)\right) \leq 4.$$

Since all terms in this sum are non-negative there exists an integer $I > 0$ such that

$$\sum_{i=I+1}^{\infty} 2||\psi_i||^2 \left(1 - \cos\left(\frac{\epsilon_i}{\hbar} t^*\right)\right) \leq \frac{\eta}{2}.$$

It is therefore enough to prove that

$$\sum_{i=0}^{I} 2||\psi_i||^2 \left(1 - \cos\left(\frac{\epsilon_i}{\hbar} t^*\right)\right) \leq \frac{\eta}{2}.$$

This however follows from a classic theorem by Bohr and Wennberg. Indeed, since the finite sum above is a quasi-periodic function, this theorem asserts that for any $\delta > 0$ there is a set of integers N_i and a dense set of values $t^*(\delta, N_i)$ such that

$$\left| \frac{\epsilon_i t^*}{\hbar} - N_i 2\pi \right| < \delta.$$

This then implies that

$$\sum_{i=0}^{I} 2||\psi_i||^2 \left(1 - \cos\left(\frac{\epsilon_i}{\hbar} t^*\right)\right) \leq \sum_{i=0}^{I} 2||\psi_i||^2 \frac{\delta^2}{2}$$

which gives the desired bound for $\delta^2 \leq \frac{\eta}{2}$. □

In spite of this negative result attempts to put the foundations of statistical mechanics on a molecular basis have persisted and have led to the emergence of two key concepts. These are the concepts of ergodicity and of mixing to which we now turn.

12.3 Ergodicity

Consider the following simple example. We have a particle moving on the surface of a torus. The torus can be thought of as the constant energy surface in phase space. The torus is described in terms of the phase space variables p, q of the particle with $0 < p < 1$ and $0 < q < 1$ and by imposing periodic boundary conditions. Now suppose the equations of motion of the particle are given by

$$\frac{\mathrm{d}p}{\mathrm{d}t} = \alpha$$
$$\frac{\mathrm{d}q}{\mathrm{d}t} = 1$$

Solving these equations, the phase trajectory of the particle is given by

$$p = p_0 + \alpha(q - q_0).$$

If α is a rational number, $\alpha = m/n$, the trajectory will be periodic and repeat itself after a period of time $T = n$. If α is irrational the trajectory will never close. Let us examine the case when α is irrational more closely. Since the phase space is periodic any integrable function $f(p, q)$ can be expanded in a Fourier series as

$$f(p, q) = \sum_{l,m} f_{lm}\, \mathrm{e}^{2\pi \mathrm{i}(pl+qm)}.$$

We now claim that the time average of $f(p, q)$ is the same as its phase space average for this system. The time average is given by

$$< f >_T = \lim_{T\to\infty} \frac{1}{T} \int_{t_0}^{t_0+T} \mathrm{d}t \sum_{l,m} f_{lm}\, \mathrm{e}^{2\pi \mathrm{i}(p(t)l+q(t)m)}.$$

From the equations of motion we have

$$q(t) = q_0 + t$$
$$p(t) = p_0 + \alpha t$$

For α irrational this gives

$$< f >_T = f_{00}$$

On the other hand the phase space average of $f(p, q)$ is given by

$$< f >_s = \int_0^1 \int_0^1 \mathrm{d}p\mathrm{d}q f(p, q).$$

Evaluating the integral using the Fourier series representation for $f(p, q)$ we again get

$$< f >_s = f_{00}$$

which establishes the claim.

A dynamical system which has the property that the time average of any function of the phase space variables is equal to its phase space average, is said to be *ergodic*. The example we have considered of a particle on a torus is thus an ergodic system.

The phase space trajectory of an ergodic system covers all of the available phase space over time. It would seem that a dynamical system with such a property could provide a basis for the postulates of statistical mechanics. For such a system, if a probability distribution function over phase space can be chosen so that it is constant on a subspace of constant energy then the phase space average of any function using this probability distribution would be equal to the time average of the same function. Such a picture would justify the postulates formulated for the micro canonical ensemble. There would remain the problem of justifying why a result involving an infinite time average is relevant for real systems studied in statistical mechanics since statistical mechanics is only valid for systems over a limited time range. For real systems we know that the time of observation must be much greater than some microscopic timescale of the system and at the same time be smaller than a timescale set by the macroscopic properties of the system and its environment. For instance water in a container in a room can be thought of as being in equilibrium with its environment for a few hours but after a few days the container will be empty. The water will have evaporated into the room. To treat this problem by taking the limit as $T \to \infty$ can be justified only if the ratio of the appropriate microscopic timescale τ of the system is much smaller than the observation time T. In this case $T = N\tau$ with N a very large number.

It is important to note that for ergodic systems a distribution in phase space will not automatically tend to the uniform distribution assumed in statistical mechanics. An example will make this clear. Consider the distribution in phase space

$$\rho(p_0, q_0, 0) = \sin(\pi p_0)\sin(\pi q_0)$$

in the example considered earlier. After time t this distribution will become

$$\rho(p_0, q_0, t) = \sin(\pi(p_0 - \alpha t))\sin(\pi(q_0 - t)).$$

The distribution has not spread out. Its shape has not changed. All that has happened is that it has been displaced in phase space. The example shows that an arbitrary phase space distribution in an ergodic system does not always tend to a uniform distribution in phase.

In order to get round this difficulty the concept of *mixing* has been suggested by some authors. This property corresponds intuitively to requiring that any initial distribution in phase space must ultimately spread throughout phase space as time progresses. Systems with the mixing property are ergodic but as we saw in our example the converse is not true.

12.4 Equipartition law and temperature

In Chapter 1 the argument was made that collisions between molecules would result in the system reaching a state of equilibrium in which the entire system of molecules would have a well-defined average kinetic energy. The fast molecules, it was suggested, as a result of collisions would on average be slowed down while the slow molecules would similarly be speeded up. This expectation provided a molecular basis for the equipartition law of energy which in turn led to the concept of thermal equilibrium formalized in the zeroth law of thermodynamics. In this section we reconsider this argument critically. We also point out some peculiarities about negative temperatures.

12.4.1 Equipartition law

Adapting the argument of Chapter 1 to a solid, one assumes that the equipartition law of energy will hold for a system of harmonic oscillators, regarded as a model for a solid. Here the molecular basis for equipartition was expected to follow from the interaction between the harmonic oscillators. These interactions could be modeled by including non-linear terms. (Without such non-linear terms the harmonic oscillators could never share their energies.) A redistribution of energy between the different modes could only result if non-linearities, representing interactions between different modes, are present in the system. These "interaction terms" would lead to "scattering" between different modes and thus lead to the equipartition law of energy. A justification for the zeroth law of thermodynamics for a solid would then have been achieved.

In 1955 Fermi, Pasta, and Ulam (FPU) proceeded to check if this was the case by studying the vibrations of particles connected by non-linear springs in one dimension. The solution of the non-linear problem was carried out on an electronic computer which had just been developed.

Much to their surprise FPU found that the system did not approach an equilibrium state with all modes of the system sharing energy. The energy was shared only among a few low-energy modes of the system. This conclusion was tested by varying the type of non-linearity introduced but no qualitative difference in the result occurred.

The behavior discovered by FPU is now well understood in terms of a theorem due to Kolmogorov, Arnold, and Moser (KAM). This theorem states that for a system with weak anharmonic couplings most of the constant energy surface in phase space will consist of confined regions known as invariant tori. The system will remain in these regions even under small perturbations. As the perturbations increase these invariant regions break down and, at some point, a transition to chaotic behavior occurs. At this stage something similar to the equipartition law emerges. Thus the justification for the zeroth law of thermodynamics on a molecular basis is more involved than had originally been suspected.

12.4.2 Negative temperatures

We now turn to the curious possibility of the temperature being negative. As an example of how this can happen let us consider the quantum version of the paramagnetic model considered in Chapter 2. Each dipole can now be in one of two energy states, namely $E_- = -\mu B$ and $E_+ = \mu B$ ($\mu > 0$) so that the partition function is simply

$$Z(\beta) = (\mathrm{e}^{-\mu\beta B} + \mathrm{e}^{\mu\beta B})^N.$$

The probability that the system is in the energy state $E_+ = \mu B$ is

$$P(E_+) = \frac{\mathrm{e}^{-\mu\beta B}}{Z(\beta)}$$

while the probability of the dipole being in the energy state $E_- = -\mu\beta$ is

$$P(E_-) = \frac{e^{\mu\beta B}}{Z(\beta)}.$$

Thus

$$\frac{P(E_+)}{P(E_-)} = e^{-2\mu\beta B}.$$

This expression implies that the energy $E_+ > E_-$ is less probable as the temperature is lowered. In particular, $P(E_+)/P(E_-) \to 0$ as $\beta \to \infty$. This is the normal situation.

There are situations, however, where population inversions can occur, that is, where the higher energy states are more populated than the lower energy states. An example of such a situation is a crystalline system with special nuclear spin properties. The nuclear moments of the crystal have a relaxation time t_1 for mutual interaction of nuclear spins which is very small compared to the relaxation time t_2 for interaction between the nuclear spins and the lattice. If such a crystal is placed in a strong magnetic field which is then quickly reversed then the spins are unable to follow the switch over. This leaves the system in a state in which the nuclear spin states of high energy are more numerous than those of lower energy. The system will remain in this state for a time of the order of t_1. During this time, we can say the system is in a negative temperature state since the anomalous population inversion can be obtained by reversing the sign of temperature in the expression above. Experimentally such systems have been observed.

It is amusing to work out the thermodynamic properties of our simple model for both positive and negative temperatures. The free energy of the system is

$$\begin{aligned} F &= -\frac{1}{\beta}\ln Z(\beta) \\ &= -NkT\ln\left[2\cosh\left(\frac{\mu B}{kT}\right)\right] \end{aligned}$$

from which it follows that

$$\begin{aligned} S &= Nk\left[\ln\left[2\cosh\left(\frac{\mu B}{kT}\right)\right] - \frac{\mu B}{kT}\tanh\left(\frac{\mu B}{kT}\right)\right] \\ U &= F + TS = -N\mu B\tanh\left(\frac{\mu B}{kT}\right) \\ M &= -\left(\frac{\partial F}{\partial B}\right)_T = N\mu\tanh\left(\frac{\mu B}{kT}\right) \end{aligned}$$

where S, U and M represent the entropy, internal energy, and magnetization of the system. Now observe at $T = 0$ the entropy is equal to zero and $U = -N\mu B$. As the temperature increases, both S and U increase monotonically. At $T = \infty$ $U = 0$ and S reaches its maximum value $Nk \ln 2$.

The temperature $T = -\infty$ is physically identical to $T = \infty$. Both give identical values for thermodynamic quantities. Finally, at $T = 0_{-}$, we have $U = N\mu B$, its maximum value. Thus the negative temperature region is not below absolute zero but above infinite temperature!

Let us now review different arguments for allowing only positive temperatures in statistical mechanics and thermodynamics. We will find that for systems with a finite number of finite energy configurations none of these arguments apply. We first consider the thermodynamic argument.

We recall that in the absolute temperature scale the relation

$$\frac{Q_1}{T_1} = \frac{Q_2}{T_2}$$

holds for a reversible Carnot engine which absorbs heat Q_1 from a reservoir of temperature T_1 and rejects heat Q_2 to a reservoir of temperature T_2 where T_1 is bigger than T_2. The efficiency, η, of such an engine was defined to be

$$\eta = \frac{Q_1 - Q_2}{Q_1} = 1 - \frac{T_2}{T_1}.$$

If we choose T_1 positive then we must have T_2 strictly positive. Otherwise the second law of thermodynamics would be violated. This shows that the temperature of a reservoir cannot be negative. It has nothing to say about a system with a finite number of finite energy configurations.

Next we consider the statistical mechanics argument for positive temperature using the canonical ensemble. We note that the partition function is defined to be

$$Z(\beta) = \sum_{states} e^{-\beta E}.$$

For any system where the sum over energy configurations is infinite, $\beta < 0$ would lead to $Z(\beta)$ diverging. This is not acceptable, hence we must have β strictly positive for such systems.

Our final argument for positive temperature considers the entropy, S, of a closed system in the micro canonical ensemble. We will show that the condition for equilibrium does not allow the temperature T to be negative. However, as we saw states of partial equilibrium of a system involving only a finite number of degrees of freedom can be in a state of negative temperature. Let us now present the argument.

We divide a system in equilibrium into a large number of small (but macroscopic) parts and let M_k, E_k, $\mathbf{P}_k$ denote the mass, energy, and the momentum of the kth part. The entropy of this system is a function of the internal energy. The internal energy depends only on the microscopic energy of the state. It cannot depend on the net macroscopic motion of the system since such motions can be eliminated by a suitable coordinate transformation which takes the system from one inertial coordinate system to another. Thus in the micro canonical ensemble the internal energy of the kth part is given by

$$U_k = E_k - \left(\frac{|\mathbf{P}_k|^2}{2M_k}\right)$$

and

$$S = \sum_k S(U_k).$$

If the system is taken to be closed with the total momentum, angular momentum, and energy conserved then we must have

$$\sum_k \mathbf{P}_k = const$$

$$\sum_k \mathbf{r}_k \times \mathbf{P}_k = const$$

with r_k the coordinate of the center of mass. In the equilibrium state the entropy as a function of $\mathbf{P}_k$ must be a maximum subject to these conservation laws. Using the method of Lagrange multipliers the condition for the entropy to be a maximum can be obtained by maximizing

$$\hat{S} = \sum_k [S_k + \boldsymbol{\lambda} \cdot \mathbf{P}_k + \boldsymbol{\mu} \cdot (\mathbf{r}_k \times \mathbf{P}_k)]$$

with respect to $\mathbf{P}_k$. Here λ and μ are the Lagrange multipliers. Using the thermodynamic identity

$$\frac{\partial S}{\partial U} = \frac{1}{T}$$

we get

$$\mathbf{v}_k = \frac{\mathbf{P}_k}{M_k} = T\boldsymbol{\lambda} + T\boldsymbol{\mu} \times \mathbf{r}_k .$$

Thus a system in equilibrium characterized by the entropy being maximum cannot have internal macroscopic motion. The entire system can have an overall translation velocity $T\lambda$ and a velocity due to the entire system rotating with constant angular velocity $T\mu$.

If we now impose the sufficient condition for $\hat{S}$ to be a maximum with respect to variations of $\mathbf{P}_k$ we must have

$$\frac{\partial^2 \hat{S}}{\partial^2 \mathbf{P}_k} < 0$$

but

$$\frac{\partial^2 \hat{S}}{\partial^2 \mathbf{P}_k} = -\frac{1}{T M_k},$$

and hence T is positive. Therefore a system with negative temperature cannot be in thermodynamic equilibrium. However, negative temperatures can arise at intermediate times when two timescales are involved in the approach to equilibrium. As an illustration we describe, in the next subsection, a system in which mass motion occurs at intermediate times and which results in negative temperature.

12.4.3 Negative temperatures and turbulence in two dimensions

In our presentation of statistical mechanics we have assumed throughout that the system is governed by Hamiltonian dynamics. Turbulence on the other hand is dissipative. However, in two dimensions, if the viscosity is small then so is the energy dissipation. In this regime a statistical mechanics approach may be reasonable. In this subsection we consider a simple model with negative temperature which could represent a turbulent fluid in two dimensions. Turbulence is characterized by the appearance of a large number of eddies or vortices with random chaotic properties.

In fluid dynamics the relevant equation is the Navier–Stokes equation. In the absence of viscosity this equation reduces to Euler's equation which, for an incompressible fluid moving with a velocity field $\mathbf{v}(\mathbf{x}, t)$ is given by

$$\frac{\partial \mathbf{v}}{\partial t} + (\mathbf{v} \cdot \nabla)\mathbf{v} = \mathbf{f}, \qquad \nabla \cdot \mathbf{v} = 0,$$

where $\mathbf{f}$ is the net force acting on the fluid of particles. We have also assumed that the fluid has unit density. In the statistical approach to turbulence $\mathbf{f}$ is a random force and the velocities are assumed to have statistical properties. A simple model would be to replace $\mathbf{v}(\mathbf{x}, t)$ by $\langle \mathbf{v}(\mathbf{x}, t) \rangle$ and similarly,

$$\frac{\partial \langle \mathbf{v} \rangle}{\partial t} + (\langle \mathbf{v} \rangle \cdot \nabla)\langle \mathbf{v} \rangle = 0.$$

In this crude approximation correlations between velocities have been ignored. Such correlations are an important feature of turbulent flows but the correlations between velocities of neighboring points is weak and hence this approximation seems reasonable. It simply states that the average velocity satisfies Euler's equation. We have also set $\langle f \rangle = 0$.

Motivated by the fact that a turbulent flow contains a large number of vortices we consider a two-dimensional region D of finite area A containing N localized vortices of strength ± 1. Such a simple characterization of a vortex is possible in two dimensions. The vorticity strength $\xi(\mathbf{x}, t)$ of a fluid of particles moving with velocity $\mathbf{v}(\mathbf{x}, t)$ is defined by

$$\xi(\mathbf{x}, t) = \nabla \wedge \mathbf{v}(\mathbf{x}, t) .$$

The kinetic energy of a fluid in two dimensions can also be written in terms of vorticities since

$$\begin{aligned} E &= \frac{1}{2}\int_D (\mathbf{v}\cdot\mathbf{v})\mathrm{d}^2x \\ &= \frac{1}{4\pi}\int\int \xi(\mathbf{x})\xi(\mathbf{x}')\log|\mathbf{x}-\mathbf{x}'|\,\mathrm{d}^2x\,\mathrm{d}^2x' + C , \end{aligned}$$

where C is a boundary term. The vorticity distribution for N localized vortices in this model can be written as

$$\xi(\mathbf{x}, t) = \sum_{i=1}^{N} \Gamma_i\, \delta^{(2)}(\mathbf{x} - \mathbf{x}_i(t)) , \qquad \Gamma_i = \pm 1 .$$

We now describe this system using ideas from statistical mechanics. For this we recall that in the canonical ensemble the probability for the occurrence of a state with energy E_s is given by

$$P_s = \frac{\mathrm{e}^{-\frac{E_s}{T}}}{Z} ,$$

where

$$Z = \sum_s \mathrm{e}^{-\frac{E_s}{T}} = \mathrm{e}^{-\frac{U-TS}{T}} , \qquad U = \sum_s P_s E_s$$

is the partition sum. It should also be stressed that the parameter T introduced here is not related to the molecular temperature of the underlying fluid. To emphasize this we have not introduced the Boltzmann constant k in the formula for the partition function. The entropy is then given by

$$\begin{aligned} S &= -\left(\log Z + \frac{U}{T}\right) = -\sum_s P_s\left(\log Z + \frac{E_s}{T}\right) , \qquad \text{or} \\ S &= -\sum_s P_s \log P_s . \end{aligned}$$

To model N vortices of strength $\Gamma_i = \pm 1$, placed in a region D of finite area A, we write $P_s = f(\mathbf{x}_1, \ldots, \mathbf{x}_N)$, where $\mathbf{x}_i$ is the location of the ith vortex. Then

$$S = \int \mathrm{d}^2x_1 \cdots \int \mathrm{d}^2x_N\, f(\mathbf{x}_1, \ldots, \mathbf{x}_N) \log f(\mathbf{x}_1, \ldots, \mathbf{x}_N) .$$

The entropy is maximal for constant $f(\mathbf{x}_1, \ldots, \mathbf{x}_N) = A^{-N}$, where A is the area of the region D. Using this distribution the energy of the system becomes

$$\langle E_c \rangle = -\frac{1}{4\pi} N(N-1) \int \mathrm{d}^2x \int \mathrm{d}^2x' \log|\mathbf{x} - \mathbf{x}'| + C.$$

It is clear that higher values of $\langle E_c \rangle$ are possible simply by pushing the vortices closer together. Hovever, this would mean that the distribution function $f(\mathbf{x}_1, \ldots, \mathbf{x}_N)$ is not uniform and hence represents a configuration with lower entropy. Thus

$$\frac{\mathrm{d}S}{\mathrm{d}\langle E \rangle} < 0$$

for $\langle E \rangle > \langle E_c \rangle$. But, on the other hand, $\mathrm{d}S/\mathrm{d}\langle E \rangle = 1/T$. Thus in this model we again encounter a negative temperature. This is, however, not in contradiction with the general arguments presented above since this turbulent system is not in equilibrium.

12.5 Density of states and surface effects

As we saw in Chapter 7, the wave function for a free non-relativistic quantum particle which satisfies periodic boundary conditions is given by

$$\Psi_{lmn}(x, y, z) = \mathcal{N} \mathrm{e}^{\frac{2\pi i l}{L} x} \mathrm{e}^{\frac{2\pi i m}{L} y} \mathrm{e}^{\frac{2\pi i n}{L} z}$$

where L is the period and $\mathcal{N}$ is a normalization constant. From this solution the components of the momentum vector of the particle are found to be

$$k_1 = \frac{2\pi \hbar l}{L}$$
$$k_2 = \frac{2\pi \hbar m}{L}$$
$$k_3 = \frac{2\pi \hbar n}{L}.$$

The density of states in terms of the momentum variables $k_i = 2\pi \hbar n_i / L$ is then given by

$$\Delta l \Delta m \Delta n = \frac{V}{(2\pi \hbar)^3} \Delta k_1 \Delta k_2 \Delta k_3$$

where V is taken to be the volume within which the particle is confined. We used this expression to convert sums over states to integrals over momentum variables. We now reexamine this prescription more carefully. This will lead us to the conclusion that the density of states for this system includes in it contributions from the surface of the volume V within which the particle is confined. First we clear up a conceptual point. In order to identify V as the volume in which the particle is confined it seems in line with intuition to set the wave function $\Psi(x, y, z) = 0$ for $x = y = z = L$ or 0 rather than require $\Psi(x, y, z)$ be a periodic function of its arguments. Such boundary conditions correspond to a particle confined within a region by an infinite potential and hence represent the physical system more accurately. We will call this the box problem.

We now consider a particle in a box of sides a, b, c. The wave function for the particle is

$$\Psi_{lmn}(x, y, z) = \mathcal{N} \sin\left(\frac{l\pi z}{a}\right) \sin\left(\frac{m\pi z}{b}\right) \sin\left(\frac{n\pi z}{c}\right).$$

The magnitude of the square of the momentum of this wave function can be determined by acting on it with the operator $-\hbar^2\nabla^2$. This gives

$$k^2 = |\,\mathbf{k}^2\,| = \pi^2\hbar^2 \left[\frac{l^2}{a^2} + \frac{m^2}{b^2} + \frac{n^2}{c^2}\right]$$

with l, m, n restricted to positive non-zero integer values.

To determine the density of states $\rho(\mathbf{k})$ for this system we proceed indirectly. We first calculate $g(\mathbf{k})$ which is the function that gives the number of states with momentum less than or equal to k. Once $g(\mathbf{k})$ is known, $\rho(\mathbf{k})$ can be determined from the equation $\rho(\mathbf{k}) = \mathrm{d}g(\mathbf{k})/\mathrm{d}k$. Now

$$g(\mathbf{k}) = \sum_{l,m,n}{}' f(l, m, n)$$

where f is the characteristic function on the set of positive integers,

$$f(l, m, n) = \begin{cases} 1 & \text{where } l, m, n = 1, 2, 3, \ldots \\ 0 & \text{otherwise}, \end{cases}$$

and $\sum'$ signifies that the natural numbers l, m, n over which the summation is carried out must satisfy the inequality

$$\frac{l^2}{a^2} + \frac{m^2}{b^2} + \frac{n^2}{c^2} \leq \frac{k^2}{\pi^2\hbar^2}.$$

We note that $g(k)$ is defined over the first quadrant in the space of integers excluding zero. This restriction can be removed by writing

$$g(k) = \frac{1}{8}\Bigg[\sum_{l,m,n}{}' f^*(l, m, n) - \left(\sum_{l,m}{}' f^*(l, m, 0) + \sum_{l,n}{}' f^*(l, 0, n) + \sum_{m,n}{}' f^*(0, m, n)\right)$$
$$+ \left(\sum_{l} f^*(l, 0, 0) + \sum_{m} f^*(0, m, 0) + \sum_{n} f^*(0, 0, n)\right) - 1\Bigg]$$

where $f^*(l, m, n) = 1$ for integral values of l, m, n and is zero otherwise.

The first term is an overcount of the states in $g(\mathbf{k})$ as it includes states with any one of the labels l, m, n equal to zero. The second factor corrects for this but in the process overcompensates since each factor in the second term contains terms in which more than one of the labels is zero. The third and fourth terms can be similarly understood.

The advantage of writing $g(\mathbf{k})$ in this form is clear. The first term represents the number of lattice points contained within and on an ellipsoid whose defining equation is

$$\frac{x^2}{a^2}+\frac{y^2}{b^2}+\frac{z^2}{c^2}=\frac{k^2}{\pi^2\hbar^2}.$$

Similarly the second set of terms counts the number of lattice points on or within an ellipse corresponding to the respective cross-sections of the ellipsoid. These are all geometrical objects. For large values of $ak/\pi\hbar, bk/\pi\hbar, ck/\pi\hbar$ we can make the asymptotic replacement

$$\frac{1}{8}{\sum}' \to \frac{1}{8}V\int' \mathrm{d}x\mathrm{d}y\mathrm{d}z = V\frac{k^3}{6\pi^2\hbar^3}$$

where the standard expression for the volume of the ellipsoid has been used. In a similar manner the second set of terms become surface integrals over the area of an ellipse and the third set of terms are approximated by line integrals. Thus for large k values we have

$$g(\mathbf{k}) = V\frac{k^3}{6\pi^2\hbar^3} - S\frac{k^2}{16\pi\hbar^2} + L\frac{k}{16\pi\hbar} - \frac{1}{8} + E(\mathbf{k})$$

where $E(\mathbf{k})$ is the error made by replacing the sums by integrals. It is now a simple matter to derive the density of states function $\rho(\mathbf{k})$. We have

$$\rho(\mathbf{k}) = V\frac{k^2}{2\pi^2\hbar^3} - S\frac{k}{8\pi\hbar^2} + \cdots$$

The first term is precisely the expression for the density of states we had obtained with periodic boundary conditions since $V/(2\pi\hbar)^3\mathrm{d}^3k = V/(2\pi\hbar)^3 4\pi k^2\mathrm{d}k = \rho(\mathbf{k})\mathrm{d}k$ gives $\rho(\mathbf{k}) = Vk^2/2\pi^2$. Thus when the dimensions of the box are large both types of boundary conditions lead to the same density of states.

12.6 Zero-point energy

Before the discovery of quantum mechanics Planck suggested on the basis of semi-classical arguments to modify the expression for the energy of a harmonic oscillator from $nh\nu$ to $(n+\frac{1}{2})h\nu$. It was subsequently realized by Stern and Pauli that the presence of this extra term for a collection of molecules would have experimental consequences. The extra term $\frac{1}{2}h\nu$ is the zero-point energy of the oscillator. It is present even when the system is in its lowest energy configuration. If this term was not present in the vibrational energy of a complex molecule then Pauli and Stern showed that different isotopes of the molecule would have different high-temperature vapor pressures. Indeed the difference in vapor pressures would be so significant that a separation of the isotopes could be carried out in this way. Experimentally there was no noticeable difference in the high-temperature vapor pressures of isotopes. Thus experiments suggested that the zero-point energy term should be introduced. With the discovery of quantum mechanics it was found that the zero-point energy term was required by the theory. A particle confined to a region of space Δx was expected to have a momentum spread Δp proportional to $\hbar/\Delta x$ as a consequence of the uncertainty principle of quantum mechanics. This momentum fluctuation term contributes to the energy giving rise to the zero-point energy term. For a harmonic oscillator, this mechanism is responsible for the presence of the zero-point energy term $\frac{1}{2}h\nu$.

The presence of a zero-point energy term for radiation is a more delicate matter. If we consider radiation confined within a box of sides (L_1, L_2, L_3) then we can think of the radiation as consisting of a superposition of oscillators with frequencies given by

$$\omega(\mathbf{n}) = \pi c \sqrt{\frac{n_1^2}{L_1^2} + \frac{n_2^2}{L_2^2} + \frac{n_3^2}{L_3^2}}$$

with n_i taking values from the natural numbers. The zero-point energy of this system is then expected to be

$$E_0(\mathbf{L}) = \sum_{\mathbf{n}} \frac{1}{2} \hbar\omega(\mathbf{n})$$

which is a divergent object. This result suggests that great care is needed in applying ideas taken from the quantum theory of system with a finite number of degrees of freedom to systems, such as radiation, which contain an infinite number of degrees of freedom.

Let us, however, pursue the idea introduced at a qualitative level. We note that E_0 depends on the parameters (L_1, L_2, L_3), which represent the sides of the box within which the radiation is confined. This means that if these parameter values are changed the energy will change. Such a change would represent the presence of a force

$$\mathbf{F} = -\frac{\partial E_0}{\partial \mathbf{L}}.$$

Let us simplify the problem in order to understand the nature of this force. Suppose our box is one-dimensional so that

$$\omega(n) = \frac{\pi c n}{L}$$

then

$$E_0(L) = \frac{1}{2}\frac{\hbar\pi c}{L}\sum_{n=1}^{\infty} n.$$

The divergent nature of this expression is now very clear. To continue we need to regularize this divergent expression. For this we regard $E_0(L)$ as the analytic continuation of $E_0(L; s)$ defined by

$$E_0(L) = \frac{1}{2}\frac{\hbar\pi c}{L}\sum_{l=1}^{\infty} l^{-s}$$

to the point $s = -1$. This seemingly ad hoc procedure is more natural in a complete treatment of radiation. For the moment we continue with our calculation. We note that

$$E_0(L; s) = \frac{1}{2}\frac{\hbar\pi c}{L}\zeta(s)$$

where $\zeta(s)$ is the Riemann zeta function. This function can be analytically continued to the point $s = -1$ where $\zeta(-1) = -1/12$. We can thus define the regularized expression for $E_0(L)$ to be

$$E_0(L) = -\frac{1}{24}\left(\frac{\hbar\pi c}{L}\right).$$

From this well-defined expression we calculate the force between the ends of the box due to quantum zero-point fluctuations to be

$$F = -\left(\frac{\partial E_0}{\partial L}\right) = -\frac{1}{24}\frac{\hbar\pi c}{L^2}.$$

Such a force was first predicted by Casimir to exist between two parallel plates and was subsequently experimentally measured. Nowadays the Casimir force is used in microscopic electromechanical devices.

12.7 Internal degrees of freedom

In this section we sketch how internal degrees of freedom of a molecule are treated. In the days before quantum theory, the absence of an equipartition law for distributing energy among the internal degrees of freedom of a molecule was a puzzle. With the advent of quantum theory it was realized that internal degrees of freedom only showed up after certain energy thresholds were crossed. As an example let us consider a diatomic molecule. Besides the kinetic energy of the center of mass of the molecule, we expect the two atoms of the molecules to vibrate with respect to each other and to rotate. In principle the corresponding energy eigenvalues of the system have to be determined before the partition function can be calculated. In practice approximate values for the eigenvalues have to be used. These approximate eigenvalues contain terms which can be identified with the different motions such as relative vibrations and rotations of the constituent atoms of the molecule. One way of obtaining such a structure is by using the *Born–Oppenheimer approximation*, which we briefly summarize. For more complicated molecules appropriate approximation schemes must be introduced which allow us to determine the important degrees of freedom of the system and the energies associated with them. The energy eigenvalues of a diatomic molecule have the structure

$$E_{n,j} = \left(n + \frac{1}{2}\right)\hbar\omega + \frac{j(j+1)}{2I}\hbar^2 + E_e + E_{\text{center of mass}}$$

where $n = 0, 1, \ldots, \infty$ and $j = 0, 1, \ldots, \infty$. The first term represents the vibrational energy of the two nuclei with ω the vibrational energy. The second term represents the rotational energy of the nuclei with I, the moment of inertia of the two nuclei about the rotation axis. The third term represents the energy of the electrons in the diatomic molecule. The structure presented is for illustration only. We have neglected effects of anharmonicity and the interaction between the vibrational and rotational modes of the molecule.

In this approximation, the canonical partition function can be written as

$$Z = Z_v Z_r Z_e Z_{cm}$$

with

$$\begin{aligned}
Z_{cm} &= \frac{1}{N!}\left(\frac{2\pi M}{\hbar^2\beta}\right)^{\frac{3N}{2}} V^N \\
Z_e &= \sum_{E_e} \mathrm{d}(E_e)\,\mathrm{e}^{-\beta E_e} \\
Z_r &= \sum_j \mathrm{e}^{-\beta\frac{j(j+1)\hbar^2}{2I}} \\
Z_v &= \sum_n \mathrm{e}^{-\beta(n+\frac{1}{2})\hbar\omega}.
\end{aligned}$$

Once Z is determined, the thermal properties of the system can be calculated.

At this stage we would like to draw attention to an apparent paradox involved in taking the internal structure into account. Let us consider the hydrogen atom and take the binding energy of the electron into account, i.e. we take the energy eigenvalues of the system as

$$E_n = \frac{|\mathbf{P}|^2}{2(M+m)} - \frac{R}{n^2},$$
$$n = 1, 2, \ldots, \infty.$$

These eigenvalues are obtained by solving Schrödinger's equation for the hydrogen atom. The first term represents the center of mass energy of the proton–electron system with M the proton mass and m the electron mass. The second term represents the energy eigenstates of the bound electron. If we want to determine the partition function for a collection of hydrogen atoms in volume V then we have to calculate

$$Z_N = \sum_{n=1}^{\infty} \frac{1}{h^{3N} N!} \int \mathrm{d}^{3N} x \int \mathrm{d}^{3N} p \, \mathrm{e}^{-\beta\left(\frac{|\mathbf{P}|^2}{2(M+m)} - \frac{R}{n^2}\right)} \mathrm{d}_n$$

where $\mathrm{d}_n = 2n^2$ is the number of levels in hydrogen which all have binding energy $-R/n^2$. Thus

$$Z_N = \frac{V^N}{N!} \left(\frac{2\pi (M+m)}{\beta h^2} \right)^{\frac{3N}{2}} \sum_{n=1}^{\infty} 2n^2 \mathrm{e}^{\beta \frac{R}{n^2}}$$

with probability that at temperature T, the system is in a specific energy level E_n, given by

$$P_\beta(E_n) = \frac{1}{Z_N} \mathrm{e}^{-\beta E_n}.$$

Now we have a problem. Consider $\sum_{n=1}^{\infty} 2n^2 \mathrm{e}^{\beta \frac{R}{n^2}}$. This sum diverges which means that the probability of the atom being in any one of its bound states is zero no matter what the value of the temperature! One way of avoiding this difficulty is to observe that for a system in volume V the sum over n cannot extend to infinity. The reason is the Bohr radius R_n of the orbit associated with the state n is $R_n = n^2 r_0$ where r_0 is the Bohr orbit. If we have the system in volume $V \sim L^3$ then $R_n \leq L$, i.e. $n^2 r_0 \leq L$, i.e. there is a maximum value n can take, namely $n = \bar{n} \cong \sqrt{L/r_0}$. This resolution is experimentally ruled out! If we allow β to be reasonably small then

$$\sum_{n=1}^{\bar{n}} 2n^2 \mathrm{e}^{\beta \frac{R}{n^2}} \sim \sum_{n=1}^{\bar{n}} 2n^2 \sim \int_1^{\bar{n}} 2n^2 \mathrm{d}n \sim N_{\max}^3 \sim V^{\frac{1}{2}}.$$

This volume factor implies that the probability of the atom being in any one of its bound states is still zero in the thermodynamic limit ($V \to \infty$, $N \to \infty$, V/N finite) which is not acceptable. We then realize that the cutoff in the sum over the free hydrogen atom eigenstates should be determined not by the size of the container but by the mean free path of the atom. Roughly if λ is the mean free path then $N\lambda^3 \sim L^3$,

$\lambda \sim (V/N)^{\frac{1}{3}}$ and setting $\bar{n} \sim \sqrt{\lambda/r_0}$ gives $\sum_{n=1}^{\bar{n}} 2n^2 e^{\beta \frac{R}{n^2}} \sim (V/N)^{\frac{1}{2}}$. This modifies the equation of state from $PV = NkT$ to $PV = N(1 + 1/2N)kT = NkT$, for large N and gives a non-zero value for the probability of the atom to be in one of its bound states even in the thermodynamic limit. The paradox is thus resolved.

When we discuss chemical reactions we will suppose that it is sufficient to consider atoms to be either in their ground state or to have been converted to an ion and a free electron; the excitation probability is negligible.

Problems

Problem 12.1 Consider a container consisting of two compartments, separated by a wall with a hole, and containing N particles with positive total kinetic energy. Assume that at some initial time, t_0, all N particles are in one of the two compartments. Using Poincaré recurrence, show that there exists a time $t_1 \gg t_0$ such that at $t = t_1$ all particles are again found in the same compartment. Why is this behavior not observed experimentally?

Problem 12.2 As an illustration of the problem concerning the approach to equilibrium, consider the following simple dynamical system proposed by Kac (1956), consisting of a ring with equally spaced points $s_1, \ldots, s_q$. Assume that $p < q$ of these points are marked. Between each two points there is a ball that can be either black or white. We now consider a discrete time evolution by which each ball moves clockwise past one of the points s_i with the ball changing its color if and only if it passes a marked point.

(1) Write down the evolution equations for the number of black and white balls, $N_b(t_k)$ and $N_w(t_k)$ respectively.
(2) Solve these equations making the assumption that the color of the ball is not correlated with the fact of having a marked or unmarked point ahead of it. Does the number of black and white balls converge to an equilibrium configuration?
(3) Discuss the assumption made in (2) and give it a statistical interpretation.

Problem 12.3 A realistic application of statistical mechanical description of two-dimensional turbulence is superfluid helium in a vertical cylinder (see Figure 12.1). If angular momentum along the z-axis is injected into the system, a normal fluid will undergo rigid body rotation due to its viscosity. A superfluid, on the other hand, develops an array of quantized vortices.

We model this system by a collection of vortices subject to a central potential $V(r)$. The central potential, which replaces the rigid walls of the cylinder, acts to contain the vortices in a finite volume. Rewrite the kinetic energy for the fluid in

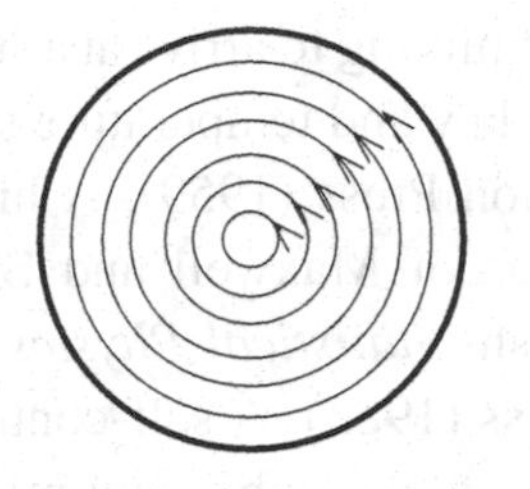
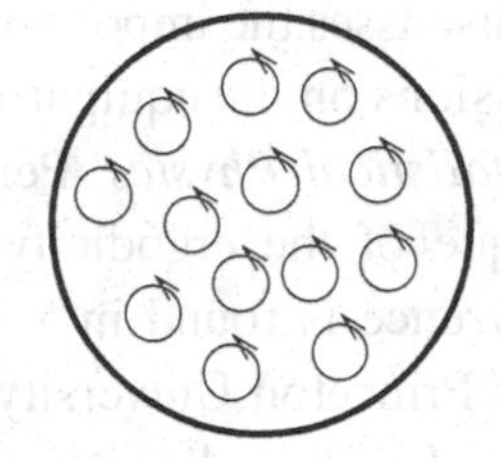

Figure 12.1 Rigid rotation of viscous fluid (left) and quantized vortices in superfluid (right).

terms of a two-dimensional effective potential between the vortices and write a program to find the configuration of lowest energy for this system. Describe the geometry of this configuration.

Problem 12.4 Compute the density of states function, $\rho(\mathbf{k})$, for a box in two dimensions and compare with periodic boundary conditions.

Historical notes

The connection between the density of states and the geometry of a cavity which we discussed in Section 12.5 is part of an extensive mathematical theory which goes under the name of spectral geometry. Mathematical interest in this problem can be traced to Lorentz who, in 1910 based on examining a few special cases, conjectured that the density of state function's dependence on volume was true for a cavity of arbitrary shape. Lorentz challenged mathematicians in a lecture given in Göttingen to prove this conjecture. It is said that the famous mathematician Hilbert who attended Lorentz's lecture predicted that the conjecture would not be proved in his lifetime. Less than two years later Hilbert's student Weyl provided a proof, thereby opening a new and important chapter in mathematical analysis. An interesting and readable account of this problem is contained in Kac's article "Can one hear the shape of a drum?" (*Amer. Math. Monthly* **73** 1–23, 1966). This question was resolved in 1991 when C. Gordon, D. Webb, and S. Wolpert showed that two drums with different geometric shapes can have identical normal modes. ("You cannot hear the shape of a drum", *Bull. Amer. Math. Soc.* **27**, 134–138, 1992.)

Further reading

A good text on the relation between ergodicity and statistical mechanics can be found in L. Reichl, *The Transition to Chaos*, Springer-Verlag (1992) and N. S. Krylov, *Works on the Foundations of Statistical Physics*, Princeton University Press

(1979) who also discusses the importance of mixing to arrive at a statistical system. For further discussions on the equipartition law and temperature see L. D. Landau and E.Lifshitz, *Statistical Physics*, Pergamon Press (1959). A historical account including a critique of the ergodicity ideas of Maxwell and Boltzmann based on Poincaré recurrence is found in S. Brush, *Statistical Physics and the Atomic Theory of Matter*, Princeton University Press (1983). A self-contained discussion of turbulence including two-dimensional systems with many relevant references can be found in U. Frisch, *Turbulence*, Cambridge University Press (1995). R. Peierls, Surprises *in Physics*, Princeton University Press (1979) discusses aspects of internal degrees of freedom in thermodynamic systems.

13

Phase transitions and the renormalization group

13.1 Basic problem

As the external conditions of a macroscopic system are changed the properties of the system can sometimes change dramatically. A good example of such a phenomenon is provided by a ferromagnet. When the temperature of a ferromagnet is increased above a certain temperature, called the *Curie temperature*, then the ferromagnet loses its magnetism and changes into a paramagnet. Furthermore in the neighborhood of the Curie temperature T_C the *susceptibility* χ diverges. We recall that if a ferromagnet is placed in an external magnetic field B its magnetization M changes. If B is now changed to $B + \delta B$ then M changes to $M + \delta M$. The susceptibility χ is defined as $\chi = (\partial M/\partial B)$. The manner in which the magnetization of a ferromagnet approaches zero as the temperature T of the system is increased to the Curie temperature of the system can be studied experimentally. It is found that in the absence of an external magnetic field and for T close to T_C

$$M = M_0 \left| \frac{T - T_C}{T_C} \right|^{\beta}$$

with $\beta \approx 0.33$ for many different ferromagnets. Similarly it has been found that $\chi(T)$ near $T \sim T_C$ behaves as

$$\chi(T) \propto \frac{1}{|T - T_C|^{\gamma}}$$

with $\gamma \approx 1.25$. Such behavior constitutes a *second-order phase transition*. The transition is from the ferromagnetic phase at $T < T_C$, characterized by $M \neq 0$ for $B = 0$ to the paramagnetic phase at $T > T_C$ where $M = 0$ for $B = 0$.

Let us now give a concrete definition of a phase transition. A given equilibrium state of a macroscopic system can be described by an *order parameter* field. For a ferromagnet the order parameter field is the magnetization density. The order parameter field can be regarded as a mapping from the system (with coordinate $\mathbf{x}$)

to an order parameter space. In a model where the magnetization density is given by a scalar function $M_z(\mathbf{x})$ this is the space of real numbers $\Re$. In general the order parameter space can be more complicated and the order parameter field need not be a scalar function.

Definition 13.1 A phase transition corresponds to the order parameter field changing qualitatively together with the emergence of singular behavior in the system.

For instance the order parameter field in the case of a ferromagnet is non-zero in the ferromagnetic phase, is zero in the paramagnetic phase, and the susceptibility of the system diverges at the phase transition temperature. Determining a suitable order parameter field to characterize a phase is part of the task of a theory of phase transitions. If the order parameter field changes continuously from one phase to another, as in the case of a ferromagnet, the transition is said to be a continuous or second-order phase transition. If it is discontinuous the transition is said to be *first order*. An example of a first order transition is when a solid melts to a liquid. The density of the system, which can be taken as the order parameter, changes discontinuously. A phase transition is a striking example of an emergent phenomenon. Starting off with only short-range interactions between its microscopic magnetic moments, the system realizes long-range correlations below T_C. We will now give an argument, based on a simple model of a ferromagnet, to show that it is impossible to understand the singular behavior of the susceptibility as a function of temperature. We will then give a second argument based on a simple model for the susceptibility χ to conclude that it is easy to understand the physical origin of the singular behavior of χ. Reconciling these two points of view will lead us to understand that certain infinite limits are important in statistical mechanics and to the *renormalization group* approach. On the way we will discuss an approach to phase transitions due to Landau.

We start with a model for a ferromagnet. We regard a ferromagnetic solid as being made out of a finite number of elementary magnets placed at locations throughout the solid. We simplify our model by assuming that each of these elementary magnets m can either point up ($m = +1$) or down ($m = -1$). Finally each elementary magnet interacts only with its nearest neighbor. A Hamiltonian for this model could be

$$\mathsf{H} = -g \sum_{n,i} m_i m_{i+n} - B \sum_i m_i$$

where the first sum is over i as well as the nearest neighbors of i. Note H decreases if m_i, m_{i+n} have the same sign for $g > 0$. There are altogether a large but finite number of magnets in a ferromagnet. We are now ready to prove our theorem.

Theorem 13.1 In the model of a ferromagnet proposed the susceptibility cannot diverge.

Proof. In any statistical mechanics the partition function of a system once calculated determines the thermal properties of the system. For a magnetic system the macroscopic variables are the magnetization of the system M, the external magnetic field B and the temperature T of the system. The canonical ensemble for this system is defined by the partition function

$$Z = \sum_{\{c\}} e^{-\beta H}, \qquad \beta = \frac{1}{kT}$$

where $\{c\}$ denotes the set of all configurations of the individual magnetic moments. We assume $\beta \neq \infty$ and define the free energy as usual by $F = -1/\beta \ln Z$. The susceptibility χ of the system is defined as in Chapter 2. We now show that χ cannot be a singular function of temperature. The proof is straightforward. Since our model involves a finite number of elementary magnets each of which interacts with a finite number of its neighbors and can exist in only two states, it follows that the number of configurations which have to be summed over to determine the partition function Z is finite. Each term of the sum is an analytic function of temperature and B. Since a finite sum of analytic functions is again an analytic function, Z is an analytic function of the temperature and B. Furthermore each term $e^{-\beta H}$ is strictly positive. So Z is also strictly positive. Thus $F = -1/\beta \ln Z$ is an analytic function of the temperature and B. An analytic function of the temperature and the external magnetic field will continue to be analytic in those variables no matter how many times the function is differentiated. That is the definition of an analytic function. Therefore $\chi = -\partial^2 F/\partial B^2$ is a non-singular function of temperature, which concludes the proof. □

One response to this theorem might be to suggest that the theorem fails if we allow the number of elementary magnets to tend to infinity. This is because an infinite sum of analytic functions need not be analytic. An elementary example of this is the series $1 + x + x^2 + \cdots + x^N$. If we let $N \to \infty$ then the series tends to $1/1 - x$ which has a singularity at $x = 1$. In order to analyze this possibility we will need to consider the statistical mechanics partition function in the limit in which the number of configurations is infinite. It is only in this limit that the phase transitions might be understood from this point of view.

Now for our second approach. This time we suppose that the external magnetic field B is changed to $B + \delta B(\mathbf{x})$, i.e. the change δB is position dependent. We expect that a change at $\mathbf{x}$, $\delta B(\mathbf{x})$, will produce a change in the magnetization δM not just at the point $\mathbf{x}$ but at other points as well. Indeed we might expect

$$\delta M(\mathbf{y}) \propto C_T(|\mathbf{x} - \mathbf{y}|)\, \delta B(\mathbf{x})$$

where $C_T(|\mathbf{x} - \mathbf{y}|)$ is a "correlation function" which determines the effect at $\mathbf{y}$ on the magnetization due to a change in the external field δB at $\mathbf{x}$. The total change

$\delta M(\mathbf{y})$ is then expected to be

$$\delta M(\mathbf{y}) = \int \mathrm{d}^3\mathbf{x}\, C_T(|\mathbf{x}-\mathbf{y}|)\, \delta B(\mathbf{x}).$$

We have assumed that the correlation function depends only on temperature and on the distance between the points $\mathbf{x}$ and $\mathbf{y}$. This is our simple model. The formula we have just guessed can be derived as a mathematical result within the framework of *linear response theory* and is known as *Kubo's formula*. For our own purpose the equation provides a convenient starting point for understanding the physical origin of the divergence of the susceptibility χ. Let us now suppose that δB is independent of $\mathbf{x}$ and let us set $\mathbf{y} = 0$. Then we have

$$\chi(0) = \frac{\delta M(0)}{\delta B} = \int \mathrm{d}^3x\, C_T(|\mathbf{x}|).$$

If we suppose

$$C_T(|\mathbf{x}|) = \begin{cases} \alpha, & \text{for } |\mathbf{x}| \leq a(T), \\ 0, & \text{for } |\mathbf{x}| > a(T), \end{cases}$$

that is, a disturbance only propagates a distance $a(T)$, then

$$\chi(0) = \frac{4\pi\alpha}{3} a^3(T).$$

Thus $\chi(0)$ will diverge if $a(T)$ diverges, that is, if correlations in the system become infinite. From this point of view the divergence in this susceptibility is due to the fact that near a phase transition disturbances propagate over large distances, $a(T) \to \infty$. Our task is to reconcile the two approaches described. This will lead us to what is known as the renormalization group approach to phase transitions.

Let us start by verifying if, by allowing the number of configurations to go to infinity, we can indeed recover ferromagnetism in our model with the associated singular behavior. To check this we consider a situation where the partition function can be exactly determined. The model we consider is the one-dimensional Ising model discussed in Chapter 2. We recall that the model was defined by a Hamiltonian

$$H = -g \sum_{i=1}^{N} S_i S_{i+1} - B \sum_{i=1}^{N} S_i.$$

In our calculation in Chapter 2 we found that in the $N \to \infty$ limit the free energy F for this system is given by

$$F = -\frac{N}{\beta} \ln \left(\mathrm{e}^{\beta g} \left[\cosh \beta B + \sqrt{\cosh^2(\beta B) - 2\mathrm{e}^{-2\beta g} \sinh(2\beta g)} \right] \right).$$

The magnetization M and the susceptibility χ are then given by

$$M = -\frac{\partial F}{\partial B}, \quad \chi = -\frac{\partial^2 F}{\partial B^2}.$$

Using this expression for F it is clear that $M = 0$ if $B = 0$ for $T = 0$ and χ is not singular. Thus our simple model fails to provide a model for a ferromagnet even in the $N \to \infty$ limit, i.e. a system for which $M \neq 0$, if $B = 0$, for some T. Our hope of getting a system to become ferromagnetic for $T < T_C$ with singular behavior for χ in the limit where the number of configurations was allowed to go to infinity has failed for this model. What went wrong? We now give an argument due to Peierls which explains why the model does not lead to ferromagnetism and why if the model is extended to two dimensions it should work!

13.2 Peierls argument

The idea of Peierls was to start with an ordered phase of the system, that is, a phase in which all the spins point in the same direction, say, $S = +1$, even in the absence of an external magnetic field, and then to introduce a length L of "disordered" (i.e. $S = -1$) spins. Peierls considered the change of the free energy F of the system for different temperatures when this was done. If F increases it means that the original ordered phase was stable so that the system could exist in a ferromagnetic phase while if F decreases it means the ordered phase was thermodynamically unstable. Indeed in Chapter 1 we established that the free energy F is minimal at equilibrium at constant temperature and under conditions when no work is done. We have by definition

$$F = U - TS.$$

Keeping T fixed if the system is changed in the way we described we have

$$\Delta F = \Delta U - T\Delta S.$$

Now observe that the change in U can be regarded as due to the change in the energy of the system. Setting $B = 0$ we note that if all $S_i = +1$ or all $S_i = -1$, the Hamiltonian has the same value. Thus if a disordered element of $S_i = -1$ of length L is introduced in a one-dimensional system with $(N - L)$ spins $S_i = +1$ then the energy of the system only changes at the boundaries. This is because it was assumed that we only have nearest neighbors interact in our model. For a one-dimensional system there are only two boundaries. Thus

$$\Delta U = 2W,$$

where W is the energy change in replacing $S_i = 1$, $S_{i+1} = 1$ by $S_i = 1$, $S_{i+1} = -1$, i.e. $W = 2g$.

Next we have to calculate ΔS. To do this we recall that the entropy is given by $k \ln N_0$ where N_0 is the number of ways the configuration could be constructed. In the case of the one-dimensional model N_0 is given by the number of ways in which a disordered length L can be introduced into the system. It is clear that the disordered length can be made to start at $i = 1, 2, \ldots, N - L$. Thus N_0, the number of ways the disordered configuration can be introduced is equal to $N - L$. Therefore

$$\Delta F = 2W - kT \ln(N - L).$$

We are to take the $N \to \infty$ limit in order to allow a phase transition to occur as we saw. Thus $\Delta F < 0$ when the disordered element is introduced. Hence no ordered phase should be expected for the model. This is a reassuring result. It tells us that on general grounds a one-dimensional model with nearest neighbor interactions cannot undergo a phase transition and this explains our failure to get a phase transition for the one-dimensional Ising model.

Let us consider a generalization of the Ising model to two dimensions and apply Peierls' ideas. We will see that a phase transition is now possible and the argument gives a value for the critical (or Curie) temperature for the system. Our discussion is not intended to be complete but only to explain clearly the essential difference between the one- and two-dimensional models.

The model we now consider is given by

$$H = -g \sum_{i,j} (S_{i,j} S_{i+1,j} + S_{i,j} S_{i,j+1}).$$

where the spins S_i are placed on a $N \times N$ periodic lattice. We start with an ordered phase of the system when all the spins $S_i = +1$. We again introduce a line of length L of disordered elements. A picture might be helpful.

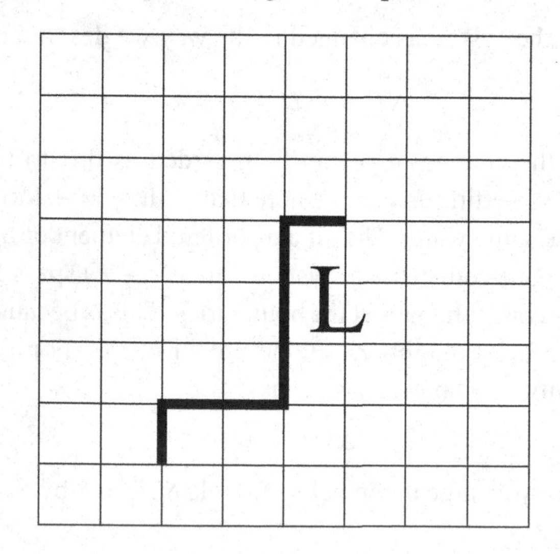

In this case $\Delta U = (2WL)$ as introducing a line L disordered element in two dimensions introduces a boundary of length L as well. Next we have to determine the number of ways, N_0, such a line can be introduced in the system. We note that

- The line can start at any one of the points on the $N \times N$ lattice.
- From the starting point the next disordered element can be chosen in three ways (for the square lattice system we consider).

Thus

$$N_0 = N^2 3^L.$$

This simple way of counting does not take into account the fact that the defect line has no orientation and we are thus over counting by a factor of two. This is, however, not relevant for our purpose. For concreteness let us count only configurations which are almost maximally disordered. For such configurations we have

$$L = fN^2$$

where f is a number between 0 and 1. Thus

$$\begin{aligned}\Delta F &= 2WfN^2 - kT \ln\left(N^2 3^{fN^2}\right)\\ &= 2WfN^2 - kT(fN^2 \ln 3 + \ln N^2).\end{aligned}$$

For large N the last term can be neglected so that

$$\Delta F \simeq fN^2(2W - kT \ln 3).$$

Thus $\Delta F > 0$ if $2W > kT \ln 3$ or $kT < (2W/\ln 3)$. In the two-dimensional case even with a nearest neighbor model the argument of Peierls suggests that a phase transition is allowed provided the temperature T of the system is less than $2W/k \ln 3$. This expectation is justified for two reasons. First, the numerical study of the two-dimensional Ising model in Chapter 5 confirms our expectations. Second, the two-dimensional Ising model was solved exactly by Onsager and a phase transition with singular behavior for its susceptibility is found in the limit where the number of configurations is allowed to go to infinity.

Now that we know that statistical mechanics can indeed be used to study phase transitions we turn to a simple phenomenological approach for understanding phase transitions due to Landau.

13.3 Landau theory of phase transitions

We have already made use of the fact that for any change of a system in which the temperature is kept fixed and no work is done by the system, the change of free energy, ΔF, is always negative so that a state of equilibrium must be a minimum of F. Landau utilized this property of the free energy in his theory of phase transitions. Let us examine this approach for the case of a ferromagnet. The basic idea is to make a model for the free energy F near the Curie temperature T_C when the system is still a ferromagnet. We know that for $T < T_C$ long-range correlations are present, that is, the spin at lattice site $\mathbf{x}$ must point in the same direction as that at site $\mathbf{y}$ even when $\mathbf{x}$ and $\mathbf{y}$ are not adjacent. Otherwise the observed macroscopic magnetic properties of the system would not exist. The basic assumption underlying Landau's theory is that, near the critical temperature T_C, the properties of a ferromagnet can be described in terms of a magnetization density function $\mathbf{M}(\mathbf{x})$. The function $\mathbf{M}(\mathbf{x})$ can be defined by considering a volume element ΔV, large compared to the lattice cell volume, but small compared to the volume of correlated spins centered around the point $\mathbf{x}$. The magnetization of the volume element ΔV is defined to be $\mathbf{M}(\mathbf{x})\,\Delta V$. For this definition of $\mathbf{M}(\mathbf{x})$ to be useful it is important that $\mathbf{M}(\mathbf{x})$ should not be a rapidly varying function of position. Near the Curie temperature T_C we also expect $\mathbf{M}(\mathbf{x})$ to be small in amplitude. Since we are implicitly assuming in this approach that a spin-spin type of interaction is responsible for the phenomenon of ferromagnetism it seems reasonable to expect the free energy density to be a function of $\mathbf{M}(\mathbf{x}) \cdot \mathbf{M}(\mathbf{x})$. On the basis of arguments of this kind Landau proposed to introduce a functional $F_L[T, \mathbf{B}, \mathbf{M}]$ of the magnetization density $\mathbf{M}(\mathbf{x})$, temperature T, and external magnetic field $\mathbf{B}(\mathbf{x})$ of the form

$$
\begin{aligned}
F_L[T, \mathbf{B}, \mathbf{M}] &= F_L[T, \mathbf{B}, \mathbf{M} = 0] \\
&\quad + \int \mathrm{d}^3x \Big[a(T)\,\mathbf{M}(\mathbf{x}) \cdot \mathbf{M}(\mathbf{x}) + b(T)\,(\mathbf{M}(\mathbf{x}) \cdot \mathbf{M}(\mathbf{x}))^2 \\
&\quad + \cdots + c(T) \sum_{i,j} (\nabla_j M_i(\mathbf{x})) \cdot (\nabla_j M_i(\mathbf{x})) + \cdots - \mathbf{B} \cdot \mathbf{M}(\mathbf{x}) \Big].
\end{aligned}
$$

The free energy $F_L(T, \mathbf{B})$ is then obtained by minimizing $F_L[T, \mathbf{B}, \mathbf{M}]$ with respect to $\mathbf{M}$. Note that the temperature dependent coefficients $a(T), b(T), c(T), \ldots$ are assumed to be smooth functions of temperature. We will simplify the model function by assuming $\mathbf{B}(\mathbf{x})$ acts along the z-direction and that $\mathbf{M}(\mathbf{x})$ only has components in the z-direction. Then we have

$$
\begin{aligned}
F_L[T, B_z, M_z] = F_L[T, B_z, M_z = 0] + \int \mathrm{d}^3x\ &[a(T)\,M_z^2(\mathbf{x}) + b(T)\,M_z^4(\mathbf{x}) \\
&+ \cdots + c(T)\,(\nabla M_z(\mathbf{x})) \cdot (\nabla M_z(\mathbf{x})) \\
&+ \cdots - B_z(\mathbf{x})\,M_z(\mathbf{x})].
\end{aligned}
$$

The expression for the Landau free energy F_L is expected to be useful when T is close to the Curie temperature T_C. In this region $M_z(\mathbf{x})$ is expected to be small and we also expect $(\nabla M_z(\mathbf{x}) \cdot \nabla M_z(\mathbf{x}))$ to be small. Because of these reasons we will from now on ignore the effect of the higher powers of $M_z(\mathbf{x})$ and higher gradient terms. To determine the equilibrium configuration of the magnetization $M_z(\mathbf{x})$ we have to minimize the free energy with respect to $M_z(\mathbf{x})$. Using

$$\delta F_L = \int d^3x \left[2a(T)\, M_z(\mathbf{x}) + 4b(T)\, M_z^3(\mathbf{x}) - 2c(T)\, \nabla^2 M_z(\mathbf{x}) - B_z(\mathbf{x})\right] \delta M_z(\mathbf{x})$$

we see that vanishing of δF for arbitrary $\delta M_z(\mathbf{x})$ requires

$$2a(T)\, M_z(\mathbf{x}) + 4b(T)\, M_z^3(\mathbf{x}) - 2(c(T)\nabla^2 M_z(\mathbf{x})) = B_z(\mathbf{x}).$$

Suppose now that $B_z(\mathbf{x})$ does not depend on $\mathbf{x}$ and let us see if a solution for $M_z(\mathbf{x})$ independent of $\mathbf{x}$ is possible. Such an $\mathbf{x}$ independent solution must satisfy

$$\left[2a(T)\, M_z + 4b(T)\, M_z^3\right] = B_z.$$

Now we ask if it is possible to construct a solution with the property that $M_z \neq 0$ when $B_z = 0$ and $T < T_C$. Setting $B_z = 0$ we find three solutions

$$M_z = 0$$

$$M_z = \pm\sqrt{\frac{-a(T)}{2b(T)}}.$$

We would also like the solution to have the property

$$M_z = 0, \quad \text{when } T > T_C$$
$$M_z \neq 0, \quad \text{when } T < T_C.$$

If such a solution is possible then the expression for F represents a model for a ferromagnet. As we have stressed this model is constructed to represent a ferromagnet near its Curie temperature. We also assume that the coefficient functions $a(T), b(T), c(T)$ are all smooth functions of temperature. We thus expect the M_z^4 term to be small compared to the M_z^2 term. It is then reasonable to replace $b(T)$ by $b(T_C) = b_0$, a constant. Finally, setting $a(T) \simeq a_0(T - T_C)$ we have as our equilibrium $\mathbf{x}$ independent solution

$$M_z = 0 \quad \text{or}$$

$$M_z = \pm\sqrt{\frac{-a_0(T - T_C)}{2b_0}}.$$

Let us now discuss the Landau free energy for arbitrary $M_z(\mathbf{x})$ independent of $\mathbf{x}$ and $a(T) = a_0(T - T_C), b(T) = b, B_z = 0$ respectively. Ignoring the M_z independent contribution in F_L, we find

$$F_L(T, M_z) = V\left[a_0(T - T_C) M_z^2 + b_0 M_z^4\right].$$

Let us now discuss the choice of the constants a_0 and b_0 in turn. We distinguish three cases, depending on the sign of the coefficients a_0 and b_0. In each case we plot F_L/V as a function of M_z.

(1) $a_0 > 0, b_0 > 0$:

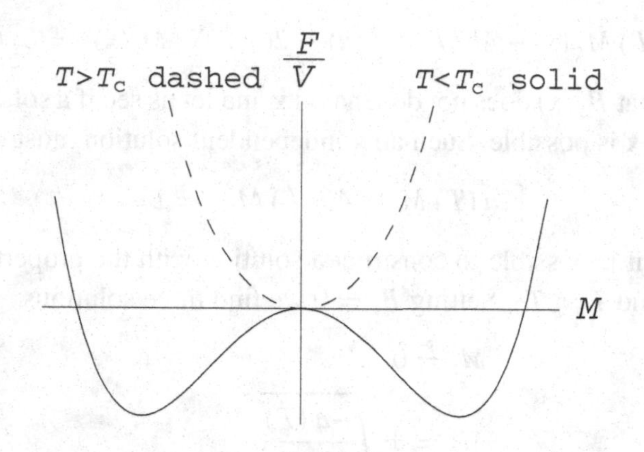

$$M_1 = -\sqrt{\frac{a_0(T_C - T)}{b_0}} \quad M_2 = +\sqrt{\frac{a_0(T_C - T)}{b_0}}$$

Note that M_1 and M_2 have lower free energy than the solution $M_z = 0$. Thus according to Landau's theory the system would settle to M_1 or M_2 as its equilibrium magnetization configuration. Observe also that for $T > T_C$ the free energy F_L/V is a convex function of M_z. Thus $M_z = 0$ is the minimum of the free energy for $T > T_C$.

(2) $a_0 > 0, b_0 < 0$:

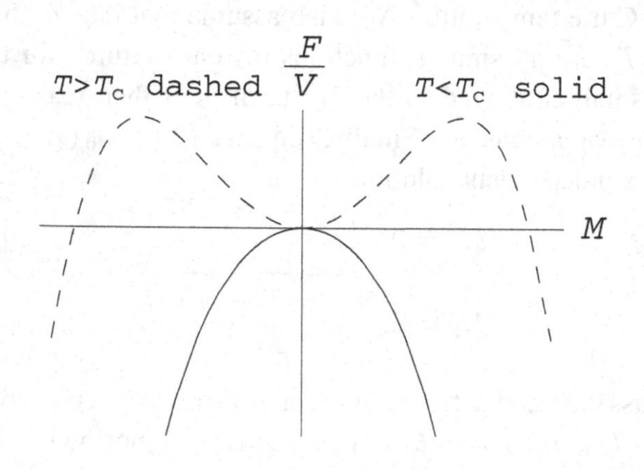

In this case the equilibrium distribution $M_z = 0$ is unstable for any temperature. The same situation arises for $a_0 < 0, b_0 < 0$ which simply amounts to interchanging the two graphs for $T < T_C$ and $T > T_C$.

(3) $a_0 < 0, b_0 > 0$:

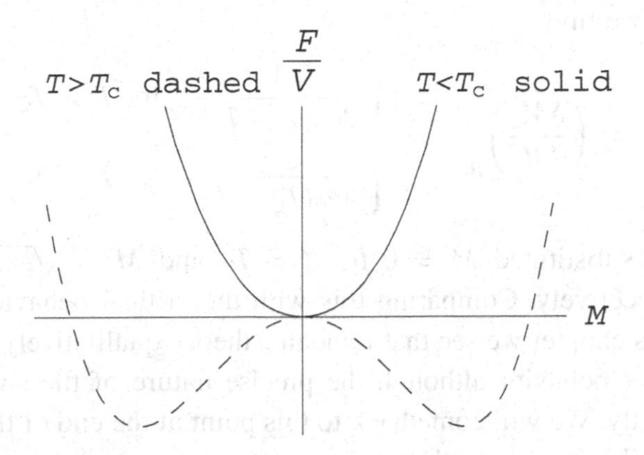

This case is not physically reasonable since it leads to spontaneous magnetization above the critical temperature!

To summarize, if Landau's expression for the free energy is to represent a ferromagnet we must choose $a_0 > 0, b_0 > 0$. The alert reader will have noticed that the free energy plotted in the above figures is not a convex function. This appears to be in contradiction with our result in Chapter 1 that the free energy F is a thermodynamic potential and as such a convex function of its extensive variables. This puzzle is resolved by noting that the true free energy is, in fact, the convex hull of the functions plotted above. This is because the equilibrium states can be mixtures of pure phases, rather than just pure phases which we have implicitly assumed by setting $M_z = constant$, independent of the position. We consider a concrete example of mixing in Problem 13.4.

Having determined the signs of a_0 and b_0 we now want to explore the predictions of the Landau theory. From the equation

$$\left(2a_0(T - T_C) + 4b_0 M_z^2\right) M_z = B_z$$

it then follows that for $B_z = 0$ and $T < T_C$

$$M_z = \pm\sqrt{\frac{a_0}{2b_0}}(T_C - T)^{\frac{1}{2}}\,.$$

Furthermore, for $T = T_C$,

$$M_z^3 = \frac{B_z}{4b_0}.$$

Finally we note that if B_z is changed to $B_z + \delta B_z$, the corresponding equilibrium distribution M_z can be written as $M_z + \delta M_z$. We thus get

$$2a_0 (T - T_C)\, \delta M_z + 12b_0 M_z^2 \delta M_z = \delta B_z.$$

Setting $B_z = 0$, we find

$$\chi = \left(\frac{\delta M_z}{\delta B_z}\right)_{B_z=0} = \begin{cases} \dfrac{1}{2a_0(T - T_C)} & \text{for } T > T_C \\ \dfrac{1}{4a_0(T_C - T)} & \text{for } T < T_C. \end{cases}$$

where we have substituted $M_z = 0$ for $T > T_C$ and $M_z = \sqrt{a_0/2b_0}\,(T_C - T)^{\frac{1}{2}}$ for $T < T_C$ respectively. Comparing this with the critical behavior stated at the beginning of this chapter we see that Landau's theory qualitatively reproduces the expected singular behavior although the precise nature of the singularity is not reproduced exactly. We will come back to this point at the end of this section.

It is also possible to get a rather precise statement regarding long-range correlations within the framework of Landau's theory. To do this we recall our linear response relation introduced in Section 13.1

$$\delta M_z (\mathbf{x}) = \int \mathrm{d}^3 y\, C_T (|\mathbf{x} - \mathbf{y}|)\, \delta B_z (\mathbf{y})$$

where we regard δM_z as the change in the equilibrium magnetization density brought about by changing the external magnetic field by $\delta B_z (\mathbf{y})$. From the equations which describe the equilibrium magnetization density we have with $C(T) = c_0$,

$$2a (T)\, M_z (\mathbf{x}) + 4b_0 M_z^3 (\mathbf{x}) - 2c_0 \nabla^2 M_z (\mathbf{x}) = B_z (\mathbf{x})$$

and

$$2a (T) (M_z + \delta M_z) + 4b_0 (M_z + \delta M_z)^3 - 2c_0 \nabla^2 (M_z + B_z) = B_z + \delta B_z.$$

From these two equations it follows that

$$\left[2a_0 (T - T_C) + 12b_0 M_z^2 - 2c_0 \nabla^2\right] \delta M_z (\mathbf{x}) = \delta B_z (\mathbf{x}).$$

Substituting the linear response relation expression for $\delta M_z (\mathbf{x})$ we get

$$\left(2a_0 (T - T_C) + 12b_0 M_z^2 - 2c_0 \nabla^2\right) C_T (|\mathbf{x} - \mathbf{y}|) = \delta^{(3)} (\mathbf{x} - \mathbf{y})$$

where we have written $\delta B_z (\mathbf{x}) = \int \mathrm{d}^3 y \delta (\mathbf{x} - \mathbf{y})\, \delta B (\mathbf{y})$. Setting $B_z = 0$ and rescaling the coordinates as $(\mathbf{u}, \mathbf{v}) = 2c_0(\mathbf{x}, \mathbf{y})$ we get

$$(2\bar{a}_0(T - T_C) - \nabla^2) C_T (|\mathbf{u} - \mathbf{v}|) = \delta^{(3)} (\mathbf{u} - \mathbf{v}),$$

where $\bar{a}_0 = a_0/(2c_0)^3$. This result is valid in the region $T > T_C$. For $T < T_C$ we get an analogous equation with $\bar{a}_0 \to -2\bar{a}_0$. The solution of this equation can be inferred from the following result.

Lemma 13.2 The differential equation

$$(m^2 - \nabla^2)C_T(|\mathbf{u}|) = \delta^3(\mathbf{u})$$

has the solution

$$C_T(|\mathbf{u}|) = \frac{1}{4\pi}\frac{\mathrm{e}^{-m|\mathbf{u}|}}{|\mathbf{u}|},$$

modulo solutions of the homogeneous equation.

Proof. To prove the lemma let us write the delta function in momentum space as

$$\delta^3(\mathbf{u}) = \int \frac{\mathrm{d}^3k}{(2\pi)^3}\,\mathrm{e}^{\mathrm{i}\mathbf{k}\cdot\mathbf{u}}$$

and similarly

$$C_T(|\mathbf{u}|) = \int \frac{\mathrm{d}^3k}{(2\pi)^3}\,e^{\mathrm{i}\mathbf{k}\cdot\mathbf{u}}\,\Delta(|\mathbf{k}|)\,.$$

Substituting these expressions in the differential equation we get

$$(m^2 + |\mathbf{k}|^2)\Delta(|\mathbf{k}|) = 1.$$

Then, performing the inverse Fourier transform we end up with

$$C_T(|\mathbf{u}|) = \int \frac{\mathrm{d}^3k}{(2\pi)^3}\,\mathrm{e}^{\mathrm{i}\mathbf{k}\cdot\mathbf{u}}\,\frac{1}{|k|^2 + m^2}.$$

In order to evaluate this integral let us choose a coordinate system in which $\mathbf{u}$ points along the z-axis. Changing to polar coordinates we have

$$\begin{aligned} C_T(|\mathbf{u}|) &= \int_0^\infty \frac{\mathrm{d}kk^2}{(2\pi)^3}\int_0^{2\pi}\mathrm{d}\phi\int_{-1}^{1}\mathrm{d}\cos(\theta)\mathrm{e}^{ik\cos\theta|\mathbf{u}|}\left(\frac{1}{k^2+m^2}\right) \\ &= \frac{1}{2\pi^2}\frac{1}{|\mathbf{u}|}\int_0^\infty \frac{\mathrm{d}kk}{k^2+m^2}\sin k|\mathbf{u}| \\ &= \frac{1}{4\pi^2}\frac{1}{|\mathbf{u}|}\mathrm{Im}\left(\int_{-\infty}^{\infty}\frac{\mathrm{d}kk}{k^2+m^2}\mathrm{e}^{ik|\mathbf{u}|}\right). \end{aligned}$$

This integral can be evaluated by closing the contour of integration above and below the real line respectively leading to

$$C_T(|\mathbf{u}|) = \frac{1}{4\pi}\frac{\mathrm{e}^{-m|\mathbf{u}|}}{|\mathbf{u}|}$$

which is the claimed solution. □

To continue we restore the original coordinates $(\mathbf{x}, \mathbf{y})$ and introduce the *correlation length* $\xi^2 = c_0/a(T)$. Then

$$C_T(|\mathbf{x}-\mathbf{y}|) = \frac{1}{4\pi}\frac{e^{-\frac{|\mathbf{x}-\mathbf{y}|}{\xi}}}{|\mathbf{x}-\mathbf{y}|}.$$

The corresponding result for $T < T_C$ is obtained simply by replacing $a(T) \to -2a(T)$ in the correlation length. We notice that $\xi \to \infty$ as $T \to T_C$. Thus Landau's theory is in qualitative agreement with the intuitive idea, introduced earlier in Section 13.1 that long-range correlations are generated in a ferromagnet as $T \to T_C$. Another point to note is that if δB_z were $\mathbf{x}$ independent, then as we saw before

$$\chi(0) = \frac{\delta M_z(0)}{\delta B_z} = \int \mathrm{d}^3y \, C_T(|\mathbf{y}|).$$

From the expression for $C_T(|\mathbf{y}|)$ it then follows that

$$\chi \sim \xi^2 \to \infty \quad \text{as } T \to T_C.$$

Let us summarize the results obtained from Landau's approach. The approach focused on long-range correlations and suggested that the singular behavior of the susceptibility was due to such correlations when $T \to T_C$. The approach also predicts that the relation between different macroscopic parameters involves power laws,

$$\begin{aligned} M_z &\sim (T_C - T)^\beta, \quad T \to T_C \\ M_z &\sim B_z^{\frac{1}{\delta}}, \quad T = T_C \\ \chi &\sim \frac{1}{(T_C - T)^\gamma}, \quad T \to T_C, \end{aligned}$$

with $\beta = \frac{1}{2}$, $\gamma = 1$, and $\delta = 3$. The parameters β, δ, γ are called *critical exponents* and are measured experimentally. The experimental values for these parameters $\beta \simeq 0.33$, $\delta \simeq 4.5$, and $\gamma \simeq 1.2$ are found for different ferromagnets with different lattice structures and widely differing values for the Curie temperature T_C. These parameters thus are a *universal* property of the ferromagnetic phase transition. This is also a feature of Landau's theory. Thus Landau's theory is in qualitative agreement with experiment.

13.4 Renormalization group

Although Landau's theory is in good qualitative agreement with experiment there is room for improvement on the quantitative level concerning the critical exponents.

This will lead us to the renormalization group. Our treatment will be rather brief as our aim is to explain the basic ideas involved in the renormalization group approach and not explore technical issues. The reader is referred to some of the books listed at the end of the chapter for a more thorough treatment of this important topic.

The renormalization group can be approached in two different ways. There is the real space approach and there is the field theory approach. In both approaches the crucial physical input is the assumption that a certain length scale of the system approaches infinity near a phase transition.

Let us start by looking at the real space approach. Instead of considering a model for the free energy we now consider the partition function directly. Since the dynamics of a physical system is completely determined by the Hamiltonian we can think of the partition function as a function of the Hamiltonian H and the temperature. Thus

$$\begin{aligned} Z &= Z(\beta, \mathsf{H}) \\ &= Z(\mathsf{H}), \quad \beta = \frac{1}{kT} \text{ fixed }. \end{aligned}$$

If we now assume that there is a natural length scale ξ, the correlation length, which should be used as the unit of length and that $\xi \to \infty$ as $T \to T_C$ then Z better be *scale invariant*, since in this limit there is no physical length left to set the scale. To illustrate this we may consider a model for a ferromagnet in which elementary magnets are present in a cubic lattice of size a_0. Scale invariance then means that the physics at scales of the order ξ should not change if the microscopic a_0 is replaced by $a_L = La_0$ as long as $La_0 \ll \xi$. If the Hamiltonian H represents the dynamics of the system when the scale of the system is a_0 and H_L represents the dynamics when the scale is a_L then the statement that the physical properties of the system are scale invariant means in particular that

$$Z(\mathsf{H}) = Z(\mathsf{H}_L).$$

This is the key assumption of the real space renormalization group approach. Now, since the correlation length ξ depends solely on the Hamiltonian of the system we can write $\xi = \xi(\mathsf{H})$. On the other hand, since ξ is a length, dimensional analysis implies

$$\xi(\mathsf{H}_L) = \frac{1}{L}\xi(\mathsf{H}).$$

The intuitive picture behind this equation is simply that the same correlation length is measured in different units namely a_0 and $a_L = La_0$. As for the free energy *density* for the system (which we will denote a gain by F) it is given by

$$F = -\frac{1}{\beta V} \ln Z(\mathsf{H}),$$

and similarly

$$F(\mathsf{H}_L) = -\frac{1}{\beta V_L} \ln Z(\mathsf{H}_L).$$

We can repeat the argument we used for the correlation length. Since the volume V is fixed and only the length scale is changed we must have $V_L = L^{-D} V$ and therefore

$$F(\mathsf{H}_L) = L^D F(\mathsf{H})$$

for a D-dimensional system. Of course, these equations acquire content only if we know how H_L and H are related or, in mathematical terms, if the transformation function τ

$$\tau(\mathsf{H}) = \mathsf{H}_L,$$

which relates H and H_L is known. The function τ is the *real space renormalization group transformation*. We will now sketch how in principle τ can be obtained given a certain Hamiltonian and how once τ is known it is possible to calculate the actual critical exponents. We shall find that our qualitative argument leads to the statement that the free energy is a *generalized homogeneous function*. We will explain what this means and point out that this has experimentally testable implications. Our aim is to make the general strategy of the renormalization group approach clear. We will not go into the details of how the critical exponents are actually determined in this approach. We will learn how to calculate the critical exponents when we discuss the field theory approach to the renormalization group. For concreteness let us consider a two-dimensional ferromagnetic system described in terms of an Ising model. Let

$$\begin{aligned}\tilde{\mathsf{H}} &\equiv \beta \mathsf{H} \\ &= -g\,(T) \sum_{n,i} S_i S_{i+n} - B\,(T) \sum_i S_i\end{aligned}$$

where $g(T) = \beta g_0$ and $h(T) = \beta B_0$ are the "coupling constants" and $S_n = \pm 1$. The summation in the first term is over nearest neighbors. When the length scale of the system is a_0, the nearest neighbors of a given spin variable S_n are located at a distance a_0 from S_n. If the length scale is changed from a_0 to La_0 the nearest neighbors of S_n^L will be at a distance La_0. This assumes, of course, that H and H_L have the same structure. In H, S_n took the value ± 1. For H_L we have to determine the corresponding range of values of S_n^L.

Since the total number of spin variables N is fixed by changing length scales from a_0 to La_0 the number of degrees of freedom of the system are reduced. We can then interpret S_n^L as the value of the block of spins surrounding S_n and contained within a cube of side La_0. For $T \sim T_C$ if $\xi \gg La_0$ we might expect all the spins within the block to be correlated, that is, if one of the spins within the block takes on the value $+1$ then so do all the other spins in the block. If there are N_L spins in the block then S_n^L can take values $\pm N_L$. The picture leads us to expect that H_L differs from H in three ways. First the length scale a_L is different from a_0. Second, the magnitude of the spin variables differs and finally the coupling constants differ. Thus

$$\tilde{\mathsf{H}}_L = -g_L(T)\sum_{n,i} S_n^L S_{n+i}^L - B_L(T)\sum_i S_i^L .$$

The renormalization group transformation τ in this case corresponds to the set of equations:

$$g_L = U_L(g, B)$$

$$B_L = V_L(g, B) .$$

Clearly this process can be repeated; we can start with H_L and generate H_{L^2}, from H_{L^2} generate H_{L^3} etc. All of these Hamiltonians form a class of "equivalent" Hamiltonians. In particular, they lead to the same partition function. Note for this argument to work it is necessary for the correlation length ξ to be such that $\xi \gg L^{N_0}$ ($N_0 = 1, 2, 3, \ldots$). It is possible that the sequence $\mathsf{H}, \mathsf{H}_L, \mathsf{H}_{L^2}, \ldots$ converges to a H^* such that

$$\tau(\mathsf{H}^*) = \mathsf{H}^* .$$

The Hamiltonian H^* is then a *fixed point* of the renormalization group transformation τ. Note that the existence of a fixed point is a property of τ and hence will not depend greatly on the starting Hamiltonian H. Of course, there could be several fixed points so that one class of Hamiltonian which acted on by τ will converge to H_1^*, another class which converges to some other fixed point H_2^* and so on.

Let us study qualitatively the implication of assuming that H^* describes the behavior of a physical system near the critical temperature. To keep things simple we will first assume a vanishing external field, $B = 0$. The renormalization group transformation of g_L is then

$$g_L = U_L(g) .$$

We want to investigate the effect of repeated applications of the transformation τ. For this we consider

$$g_{L'L} = U_{L'}(g_L) .$$

We assume that $U_{L'}(g_L)$ is a smooth function of L' and g_L. We then have

$$\frac{\mathrm{d}g_L}{\mathrm{d}L} = \frac{1}{L}\lim_{\delta\to 0}\frac{g_{(1+\delta)L} - g_L}{\Delta L} \equiv \frac{1}{L}u(g_L).$$

The characteristic property of H^* is $\tau(\mathsf{H}^*) = \mathsf{H}^*$. In terms of g_L this implies $g^* = U_{L'}(g^*)$ or

$$L\frac{\mathrm{d}g^*}{\mathrm{d}L} = 0$$

i.e. $u(g^*) = 0$. Furthermore, since we assume $u(g_L)$ is a smooth function of g_L we have, for g_L close to g^*,

$$u(g_L) = (g_L - g^*)y\,, \quad y = \left(\frac{\partial u_L}{\partial g_L}\right)_{g_L = g^*}.$$

Near the critical point we therefore have

$$L\frac{\mathrm{d}g_L}{\mathrm{d}L} = (g_L - g^*)y.$$

What we have just done is known as the *linearization* of the renormalization group equation near a critical point. Note that the right-hand side of this equation does not explicitly depend on L.

Next we study the correlation length ξ. We have seen already that

$$L\xi\,(g_L) = \xi(g)$$

Since this relation is true for all L we have in particular, that $(L + \Delta L)\xi\,(g_{L+\Delta L}) = \xi\,(g) = L\xi\,(g_L)$ or, letting $\Delta L \to 0$,

$$\xi\,(g_L) + \left(L\frac{\mathrm{d}g_L}{\mathrm{d}L}\right)\frac{\partial \xi}{\partial g_L} = 0.$$

Substituting the expression for g_L close to g^* we then end up with

$$\xi(g_L) + (g_L - g^*)y\frac{\partial \xi}{\partial g_L} = 0.$$

The solution to this differential equation is easily found to be

$$\xi(g_L) \sim (g_L - g^*)^{-\frac{1}{y}}.$$

Since we identify scale invariance with the system being at critical temperature we identify $g^* = g_0^*/kT_{\mathrm{C}}$ and similarly, $g_L = g/kT$. To continue it is convenient to introduce the dimensionless variable

$$t = \left(\frac{T_{\mathrm{C}} - T}{T_{\mathrm{C}}}\right).$$

Substitution into the equation for ξ then gives $\xi(t) \sim t^{-1/y}$. Thus, $\xi \to \infty$, provided $y > 0$ as $t \to 0$ with critical exponent $\gamma = 2/y$.

We now want to study the scaling properties of the free energy $F(t_L, B_L)$, near the critical temperature. Here, $t_L = (g_L - g^*)$ measures the deviation away from the fixed point. From the renormalization group equation for g_L we then have $t_L = L^y t$. As for the external field B_L it is not hard to see that the linearization of the renormalization group equation near the fixed point leads to the scaling behaviour

$$B_L = L^x B, \quad \text{with} \quad x = \left(\frac{\partial V_L(g_L, B_L)}{\partial B_L}\right)_{g^*, B^*}.$$

Combining this with the scaling behavior of the free energy, $F(\mathsf{H}_L) = L^D F(\mathsf{H})$, we get

$$F(t_L, B_L) = L^D F(t, B),$$

or,

$$F(L^y t, L^x B) = L^D F(t, B).$$

If we define $\lambda = L^D$, so that $L = \lambda^{\frac{1}{D}}$ we have

$$F\left(\lambda^{\frac{y}{D}} t, \lambda^{\frac{x}{D}} B\right) = \lambda F(t, B)$$

This is the definition of a *generalized homogeneous function.* If the function $U_{L'}(g_L, B_L)$ and $V_{L'}(g_L, B_L)$ are known, then the coefficients x and y can be calculated. These would determine the coefficients $a_t = y/D$, $a_B = x/D$. Hence all the critical properties of the system could be calculated as we now demonstrate. We have

Theorem 13.3 If $F(\lambda^{a_t} t, \lambda^{a_B} B) = \lambda F(t, B)$ with λ, a constant and a_t, a_B two parameters then:

$$M(t, 0) = t^{\frac{1-a_B}{a_t}} M(1, 0)$$
$$M(0, B) = B^{\frac{1-a_B}{a_B}} M(0, 1)$$
$$\chi = t^{\frac{1-2a_B}{a_t}} \chi(1, 0)$$

that is, we have the following expressions for the critical exponents $\beta - (1 - a_B)/a_t$, $\delta = a_B/1 - a_B$ and $\gamma = (2a_B - 1)/a_t$.

Proof. We start with the thermodynamic identities:

$$M(t, B) = -\left(\frac{\partial F}{\partial B}\right)_t \quad \text{and} \quad \chi(t, B) = \left(\frac{\partial M}{\partial B}\right)_t$$

Using the generalized homogeneous function structure of the free energy we have

$$\begin{aligned}
M(\lambda^{a_t} t, \lambda^{a_B} B) &= \frac{\partial F(\lambda^{a_t} t, \lambda^{a_B} B)}{\partial(\lambda^{a_B} B)} \\
&= \lambda^{-a_B} \frac{\partial F(\lambda^{a_t} t, \lambda^{a_B} B)}{\partial B} \\
&= \lambda^{-a_B} \frac{\partial(\lambda F(t, B))}{\partial B} \\
&= \lambda^{1-a_B} M(t, B).
\end{aligned}$$

Thus

$$M(\lambda^{a_t} t, \lambda^{a_B} B) = \lambda^{1-a_B} M(t, B).$$

If we set $B = 0$ and $\lambda^{a_t} \cdot t = 1$ we obtain the first identity in the theorem:

$$M(t, 0) = t^{\frac{1-a_B}{a_t}} M(1, 0).$$

In order to prove the second identity we set $t = 0$ and $\lambda^{a_B} B = 1$ which gives

$$M(0, B) = B^{\frac{1-a_B}{a_B}} M(0, 1).$$

Finally, using $\chi(t, B) = \partial M / \partial B$ leads to

$$\chi(\lambda^{a_t} t, \lambda^{a_B} B) = \lambda^{1-2a_B} \chi(t, B).$$

Setting $B = 0$ and $\lambda^{a_t} t = 1$ gives

$$\chi(t, 0) = t^{\frac{1-2a_B}{a_t}} \chi(1, 0)$$

which completes the proof. □

To summarize, we have three experimental parameters β, γ, δ expressed in terms of two theoretical parameters, a_B, a_t. Thus, an immediate corollary is the *scaling law* $\gamma = \beta(\delta - 1)$. Putting in numbers $\gamma = 1.25$, $\beta = 0.33$, $\delta = 4.5$, we see it is approximately correct. This scaling law is just one of several relations that can be obtained in this way. We leave the derivation of the other scaling laws as a guided problem at the end of this chapter.

13.5 Critical exponent calculations

We now turn to the problem of calculating the various critical exponents that appeared in the last section borrowing techniques developed in our exposé on quantum field theory. The method presented here will improve on Landau's theory while reducing to the latter in a certain limit. The idea is to represent Landau's model as a certain approximation to the partition function of some quantum field theory for the order parameter. In analogy with our treatment of the path integral approach to quantum field theory we would like to write the partition function Z as a functional integral. As a guide to our intuition we use Landau's ansatz for the free energy. We thus write

$$Z = \int D\Phi \, \mathrm{e}^{-\frac{1}{\hbar}S[\Phi]}$$

where

$$S[\Phi(\mathbf{x})] = \int [\Phi(\mathbf{x})(-\nabla^2)\Phi(\mathbf{x}) + V(\Phi)]\mathrm{d}^D x$$

is a functional of the order parameter field with $V(\Phi)$ yet to be determined. As we will see below, this ansatz reproduces Landau's ansatz for the free energy in a saddle point approximation. A few comments are in order to justify these manipulations.

(1) The functional integral above is *not* derived from a concrete Hamiltonian as we did in Chapter 11. The path integral should be understood rather as a *statistical averaging* over different configurations of the order parameter $\Phi(\mathbf{x})$.
(2) We have redefined the order parameter field as $\sqrt{\beta\hbar}\Phi(\mathbf{x}) \to \Phi(\mathbf{x})$. The parameter $\hbar$ appearing in front of the action should thus not be interpreted as the fundamental Planck constant. Rather is is an auxiliary counting parameter introduced for convenience.
(3) If Φ is viewed as a fundamental field, then, as indicated in Section 11.4, the above expressions can be thought as the Euclidean path integral representation of a relativistic quantum field theory with action $S[\Phi(x)]$.

Having made these comments we will now proceed by applying the usual perturbative techniques from quantum field theory to the above functional integral, that is, we approximate $S[\Phi]$ by its functional Taylor expansion about $\Phi_0(\mathbf{x})$, which, in turn, is defined by the condition $\delta S/\delta\Phi = 0$. Thus

$$S[\Phi] = S[\Phi_0] + \frac{1}{2!}\int (\Phi - \Phi_0)_x(\Phi - \Phi_0)_y \frac{\delta^2 S}{\delta\Phi_x\delta\Phi_y}\mathrm{d}^D x\mathrm{d}^D y$$

where we use the notation $(\Phi - \Phi_0)_x = \Phi(\mathbf{x}) - \Phi_0(\mathbf{x})$. We can think of the functional derivatives introduced as generalizing the familiar notion of partial derivatives to continuous variables. Let us explain: recall if Φ was a n-tuple, namely $\boldsymbol{\Phi}$ with components $\Phi_i, i = 1, \ldots, N$ and S was a function of Φ_i then a Taylor expansion of S for $\mathbf{\Phi}$ close to $\mathbf{\Phi}_0 = \Phi_{0i}$ would be

$$S(\mathbf{\Phi}) \simeq S(\mathbf{\Phi}_0) + \sum_{i=1}^{N} (\Phi - \Phi_0)_i \left(\frac{\partial S}{\partial \Phi_i} \right)_{\mathbf{\Phi}_0} + \frac{1}{2!} \sum_{i=1}^{N} \sum_{j=1}^{N} (\Phi - \Phi_0)_i)(\Phi - \Phi_0)_j \left(\frac{\partial^2 S}{\partial \Phi_i \partial \Phi_j} \right)_{\mathbf{\Phi}_0} + \cdots$$

If we differentiate the components Φ_j with respect to Φ_i the result is simply

$$\frac{\partial \Phi_j}{\partial \Phi_i} = \delta_{ij}.$$

We now formally extend these ideas to $\Phi(\mathbf{x})$, where we think of $\Phi(\mathbf{x})$ as the components Φ_x of an "infinite-dimensional vector" Φ. Then the expansion of $S[\Phi(\mathbf{x})]$ in a functional Taylor series about $\Phi_0(\mathbf{x})$ should have the structure,

$$S[\Phi(\mathbf{x})] \simeq S[\Phi_0(\mathbf{x})] + \int_x (\Phi - \Phi_0)_x \left(\frac{\delta S}{\delta \Phi_x} \right)_{\Phi_0(\mathbf{x})} + \frac{1}{2!} \int_x (\Phi - \Phi_0)_x (\Phi - \Phi_0)_y \left(\frac{\delta^2 S}{\delta \Phi_x \delta \Phi_y} \right)_{\Phi_0(\mathbf{x})} + \cdots$$

where $\int \mathrm{d}^D x$ replaces $\sum_i$ by analogy with the discrete case. Similarly, the basic rule of "functional differentiation" should be

$$\frac{\delta \Phi(\mathbf{x})}{\delta \Phi(\mathbf{y})} = \delta^D(\mathbf{x} - \mathbf{y})$$

that is the D-dimensional Dirac delta function replaces the Kronecker delta δ_{ij} for the discrete case.

We recall that a stationary point (actually a function) of $S[\Phi]$ corresponds to finding a solution to the classical equation,

$$\delta S = 0$$

which corresponds to solving the Euler–Lagrange equations:

$$\frac{\partial L}{\partial \Phi_0} = \partial_i \left(\frac{\partial L}{\partial(\partial_i \Phi_0)} \right).$$

Such a $\Phi_0(\mathbf{x})$ is the classical solution. A classical approximation would correspond to replacing $S[\Phi]$ by $S[\Phi_0]$ in Z. A semi-classical approximation would involve replacing $S[\Phi]$ by

$$S[\Phi_0] + \frac{1}{2!} \int \mathrm{d}^D x \int \mathrm{d}^D y (\Phi - \Phi_0)_x (\Phi - \Phi_0)_y \left(\frac{\delta^2 S}{\delta \Phi_x \delta \Phi_y} \right)_{\Phi_0}.$$

First let us consider the classical approximation to the partition function Z. Approximating the functional integral by its saddle point we find

$$Z \simeq e^{-\frac{1}{\hbar}S[\Phi_0]}.$$

On the other hand we have in the canonical ensemble

$$Z = e^{-\beta F}.$$

Thus we can identify the free energy of statistical mechanics with $S[\Phi_0]$ of our field theory

$$F = \frac{1}{\beta\hbar} S(\Phi_0).$$

It is now clear how to generalize Landau's theory. Instead of the saddle point, or classical approximation, we consider the full path integral

$$Z(\beta, g, m) = \int D\Phi \, e^{-\frac{1}{\hbar}S[\Phi]}$$

with

$$S[\Phi] = \int d^D x \left[\frac{1}{2}\Phi(\mathbf{x})(-\nabla^2)\Phi(\mathbf{x}) + V(\Phi) \right]$$

$$V(\Phi) = \frac{1}{2}m^2\Phi^2(\mathbf{x}) + \frac{g}{4!}\Phi^4(\mathbf{x}),$$

that is, we take for $S[\Phi]$ Landau's free energy F_L, replacing $M_z(\mathbf{x})$ by $\Phi(\mathbf{x})$. From our discussion of the path integral approach to quantum field theory, we know that a perturbation expansion for the partition function and correlation functions in the coupling constant g can be constructed. Our aim is then to see if by choosing special values for the coupling constants m^2 and g the partition function Z can be made scale invariant. The system then will be close to a phase transition configuration. Using the Landau form for $S(\Phi)$ we will analyze the corresponding field theory and see if for special values of the coupling constants the partition function Z can be made scale invariant. We would like to study the above functional integral within the framework of quantum field theory. To carry out this program a few results about functional integrals are needed which we will present in the next section.

13.6 Correlation functions

In Section 13.1 we have emphasized the relevance of correlation functions for explaining the singular behavior of the susceptibility near the critical temperature, T_C. The following result shows how the two-point correlation functions are related to the path integral representation of the partition function. To begin with we extend Result 11.2 for finite-dimensional integrals formally to the case of a field $\Phi(\mathbf{x})$ as

Result 13.4

$$\int D\Phi\, \mathrm{e}^{-\frac{1}{2}(\Phi,\mathsf{A}\Phi)+(J,\Phi)} = (\det \mathsf{A})^{-\frac{1}{2}}\, \mathrm{e}^{\frac{1}{2}(J,\mathsf{A}^{-1}J)}$$

where

$$(\Phi, \mathsf{A}\Phi) = \int \mathrm{d}^D x\, \Phi(x)(-\nabla^2 + m^2)\Phi(\mathbf{x})$$
$$(J, \Phi) = \int \mathrm{d}^D x\, J(\mathbf{x})\Phi(\mathbf{x}).$$

The determinant of the operator $\mathsf{A} = (-\nabla^2 + m^2)$ has to be suitably defined in terms of the non-zero eigenvalues of the differential operator $(-\nabla^2 + m^2)$.

This is an important identity since it allows us to express the correlation functions of an arbitrary number of fields in terms of derivatives of the partition sum with respect to the "external field" $J(\mathbf{x})$. We will show this explicitly for the case of the two-point function.

Result 13.5 The two-point correlation function

$$C_T(|\,\mathbf{x}\,|) = \frac{\int D\Phi\, \mathrm{e}^{-(\Phi,\mathsf{A}\Phi)}\Phi(\mathbf{x})\Phi(0)}{\int D\Phi\, \mathrm{e}^{-(\Phi,\mathsf{A}\Phi)}}, \quad \mathsf{A} = (-\nabla^2 + m^2)$$

satisfies the partial differential equation

$$(-\nabla_x^2 + m^2)C_T(|\,\mathbf{x}\,|) = \delta^3(\mathbf{x}).$$

This shows, in particular that $C_T(|\,\mathbf{x}\,|)$ agrees with the correlation function introduced in Section 13.3 up to a solution of the homogeneous equation $(-\nabla_x^2 + m^2)f(\mathbf{x}) = 0$.

Proof. In order to verify this claim we consider

$$\int D\Phi\, \mathrm{e}^{-\frac{1}{2}(\Phi,\mathsf{A}\Phi)}\Phi(x)\Phi(y).$$

Using Result 13.4 this becomes

$$(\det \mathsf{A})^{-\frac{1}{2}} \frac{\delta}{\delta J(x)} \frac{\delta}{\delta J(y)} \mathrm{e}^{\frac{1}{2}(J,\mathsf{A}^{-1}J)}\,|_{J=0}$$

where

$$(J, \mathsf{A}^{-1}J) = \int \mathrm{d}^D x \int \mathrm{d}^D y\, J(\mathbf{x})\mathsf{A}^{-1}(\mathbf{x}, \mathbf{y})J(\mathbf{y}).$$

After expanding the exponential we will have to evaluate

$$\frac{\delta}{\delta J(\mathbf{x})}\frac{\delta}{\delta J(\mathbf{y})}\left[1+\frac{1}{2}(J,\mathsf{A}^{-1}J)+\frac{1}{2!}\frac{1}{4}(J,\mathsf{A}^{-1}J)^2+\cdots\right]\Bigg|_{J=0}$$

To continue we recall the functional derivatives given in Section 13.5

$$\frac{\delta J(\mathbf{z})}{\delta J(\mathbf{x})}=\delta^D(\mathbf{x}-\mathbf{z}).$$

Since J is to be set equal to zero at the end, the only term which contributes to the result is

$$\frac{\delta}{\delta J(\mathbf{x})}\frac{\delta}{\delta J(\mathbf{y})}\frac{1}{2}(J,\mathsf{A}^{-1}J)\mid_{J=0}$$
$$=\frac{1}{2}[\mathsf{A}^{-1}(\mathbf{x},\mathbf{y})+\mathsf{A}^{-1}(\mathbf{y},\mathbf{x})].$$

Since $\mathsf{A}^{-1}(\mathbf{x},\mathbf{y})=\mathsf{A}^{-1}(\mathbf{y},\mathbf{x})$ is a Hermitian operator we then have

$$\frac{\int D\Phi e^{-(\Phi,\mathsf{A}\Phi)}\Phi(\mathbf{x})\Phi(0)}{\int D\Phi e^{-(\Phi,\mathsf{A}\Phi)}}=\mathsf{A}^{-1}(\mathbf{x},\mathbf{y}).$$

To conclude the proof we have to determine $\mathsf{A}^{-1}(\mathbf{x},\mathbf{y})$ by solving the equation

$$\mathsf{A}\mathsf{A}^{-1}=\mathsf{I},\quad\text{or}\quad \mathsf{A}\mathsf{A}^{-1}(\mathbf{x},\mathbf{y})=\delta^D(\mathbf{x},\mathbf{y}).$$

Upon substitution of $\mathsf{A}=(-\nabla^2+m^2)$ the equation for $A^{-1}(\mathbf{x},\mathbf{y})$ is given by

$$\left(-\nabla_x^2+m^2\right)A^{-1}(\mathbf{x},\mathbf{y})=\delta^{(3)}(\mathbf{x}-\mathbf{y}),$$

which, in turn, establishes the result. □

13.7 Epsilon expansion

Before we consider the corrections to the Landau approach let us recall how $C_T(\mid\mathbf{x}\mid)$ determines the singular behavior of the susceptibility χ. As explained in Section 13.3 we have

$$\chi=\int \mathrm{d}^3x\, C_T(\mid\mathbf{x}\mid)=(2\pi)^3\Delta(0)$$

where $\Delta(\mathbf{k})$ is the Fourier transform of $C_T(\mid\mathbf{x}\mid)$. In particular if we ignore the interaction term, $g/4!\int \mathrm{d}^D\Phi^4(x)$, we have just seen that $\Delta^{g=0}(0)=1/m^2$ and thus $\chi=(2\pi)^3/m^2$. Setting $m^2\propto(T_C-T)$ Landau's result is recovered. The singular behavior of χ is thus determined by the way $m^2\to 0$ as $T\to T_C$.

The way to improve on Landau's result is now clear. We should study the full propagator

$$C_T^g(|\mathbf{x}|) = \frac{\int D\Phi\, \mathrm{e}^{-(\Phi,A\Phi)-\frac{g}{4!}\int \mathrm{d}^D x \Phi^4}\Phi(\mathbf{x})\Phi(0)}{\int D\Phi\, \mathrm{e}^{-(\Phi,A\Phi)-\frac{g}{4!}\int \mathrm{d}^D x\, \Phi^4}}$$

in the region where the parameters m and g of the theory are close to values which lead to a scale invariant partition function. The susceptibility χ is then determined by calculating $\Delta^g(0)$ corresponding to $\int C_T^g(|\mathbf{x}|)$.

Let us evaluate the lowest-order modification to $C_T(|\mathbf{x}|)$ brought about by the interaction term $g/4! \int \mathrm{d}^D x \Phi^4(\mathbf{x})$. To do this we expand the interaction in powers of g

$$\mathrm{e}^{-\frac{g}{4!}\int \Phi^4(\mathbf{x})\mathrm{d}^D x} \simeq 1 - \frac{g}{4!}\int \Phi^4(\mathbf{x})\mathrm{d}^D x.$$

Then to O(g) :

$$\begin{aligned} C_T(|\mathbf{x}_1 - \mathbf{x}_2|) = C_T^0(|\mathbf{x}_1 - \mathbf{x}_2|) &- \frac{g}{4!}\frac{\int D\Phi\, \mathrm{e}^{-(\Phi,A\Phi)}\Phi(\mathbf{x}_1)\Phi(\mathbf{x}_2)\int \mathrm{d}^D y\, \Phi^4(\mathbf{y})}{\int D\Phi\, \mathrm{e}^{-(\Phi,A\Phi)}} \\ &+ \frac{g}{4!}\frac{\int D\Phi\, \mathrm{e}^{-(\Phi,A\Phi)}\Phi(\mathbf{x}_1)\Phi(\mathbf{x}_2)\int D\Phi\, \mathrm{e}^{-(\Phi,A\Phi)}\int \mathrm{d}^D y\, \Phi^4(\mathbf{y})}{\left(\int D\Phi\, \mathrm{e}^{-(\Phi,A\Phi)}\right)^2}. \end{aligned}$$

The first term on the right-hand side is just our result for the two point correlator in the absence of interactions. The second and the third term can now be evaluated in terms of C_T^0 using Results 13.4 and 13.5 leading to

$$C_T(|\mathbf{x}_1 - \mathbf{x}_2|) = C_T^0(|\mathbf{x}_1|) - \frac{12g}{4!}\left(\int C_T^0(|\mathbf{x}_1 - \mathbf{y}|)C_T^0(0)C_T^0(|\mathbf{x}_2 - \mathbf{y}|)\mathrm{d}^D y\right).$$

The factor 12 is a combinatorial factor arising when applying Result 13.4. Just like in our discussion in quantum field theory in Chapter 9 the above equation has a simple graphical interpretation. The right-hand side can be represented by drawing all *connected graphs* with two external lines corresponding to $\Phi(\mathbf{x}_1)$ and $\Phi(\mathbf{x}_2)$ and one vertex corresponding to $\int \Phi^4(\mathbf{x})\mathrm{d}^D x$, that is

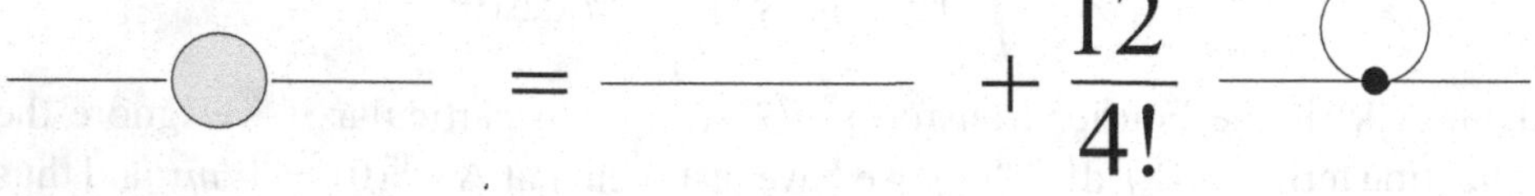

$$= \quad + \frac{12}{4!}$$

The combinatorial factor 12 then simply counts the number of ways to connect the legs of the vertex with itself and the external lines, $\Phi(\mathbf{x}_1)$ and $\Phi(\mathbf{x}_2)$. Thus we recover the Feynman rules established in Chapter 9. This is as it should be since as we have already mentioned in the remarks above we are treating the path integral like a fictitious quantum field theory with action $S[\Phi]$. The reason that we draw only connected Feynman graphs is due to the fact that the disconnected, or *vacuum graphs*, are cancelled by the last term in the equation for C_T. Let us now rewrite our result in terms of the momentum space correlators $\Delta(\mathbf{k})$ and $\Delta^0(\mathbf{k})$. Defining

$$C_T(|\mathbf{x}|) = \int \frac{\mathrm{d}^D k}{(2\pi)^D} \mathrm{e}^{\mathrm{i}\mathbf{k}\cdot\mathbf{x}} \Delta(\mathbf{k})$$

$$C_T^0(|\mathbf{x}|) = \int \frac{\mathrm{d}^D k}{(2\pi)^D} \mathrm{e}^{\mathrm{i}\mathbf{k}\cdot\mathbf{x}} \Delta^0(\mathbf{k})$$

we get

$$\begin{aligned}\Delta(\mathbf{k}) &= \Delta^0(\mathbf{k}) - 12\Delta^0(\mathbf{k}) \left(\frac{g}{4!} \int \frac{\mathrm{d}^D q}{(2\pi)^D} \frac{1}{(q^2+m^2)} \right) \Delta^0(\mathbf{k}) \\ &= \Delta^0(\mathbf{k}) - \frac{1}{2} g \Delta^0(\mathbf{k}) \Sigma(0) \Delta^0(\mathbf{k})\end{aligned}$$

where

$$\Sigma(0) = \int \frac{\mathrm{d}^D q}{(2\pi)^D} \frac{1}{(q^2+m^2)}.$$

In fact this integral is logarithmically divergent for $D = 3$. This is a common feature in quantum field theory which is addressed by adding suitable *counter terms* to the actions $S[\Phi]$. This process which is called *renormalization* is described extensively in the literature. For our purpose we can avoid this complication since we will only be interested in the m-dependent part of $\Sigma(0)$. Since we are interested in the susceptibility we may set $k = 0$. Then

$$\begin{aligned}\Delta(0) &\simeq \Delta^0(0) - \frac{g}{2} \Delta^0(0) \Sigma(0) \Delta^0(0) \\ &= \frac{1}{m^2} - \frac{g}{2m^4} \Sigma(0) \\ &\equiv \frac{1}{M^2}.\end{aligned}$$

From our result relating the susceptibility χ to $\Delta(0)$ we have $\chi = (2\pi)^3/M^2$. Hence the singular behavior of χ can be determined, as a function of temperature, once we determine the way M depends on temperature. This in turn will follow from the way M depends on m since we know that $m^2 \propto T - T_\mathrm{C}$. We shall look for a dependence of M on m of the form $M \propto m^{1+c(g)}$. This will change the singular behavior of χ from that predicted by Landau theory.

At this stage, the singular behavior of χ seems to depend on the value of the coupling constant g of the system. However, as we have emphasized at the end of Section 13.5, in order for our quantum field theory model to describe a physical system near a phase transition, the theory must be scale invariant. Now since $m^2 \propto (T_C - T)$ scale invariance can be achieved at the critical temperature by setting $g = 0$. In this case we recover the results from Landau theory. The question which we now want to address is whether taking higher-order corrections into account we will find other non-trivial values of g leading to a scale invariant theory. For this we first introduce an *effective coupling*. This can be done by considering the four-point function

$$\Gamma(\mathbf{x}_1, \mathbf{x}_2, \mathbf{x}_3, \mathbf{x}_4) = \int D\phi \mathrm{e}^{-(\phi, \mathsf{A}\phi) - \frac{g}{4!}\int \phi^4} \phi(\mathbf{x}_1)\phi(\mathbf{x}_2)\phi(\mathbf{x}_3)\phi(\mathbf{x}_4).$$

Expanding the interaction term $\mathrm{e}^{-\frac{g}{4!}\int \phi^4}$ we get

$$\int D\phi \mathrm{e}^{-(\phi,\mathsf{A}\phi)} \phi(\mathbf{x}_1)\phi(\mathbf{x}_2)\phi(\mathbf{x}_3)\phi(\mathbf{x}_4) \left[1 - \frac{g}{4!}\int \phi^4(x)\mathrm{d}^D x + \left(\frac{g}{4!}\right)^2 \left(\frac{1}{2!}\int \phi^4(x)\mathrm{d}^D x\right)^2 + \cdots \right].$$

Now we consider the term linear in the coupling constant g, that is

$$\int D\phi \mathrm{e}^{-(\phi,\mathsf{A}\phi)} \phi(\mathbf{x}_1)\phi(\mathbf{x}_2)\phi(\mathbf{x}_3)\phi(\mathbf{x}_4) \left(\frac{-g}{4!}\right) \int \phi^4(x)\mathrm{d}^D x.$$

Using Result 13.4 this gives

$$\frac{-g}{4!}\int C^0(\mathbf{x}_1 - \mathbf{x})C^0(\mathbf{x}_2 - \mathbf{x})C^0(\mathbf{x}_3 - \mathbf{x})C^0(\mathbf{x}_4 - \mathbf{x})\mathrm{d}^D x.$$

In momentum space this becomes

$$= \frac{-g}{4!}\int \prod_{i=1}^4 \mathrm{d}^D \mathbf{k}_i \left(\frac{1}{{\mathbf{k}_1}^2 + m^2}\right)\left(\frac{1}{{\mathbf{k}_2}^2 + m^2}\right)\left(\frac{1}{{\mathbf{k}_3}^2 + m^2}\right)\left(\frac{1}{{\mathbf{k}_4}^2 + m^2}\right)\delta^D\left(\sum_{i=1}^4 \mathbf{k}_i\right)$$

where $\mathbf{k}_1, \mathbf{k}_2, \mathbf{k}_3, \mathbf{k}_4$ represent the momenta of four particles. The coupling constant $-g/4!$ is obtained from this expression by picking the term from the integral with $\mathbf{k}_i = 0$ and multiplying this expression by $(m^2)^4$. This term represents the strength of the interaction between four zero-momentum particles. We will take this to be the definition of the coupling constant of the theory.

Let us now look at the term which is of second order in g. We have

$$\frac{1}{2}\int D\phi \mathrm{e}^{-(\phi,\mathsf{A}\phi)} \phi(\mathbf{x}_1)\phi(\mathbf{x}_2)\phi(\mathbf{x}_3)\phi(\mathbf{x}_4) \left(\frac{-g}{4!}\right)^2 \left(\int \phi^4(\mathbf{x})\mathrm{d}^D x\right)^2.$$

Using Result 13.4 this can be easily evaluated. A typical term obtained by this procedure is

$$\left(\frac{-g}{4!}\right)^2 \frac{1}{2!}\int \mathrm{d}^D x \int \mathrm{d}^D y C^0(\mathbf{x}_1 - \mathbf{x})C^0(\mathbf{x}_2 - \mathbf{x})(C^0(\mathbf{x} - \mathbf{y}))^2 C^0(\mathbf{x}_3 - \mathbf{y})C^0(\mathbf{x}_4 - \mathbf{y})$$

and corresponds to the Feynman diagram

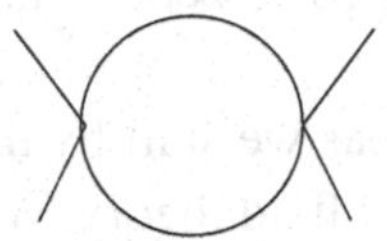

Writing this in momentum space, we get

$$\left(\frac{-g}{4!}\right)^2 \frac{1}{2!} \int \prod_{i=1}^{4} d^D\mathbf{k}_i \left(\frac{1}{\mathbf{k}_1{}^2 + m^2}\right)\left(\frac{1}{\mathbf{k}_2{}^2 + m^2}\right)\left(\frac{1}{\mathbf{k}_3{}^2 + m^2}\right)\left(\frac{1}{\mathbf{k}_4{}^2 + m^2}\right)$$
$$\times\, \delta^D\left(\sum_{i=1}^{4} \mathbf{k}_i\right)\left(\int \frac{d^Dq}{(2\pi)^D} \frac{1}{(q^2+m^2)} \frac{1}{((\mathbf{k}_1+\mathbf{k}_2-q)^2+m^2)}\right).$$

Altogether there are $2({}^4C_2)^2$ such terms. They correspond to the number of terms generated by applying Result 13.4 to evaluate the integral. From the graphical representation this factor can be understood as follows. There are $({}^4C_2)^2$ ways of selecting two lines from the two vertices and there are two ways of joining these to form the required graph. The contribution of these terms to the effective coupling constant can be obtained by again picking the term with $\mathbf{k}_i = 0$ from the integral and multiplying them by $(m^2)^4$. This gives

$$({}^4C_2)^2 2! \left(\frac{-g}{4!}\right)^2 \frac{1}{2!}\left(\int \frac{d^Dq}{(2\pi)^D} \frac{1}{(q^2+m^2)^2}\right).$$

If we ignore higher order terms we are led to define the effective coupling g_R of the field theory in terms of the linear and quadric contribution to the four-point function. Graphically we can write this as

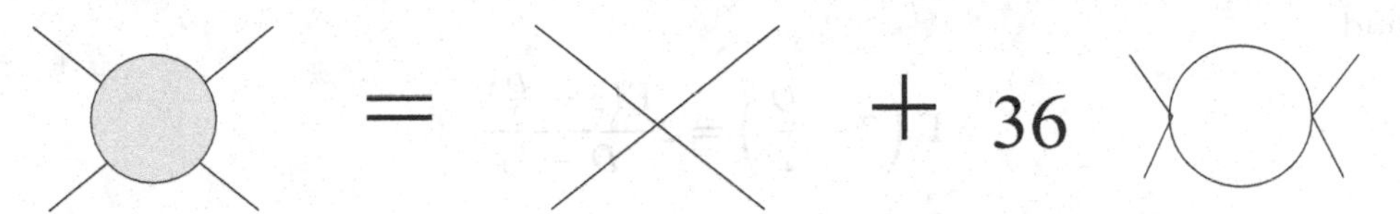

Thus we have

$$\frac{1}{4!} g_R = \frac{1}{4!} g - \left(\frac{g}{4!}\right)^2 \frac{1}{2!} ({}^4C_2)^2\, 2 \int \frac{d^Dq}{(2\pi)^D} \frac{1}{(q^2+m^2)^2}$$

i.e.

$$g_R = g - \frac{3}{2} g^2 \int \frac{d^Dq}{(2\pi)^D} \frac{1}{(q^2+m^2)^2}.$$

Observe that g_R depends on m. Near a phase transition we would like g_R to be a scale invariant quantity and m to go to zero. This would lead to a scale invariant partition function.

In order to implement these ideas we start by first determining the scale dimensions of the different objects in the field theory. We observe that in units where the Boltzmann constant and $\hbar$ are unity, $S[\Phi]$ must be a dimensionless quantity since otherwise $e^{-S[\Phi]}$ is ill-defined. By examining each term of $S[\Phi]$ we can establish the dimension "d" of the different quantities appearing in $S[\Phi]$. Introducing a length scale L, we find that

$$\begin{aligned} d_\Phi &= L^{\frac{2-D}{2}} \\ d_m &= L^{-1} \\ d_g &= L^{D-4} \end{aligned}$$

We recall that the parameter $m^2 \propto (T_C - T)$ and hence it is a variable. It is thus convenient to select m^2 as the length scale of our system. This allows us to study $\partial M^2/\partial m^2$ as a function of m^2 keeping g fixed. The parameter g is to be fixed so that the system is scale invariant. We have

$$\begin{aligned} \frac{\partial M^2}{\partial m^2} &= 1 + \frac{g}{2}(m^2)^{\left(\frac{D-4}{2}\right)}\left(\frac{A}{D-4}\right) \\ A &= \frac{2\Gamma(3-D/2)}{(2\sqrt{\pi})^D}, \end{aligned}$$

where we have used the result

$$\int \frac{d^D q}{(2\pi)^D}\frac{1}{(q^2+m^2)^n} = \frac{1}{(2\sqrt{\pi})^D} m^{2\left(\frac{D}{2}-n\right)}\frac{\Gamma\left(n-\frac{D}{2}\right)}{\Gamma(n)}$$

and

$$\Gamma\left(2-\frac{D}{2}\right) = 2\frac{\Gamma\left(3-\frac{D}{2}\right)}{(D-4)}$$

to evaluate the integral

$$\int \frac{d^D q}{(2\pi)^D}\frac{1}{(q^2+m^2)}.$$

Similarly we have for the effective coupling

$$g_R = g + \frac{3}{2}g^2(m^2)^{\frac{D-4}{2}}\frac{A}{(D-4)}.$$

The method of evaluating integrals by treating the dimension D as a parameter is known as the *dimensional regularization method*. It is very useful for regulating potentially divergent integrals and is commonly used in quantum field theory. For problems in critical phenomena it is also convenient to regard $(4 - D) = \epsilon$ as "small" and use it as an expansion parameter. This is known as the *epsilon expansion*. Writing

$$g_R = u_R m^{-2(D-4)/2} = x^{\epsilon/2} u_R$$

with $\epsilon = 4 - D$, and $x = m^2$, we have

$$x^{\epsilon/2} u_R = x^{\epsilon/2} u + \frac{3}{2} u^2 x^{\epsilon/2} \frac{A}{D-4}.$$

We now implement our requirement that g_R^* be scale invariant, that is independent of x, or $x \partial g_R^* / \partial x = 0$. This implies

$$\frac{\epsilon}{2}\left[u^* - \frac{3}{2}(u^*)^2 \frac{A}{\epsilon}\right] = 0.$$

There are two solutions $u^* = 0$, $u^* = +2\epsilon/3A$. Let us now return to the expression for

$$\left.\frac{\partial M^2}{\partial x}\right|_{g \text{ fixed}} = 1 + \frac{g}{2}(x)^{\frac{D-4}{2}}\left(\frac{A}{D-4}\right).$$

We want to determine the dependence of M^2 on x due to interactions. If we write $M^2 = Bx^{1+c(g)}$ then

$$x \frac{\mathrm{d}}{\mathrm{d}x} \ln\left(\frac{\partial M^2}{\partial x}\right) = c(g).$$

In our case

$$\ln\left(\frac{\partial M^2}{\partial x}\right) = \ln\left(1 + \frac{g}{2} x^{\frac{D-4}{2}}\left(\frac{A}{D-4}\right)\right).$$

So that

$$x \frac{\mathrm{d}}{\mathrm{d}x} \ln\left(\frac{\partial M^2}{\partial x}\right) = +A \frac{g}{4} x^{\frac{D-4}{2}} = +\frac{uA}{4}.$$

Since $g x^{\frac{D-4}{2}} = u$, we have $c = +uA/4$. We found that there were two values for u which give rise to a scale invariant g_R^*. Namely $u^* = 0$. This gives $c = 0$ and implies

$$\chi \sim \frac{1}{|\, T - T_C \,|}$$

which is Landau's result. The second solution is $u^* = +\frac{2\epsilon}{3A}$ with $\epsilon = 4 - D$. This gives $c = \epsilon/6$ and

$$\chi \sim \frac{1}{|\, T - T_C \,|^{1+\frac{\epsilon}{6}}}.$$

Table 13.1. *Critical exponents and scaling laws from ϵ-expansion (at $O(\epsilon)$), Landau theory, exact results in the 2D-Ising model, numerical results in the 3D-Ising model and a variety of experimental values to illustrate the universal behavior. The exponent α is defined in Problem 13.1.*

Exponent	ϵ-expansion	Landau	2D-Ising	3D-Ising	Experiment
α	1/6	0	0 (log. div.)	0.12	0–0.14
β	1/3	0.5	$\frac{1}{8}$	0.31	0.32–0.39
γ	7/6	1	1.75	1.25	1.3–1.4
δ	4	3	15	5	4–5
$\frac{(\beta\delta-\gamma)}{\beta}$	1/2	1	1	1	0.93 ± 0.08

In particular, for $D = 3$, we have $\epsilon = 1$ so that the critical exponent γ is changed from the Landau value of 1 to $(1 + 1/6) \simeq 1.2$. Experimentally, for many systems, as we stated earlier $\gamma \simeq 1.25$. Thus choosing $u^* = +2\epsilon/3a$ improves on Landau's original result. In Table 13.1 we compare the values for the critical exponents of different calculations with some experimental values. This gives an idea of the quality of the different approximate methods. Note that the predictions from Landau theory improve as the dimensions of the system increases. This is due to the fact that fluctuations which are important in low dimensions are neglected in the Landau theory.

Of course, $\epsilon = 1$ is not small and thus higher-order corrections should be taken into account. Our aim in this section was to emphasize the physical ideas which underlie the real space and field theory approaches to critical phenomena and to give a flavor of the way a calculation of critical exponents is actually carried out. In order to do these calculations to greater accuracy a more elaborate machinery involving renormalizations is required. The specialized books listed at the end of the section may be consulted for further details.

Problems

Problem 13.1 Consider the Landau model with

$$F_L(T, \Phi) = F_L(T, 0) + \frac{r_0(T)}{2}\Phi^2 + u_0\Phi^4 ,$$

and let $F(T) \equiv F_L(T, \Phi)|_{\Phi=\langle\Phi(T)\rangle}$ be the free energy. Here $\langle\Phi(T)\rangle$ is determined by requiring that it minimizes $F_L(T, \Phi)$. We will assume that $r_0(T) = a_0(T - T_C)/T_C$ with $a_0 > 0$. In particular, $r_0(T)$ changes sign at $T = T_C$. Show that

(1) the entropy S is continuous at $T = T_C$ and that

(2) c_V is discontinuous. Compute the critical exponent α for c_V defined through $c_V \propto ((T - T_C)/T_C)^\alpha$. Hint: use that near T_C we can approximate $c_V \simeq T_C \partial S/\partial T$ (why?)

Problem 13.2 Discuss the consequences of a cubic term in the Landau model

$$F_L(T, \Phi) = F_L(T, 0) + \frac{r_0(T)}{2}\Phi^2 + s_o\Phi^3 + u_0\Phi^4.$$

Sketch $F_L(T, \Phi)$ as a function of Φ for $T > T_C$ and $T < T_C$ and show that $\langle\Phi(T)\rangle$ is a discontinous function of T. What is the interpretation of this jump?

Problem 13.3 Revisit the Landau theory in Section 13.3 from the second viewpoint introduced in Problem 3.3. How will the corresponding formulas in Section 13.3 be modified?

Problem 13.4 Convexity of the free energy: the free energy as a function of the order parameter plotted in Section 13.3 is not convex. This is because in Section 13.3 we considered only the pure phase where the order parameter is constant. In this problem we consider the possibility of a mixture of phases. For this we start with the Landau free energy for fixed $T_0 < T_C$.

$$F_L\,[T_0, \Phi, V] = F_L(T_0, 0, V) + \int \mathrm{d}^3x \left[\frac{1}{2}(\nabla\Phi)^2 - \frac{\tau}{2}\Phi^2 + \frac{\lambda}{4!}\Phi^4\right],$$

where $\tau, \lambda > 0$, $\Phi(\mathbf{x})$ is the order parameter and $F_0(T_0, V)$ is the free energy for $\Phi = 0$. Assuming that $\Phi = \Phi_0$, a constant, the free energy as a function Φ_0 takes the form of the first plot in Section 13.3 with two minima at $\Phi_0 = \pm\sqrt{\tau 3!/\lambda}$.

(1) Show that if we allow Φ to depend on one coordinate, z say, then the free energy has another extremum for

$$\Phi_1(z) = \phi_0 \tanh\frac{z - z_0}{\xi},$$

where $\xi = \sqrt{2/\tau}$.

(2) Show that for $-L/2 \le z \le L/2$, with $L \gg \xi$ the average value for the order parameter is given by

$$\langle\Phi\rangle = -\Phi_0\frac{2z_0}{L}.$$

Thus $\langle\Phi\rangle$ as a function of z_0 interpolates between $-\Phi_0$ and Φ_0.

(3) Show that for $L \to \infty$ the function $F_L\,[T_0, \Phi_1, V]/V$ is the convex hull of $F_L\,[T_0, \Phi_0, V]/V$ between $-\Phi_0$ and Φ_0.

In this concrete example we thus see explicitly that the free energy is convex if we allow for mixing of the pure phases, $-\Phi_0$ and Φ_0.

Problem 13.5 The *mean field approximation* is a concrete procedure to implement the ideas underlying the Landau theory in a specific model. Here we consider the D-dimensional Ising model. Start with the expression for the energy of the Ising model given in Section 13.1 and expand the energy in powers of $m_i - \langle m \rangle$, where $\langle m \rangle$ is the expectation value of m_i (to be determined!). In the mean field approximation one neglects all terms of order $(m_i - \langle m \rangle)^2$ and higher. Compare the effective Hamiltonian obtained to that derived in Problem 2.5.

The partition function for the truncated expression for the energy can then be calculated exactly. Minimizing the corresponding free energy leads to an implicit equation for $\langle m \rangle$. Determine the critical temperature T_C defined as the maximal temperature for which this equation has a non-trivial solution for $\langle m \rangle$. Expand the free energy in terms of $\langle m \rangle$ for $(T - T_C)/T_C \ll 1$ up to fourth order in $\langle m \rangle$. Compare the resulting expression with the Landau free energy.

Problem 13.6 The purpose of this problem is to use dimensional arguments in order to get a quick derivation of scaling laws of the type found at the end of Section 13.4 relating the various critical exponents. The key assumption is that for $t = (T_C - T)/T_C$, the correlation length, $\xi(t) = \xi_0 t^{-\nu}$, is the only scale in the theory. Dimensional counting then implies that

$$\beta \frac{F(T, V)}{V} \propto \xi^{-D}(t),$$

where D is the space dimension. Show that this implies that $c_V \propto |t|^{D\nu - 2}$. Combining this with $c_V \propto |t|^{-\alpha}$ (see Problem 13.1) we end up with the *Josephson scaling law*

$$2 - \alpha = D\nu.$$

Similarly using that for the magnetization

$$B \langle m \rangle = \frac{F_L(T, \langle m \rangle, B) - F_L(T, 0, 0)}{V},$$

show that $\beta(1 + \delta) = D\nu$. There are in total four scaling laws and six critical exponents so that only two of them are independent.

Historical notes

The idea of the renormalization group was first introduced by Stueckelberg and Petermann in 1953 and by Gell-Mann and Low in 1954 to cure ultraviolet divergencies in the perturbative approach to relativistic quantum field theory. We have encountered an example of such a divergence for $\Sigma(0)$ in Section 13.7. Renormalization is the procedure to remove such infinities order by order in perturbation theory by adding equally infinite counter terms to the action, that is, the infinities are absorbed through a redefinition of the coupling constants and the fields. The underlying physical intuition is that the original, or "bare", theory, is merely a mathematical construct while all physical observables which are expressed solely in terms of renormalized couplings are finite. Generically this renormalization procedure introduces a new scale in the theory. The renormalization group then expresses the invariance of the physics of this new scale.

In critical phenomena the renormalization group was first formulated by Wilson in 1971. The purpose here is not to cure divergencies but to reduce the number of degrees of freedom in the effective description of critical phenomena by systematically integrating over short wavelength fluctuations. This is the philosophy that was followed in this chapter. The physics at large scales should not be affected by this procedure. Thus the sole effect should be a possible redefinition of the dimensional coupling constants to take the change of scale into account. The renormalization group transformation then relates the parameters of the theory at different steps of this process. The key point in this approach is the existence of fixed points in the renormalization group transformation. At these points the effective theory becomes scale invariant and therefore can describe a physical system at a phase transition.

Since the 1970s there has been a gradual change in philosophy also in quantum field theory as to the interpretation of the renormalized theory. Today one is more inclined to think of the renormalized theory as a long distance approximation to some yet-to-be-discovered unified theory. At long distances the details of this unknown theory are irrelevant, the low energy theory being defined by but a few coupling parameters that define the renormalized theory. The success of Wilson's interpretation of renormalization for critical phenomena surely has contributed to this shift of emphasis.

Further reading

A classic text book on the theory of phase transitions is H. E. Stanley, *Introduction to Phase Transitions and Critical Phenomena*, Oxford University Press (1971). Boundary and surface effects in phase transitions are discussed in S. K. Ma, *Statistical Mechanics*, World Scientific (1985). For further discussions of the scaling laws see e.g. K. Huang, *Statistical Mechanics*, John Wiley (1987) or M. Le Bellac, *Quantum and Statistical Field Theory*, Oxford University Press (1991).

Other useful texts on phase transitions and critical phenomena are listed below: L. Landau and E. Lifshitz, *Statistical Physics*, Pergamon Press (1959); L. E. Reichl, *A Modern Course in Statistical Physics*, Arnold (1980); S. K. Ma, *Modern Theory of Critical Phenomena*, W. A. Benjamin (1976); N. Goldenfeld, *Lectures on Phase Transitions and the Renormalisation Group*, Addison Wesley (1992), and at a more advanced level, J. Zinn-Justin, *Quantum Field Theory and Critical Phenomena 2nd ed.*, Oxford University Press (1993); G. Parisi, *Statistical Field Theory*, Addison-Wesley (1988); J. J. Binney, N. J. Dowrick, A. J. Fisher, and M. E. J. Newman, *Theory of Critical Phenomena*, Oxford University Press (1992). The experimental values for the critical exponents in Table 13.7 were taken from A. Z. Patashinskii and V. L. Pokrovskii, *Fluctuation Theory of Phase Transitions*, Pergamon Press (1979).

Index